The Bio

CM0074187ᴈ

The Bioeconomy: Delivering Sustainable Green Growth

Davide Viaggi

Department of Agricultural and Food Sciences (DISTAL), University of Bologna, Bologna, Italy

CABI is a trading name of CAB International

CABI	CABI
Nosworthy Way	745 Atlantic Avenue
Wallingford	8th Floor
Oxfordshire OX10 8DE	Boston, MA 02111
UK	USA
Tel: +44 (0)1491 832111	Tel: +1 (617)682-9015
Fax: +44 (0)1491 833508	E-mail: cabi-nao@cabi.org
E-mail: info@cabi.org	
Website: www.cabi.org	

© CAB International 2018. All rights reserved. No part of this publication may be reproduced in any form or by any means, electronically, mechanically, by photocopying, recording or otherwise, without the prior permission of the copyright owners.

A catalogue record for this book is available from the British Library, London, UK.

Library of Congress Cataloging-in-Publication Data

Names: Viaggi, Davide, 1967- author.
Title: The bioeconomy : delivering sustainable green growth / Davide Viaggi, Department of Agricultural Sciences (DipSA), University of Bologna.
Description: Oxfordshire, UK ; Boston, MA : CABI, [2018] | Includes bibliographical references and index.
Identifiers: LCCN 2018034273| ISBN 9781786392756 (hardback) | ISBN 9781786392770 (epub)
Subjects: LCSH: Sustainable development. | Biotechnology--Economic aspects. | Sustainable agriculture--Economic aspects.
Classification: LCC HC79.E5 V53 2018 | DDC 338.1--dc23
LC record available at https://lccn.loc.gov/2018034273.

ISBN-13: 978 1 78639 275 6

Commissioning editor: David Hemming
Editorial assistant: Tabitha Jay
Production editor: Ali Thompson

Typeset by SPi, Pondicherry, India
Printed and bound in the UK by Bell & Bain Ltd., Glasgow

Contents

The Author: Davide Viaggi

Davide Viaggi has a degree in Agricultural Sciences and a PhD in Agricultural Economics and Policy. He is now Associate Professor in Agricultural Economics at the University of Bologna. His research focuses primarily on: i) Agricultural policy evaluation; ii) Public goods and agri-environmental policy; iii) Economics of the bioeconomy; iv) Innovation in agriculture and food; v) Farm investment behaviour; vi) Irrigated agriculture, water resource evaluation and management; vii) Environmental impact assessment and resource economics. He has received project funding from the European Commission (22 projects altogether in fp5, fp6, fp7 and H2020), the Italian Ministry of University and a number of private organisations. Davide has coordinated two research projects in fp7 (CAP-IRE and CLAIM) and one in H2020, SC2 (PROVIDE). His work has been presented in more than 400 publications, including 70 publications indexed in Web of Science and 87 in Scopus. From 2011 to 2016, Davide was Editor-in-chief of the journal Bio-based and applied economics. Davide has provided his expertise as an advisor and evaluator for several national and international institutions, including the European Commission, the ETP Food for Life and the PPP BBI. He was a member of the European Commission expert group supporting the evaluation of the EU Bioeconomy strategy in 2011.

Preface

_____

I encountered the bioeconomy as a researcher in agricultural economics when the European Union 7th Framework programme was launched in 2007. In 2011, I participated as an expert in the evaluation of the EU bioeconomy communication subsequently published in 2012. At that time, bioenergy was already a very widely discussed issue, as well as biotechnology. However, the idea of bringing together these and other bio-based topics under a single heading was rather new. In recent years, there has been a strong political push behind the bioeconomy. In parallel, I have seen growing attempts to address the bioeconomy in economic research and policy. However, in most cases, these attempts have taken a rather narrow view, focusing on a single product or technology, and with rather disappointing results compared with more comprehensive approaches that tackle the bioeconomy in a more holistic light. Yet times have changed. Researchers and policy makers now happily use the terms 'bioeconomy' and 'bioeconomy revolution' and it is difficult to find a conference in which the Bioeconomy is not mentioned.

Over the years, I have had the opportunity to witness presentations and discuss papers, to work as chief editor of a journal dedicated to bio-based economics (_Bio-based and Applied Economics_), and have always found it very challenging to explain and indeed to fully understand the bioeconomy.

There are very good reasons for this. First, the bioeconomy includes several different sectors and fields of research that bring together pieces of expertise that have historically been far apart from each other. Moreover, given that it is a new field, there is a lack of data and clear-cut information to define and delimitate its field of investigation. Finally, people address the bioeconomy with very different views and expectations in mind (and sometimes using totally different interpretations of the concept).

In spite of this, there is clearly significant demand for economics to help understand and evaluate the potential of new technologies, to analyse new markets and for meeting societal needs linked to the bioeconomy. This is even more true as hard science and technologically focused research are quickly introducing new results and as public policy is increasingly focused on supporting the bioeconomy as the main solution to deal with resource limitations.

This book is a response to this demand and also an effort to fill a void in the literature by providing a reference that addresses the economics of the bioeconomy as a whole, while at the same time linking up its various components.

This is the main objective of this book – to present in a consistent way what the bioeconomy represents for our economies and what economists are doing in terms of economic research on the different aspects of the Bioeconomy. Also, I will try to explore the potential for future research.

But is this merely a question of 'old wine in new barrels'? This question must always be kept in mind when a new keyword suddenly appears on the scene and a lot of existing concepts and tools are re-invented or converted to analyse it, or if papers put it forward in an attempt to better sell arguments, but in fact address much of an already well-known topic.

I will seek to demonstrate that the bioeconomy has a great deal of potential that is worth investigating and that a closer look at this field will also require, and indeed push for, some innovation in economic research. Some of these ideas (e.g. related to consumer, technology change, etc.) are not specific to the bioeconomy, but are rather emphasized by the features of the bioeconomy.

In the end, I will try to show how fragile this concept may also be and how its very high potential to contribute to our society depends on actual action and tangible contributions on the part of researchers.

Davide Viaggi

Acknowledgements

_____

This book comes at the end of a journey into the bioeconomy that has entailed over two years of work and reflection, and which likely represents only just a beginning. Long preparation time implies the need to acknowledge the support of many people.

Of course, a first heart-felt thank you goes to my family for their patience and support. It includes my indebtedness for a number of Saturdays, Sundays and nights dedicated to this effort.

Thank you to the editor, David Hemming, during the long setting up of the book and to Alexandra Lainsbury, Tabitha Jay and Ali Thompson for their support during the most operational part of the book's development.

I would like to thank the initial reviewers for trusting and supporting my proposal.

I would like, in particular, to thank Robert M'Barek for his initial hints, support and comments during the development of the book.

I extend my sincere thanks to Vittorio Gallerani, Fabio Bartolini, Matteo Zavalloni, Meri Raggi and Francesco Galioto for providing suggestions and comments on earlier versions of the book.

A big thank you to David Cuming, as usual much more than a language editor.

A thank you also to my research group, sometimes left alone in the difficult waters of research funding, evaluations and publications while I was dedicating my attention to this book.

Finally, my sincere thanks to all of the many research and industry contacts, students and colleagues from whom I have picked up ideas and suggestions in the context of several projects and activities in the past few years.

Davide Viaggi

Abbreviations

ABC	anthropized biological capital
AES	agri-environmental schemes
AKIS	agricultural knowledge and innovation systems
BBI JU	bio-based industries joint undertaking
BCI	biorefinery complexity index
BIC	Bio-based Industries Consortium
CAP	common agricultural policy
CBD	convention on biological diversity
CF	carbon footprint
COREA	cumulative overall resource efficiency assessment
CSA	community supported agriculture
DNA	deoxyribonucleic acid
EIA	environmental impact assessment
EIO-LCA	economic input output - life cycle assessment
EoL	end-of-life
EPA	Environmental Protection Agency
EROI	energy return on investment
ES	ecosystem services
ESIM	European Simulation Model
EU	European Union
FOAK	first-of-a-kind
FSCW	food supply chain waste
G7	Group of 7
GBR	green biorefinery
GDP	gross domestic product
GE	genetic engineering
GHG	greenhouse gas
GM	genetically modified
GMO	genetically modified organism
GRIP	grounded innovation platform
HANPP	human appropriation of net primary production
ICT	information and communication technologies
IEA	International Energy Agency
IoT	internet of things

IPR	intellectual property rights
IRR	internal rate of return
IWF	integrative worldview framework
KBBE	knowledge-based bio-economy
KET	key enabling technologies
LCA	life cycle assessment
LCC	life cycle costing
LCOE	levelized cost of energy
LCSA	life cycle sustainability assessment
LGE	litre of gasoline equivalent
MCDM	multicritera decision-making
MOOC	massive open online course
MuSIC	modelling sustainable industrial chemicals production
NACE	Nomenclature statistique des activités économiques dans la Communauté européenne (Classification of economic activities in the European Union).
NFRA	nested-feedback risk assessment
NGO	non-governmental organization
NPV	net present value
OECD	Organisation for Economic Co-operation and Development
PE	polyethylene
PEF	product environmental footprint
PES	payments for ecosystem services
PET	polyethylene terephthalate
PHA	polyhydroxyalkanoate
PLA	polylactic acid
PO	producer organizations
PP	polypropylene
PPP	public–private partnerships
R&D	research and development
RESPONSA	REgional SPecific cONtextualized Social life cycle Assessment
RFS	renewable fuel standard
RIS	regional innovation systems
RPM	robust portfolio modelling
SAM	social accounting matrix
SDG	sustainable development goals
SE	somatic embryogenesis
SES	socio-ecological systems
SETVEW	socio-ecological technological value-enhancing web
SEVW	socio-ecological value web
SLCA	social life cycle assessment
SMAA	stochastic multi-attribute analysis
SME	small and medium-sized enterprise
SSP	shared socioeconomic pathways
SWOT	strengths, weaknesses, opportunities, threats
TFP	total factor productivity
TIM	technology and innovation management
TIS	technological innovation systems
TRL	technology readiness levels
UK	United Kingdom
UN	United Nations
US	United States
USDA	US Department of Agriculture
WTP	willingness to pay

1

Introduction and Overview

1.1 Introducing the Bioeconomy

This chapter introduces the concept of 'bioeconomy' and provides the motivation for this book. The scope and the rationale of the book are also presented, together with an overview.

The general concept of the bioeconomy is that of an economy based on the sustainable exploitation of biological resources. There are, however, differing views on this general definition. For example, the European Union (EU) defines it as a bundle of sectors including agriculture and food. The Organisation for Economic Cooperation and Development (OECD) and United States definitions, for their part, are more focused on biotechnology. Following different country definitions and strategies, there is a latent divergence between two notions of bioeconomy: one that basically uses the term as the non-food component of the economy, linked to biological resources, and another that also includes agriculture and food. Policy trends in the EU and worldwide seem to legitimize the second option, and this is the approach followed in this book.

The existence of different views is also linked to the use of the term bioeconomy in various branches of research, with different normative visions of technology. In spite of this diversity, a common feature of the existing visions of the bioeconomy, together with the foundation in biological resources, is the central role of knowledge, research and innovation, as well as the growing attention to circularity and sustainability, in particular linked to climate change concerns.

Indeed, the policy context is characterized by growing concerns for climate change, scarcity of fossil resources and growing world food needs. In parallel, there is growing recognition of the advances in technology, especially in the life sciences. The bioeconomy is envisaged as a potential transition pathway to address (at least some of) these problems, meeting society needs while respecting the world's ecosystems and natural resources.

The importance of the bioeconomy has grown considerably in policy agendas around the world. Around 50 countries now have policy agendas impacting directly on the bioeconomy. The G7 countries are becoming leaders in this strategy. Germany, the USA and Japan have produced the most ambitious bioeconomy strategies; the EU has taken a leading promotional role through its bioeconomy strategy and Horizon 2020 research framework programme (European Commission, 2012; German Bioeconomy Council, 2015a; 2015b). Most EU countries now have national strategies on issues related to the bioeconomy, and 18 out of 28 have included the bioeconomy as a priority for EU structural funds. This pathway has been marked by a number of major funding initiatives and by several major events. A recent update about country bioeconomy strategies and international initiatives on the bioeconomy has been published by the German Bioeconomy Council (2018).

© D. Viaggi 2018. *The Bioeconomy: Delivering Sustainable Green Growth* (D. Viaggi)

Steps have been made to better qualify the bioeconomy and its policy approach, with a growing emphasis on sustainability and the achievement of the United Nations Sustainable Development Goals (El-Chichakli *et al.*, 2016; Viaggi, 2016).

Another important aspect is that bioeconomy strategies are increasingly embedded in other policies, such as the EU Common Agricultural Policy, which also include measures to facilitate the supply of renewable sources of energy, by-products, wastes and residues for the bioeconomy (Viaggi, 2016). Another example concerns the integration between the bioeconomy and the circular economic strategies. For example, in the EU communication on the circular economy, the bioeconomy is expected to contribute to the circular economy, especially by providing alternatives to fossil-based products and energy. The communication includes sections on food waste, biomass and bio-based products (European Commission, 2015).

The development of the bioeconomy in different areas of the world highlights some common needs, but also shows different approaches and competing interests. In some cases the bioeconomy focuses on the exploitation of local biological resources, in others on industrial or technological progress; some countries have a strong attention to rural development (El-Chichakli *et al.*, 2016). The global interconnection of markets is growing and the effects of specific bioeconomy policies transmit much faster than in the past from one region of the world to another.

In the 2017 OECD report on the 'next production revolution', the bioeconomy is listed as one of the key sectors for technology advances (OECD, 2017) and several bioeconomy-related policy priorities are listed as being key for the future. In this sense the bioeconomy is expected to contribute to the ongoing major technology shift that will likely lead to a new technological revolution, together with other areas of technological change such as nanotechnologies and digitalization.

Indeed, much of the potential of the bioeconomy is still largely untapped, as bioeconomy technology is characterized by continuous innovation as well as more and more reactive markets and social structures. In this promising and quickly evolving sector, the interface between myth and reality, between socially constructed and actual impacts, as well as between 'industrial legends' and promising technologies is also often difficult to grasp.

1.2   Bioeconomy in Economic Research

This book is based on the idea that, in this context, economic research has an important and somewhat unique role to play. Research is devoting growing attention to the bioeconomy. This should not come as a surprise, as the bioeconomy is largely relying on research and innovation. However, the vast majority of the existing research examines individual components of the bioeconomy, rather than the bioeconomy as a whole and, in fact, only a small portion of the bioeconomy-related research specifically mentions the term 'bioeconomy'.

A search of the term 'Bioeconomy' in the Scopus database (based on title, abstract and keywords of papers) in December 2017 yielded 878 papers, with a growing trend in recent years (confirmed by additional 185 papers in the first half of 2018). Of these, 172 were classified in the category 'Social sciences' and 89 as 'Economics, econometrics and finance'. Eighty-six were in 'Business, management and accounting'. The most frequent references were the 228 papers in 'Biochemistry, genetics and molecular biology', the 225 papers in 'Environmental science' and the 217 papers in 'Agricultural and biological sciences'.

Though several papers date back to the 1990s, the number of contributions has begun to grow only recently. Moreover, there is a clear trend from very specific papers, e.g. in life sciences, which make mention of bioeconomy (with a plethora of meanings at the beginning) towards more comprehensive attempts to deal with the bioeconomy as a distinct concept or sector. Only since 2015–2016, however, has there been a clear prevalence of papers with a shared view of the bioeconomy as a comprehensive subject and that tend to identify their specific contribution in this overall concept.

Considering the bioeconomy-related papers in the subject area 'Economics, econometrics and finance' only, the number of papers per year has been (irregularly) growing over the past five years: 11 in 2012, 9 in 2013, 4 in 2014, 16 in 2015, 15 in 2016, and 22 in 2017. This also highlights that papers in economics tend to appear in the literature later than those in both the biological sciences and social sciences (the latter recording a noteworthy number of papers as early as in the period 2004–2006).

It is noteworthy that special issues have been devoted to the bioeconomy in at least two major journals dedicated to agriculture and food economics (*Agricultural Economics* and the *German Journal of Agricultural Economics*); at least one journal is entirely dedicated to bioeconomy-focused works (*AgBioForum*) and another has bioeconomy as one of its distinctive features (*Bio-based and Applied Economics*). There is also a journal called *Bioeconomics* that mostly showcases interdisciplinary collaborations in the field of biological resource management. The journal *New Biotechnology* dedicated a special issue to the bioeconomy issues in 2017 with a number of economic and technology-related papers (Aguilar *et al.*, 2018). As the subject is taking shape, some relevant books are addressing the bioeconomy with a strong or prevailing economic/social sciences perspective (Bonaccorso, 2016; Lewandowski, 2018).

Much of the bioeconomy-related economic literature is focused on the individual building blocks of the bioeconomy. Indeed, some sectors, such as agriculture and food, have been extensively studied and have a much longer history than the bioeconomy itself. Some insights into these sectors' trends can be seen from publication numbers. Economics papers related to 'agriculture' now number 13,324 in Scopus and have grown constantly from 174 per year in 2000 to 865 in 2017. This is even more evident for economics papers related to 'food', now totalling 17,265 papers in Scopus, having grown from 232 in 2000 to 1445 in 2017. Clearly, these sectors are still attracting a good deal of attention and have even become more attractive topics for research in very recent years.

More specific aspects of the bioeconomy are also the subject of a growing research effort, such as bioenergy, biorefinery and biotechnologies. Economics papers indexed in Scopus now number 1658 for the 'biotechnology' keyword (with a growing trend and average of more than 100 per year in the period 2011–2013, but decreasing afterwards) and 314 for 'bioenergy' (with similar trends, though having started much later, and a production of about 30–40 papers per year in recent years). Biorefinery, for its part,

only counts 65 economics papers, with a top production of 10 in 2011 and subsequent ups and downs. One of the most interesting components of this group, involving newly emerging products such as 'biomaterials', still only counts 22 papers in economics (compared with more than 187 in business and tens of thousands altogether).

Investigating this large bulk of literature is beyond the scope and ambition of this book. The focus is rather on the various components of this literature that point to the bioeconomy as a whole and which relate to the bioeconomy as a new concept. At the same time, some aspects of these more traditional sectors need to be considered, not only because they can be fully considered as part of the definition of bioeconomy (see the next chapters), but also because many emerging sectors are indeed largely connected to the oldest sectors (notably biorefinery or bioenergy and agriculture).

The term bioeconomy is also being integrated into the existing fields of research. Foguesatto *et al.* (2017) provide the example of bioeconomy references in agribusiness papers between the years 2000 and 2015. They found 75 papers and note the progression of publications in this 'interfacing' field, pointing to the high level of interdisciplinarity (as well as to the frequent examples of terminology confusion).

It is common for fields of study at their inception to build a lot on descriptive approaches. A number of studies address the issue of the bioeconomy by describing and analysing case studies of bioeconomy development. These may be related to specific policies, country strategies (e.g. Kamal and Che Dir, 2015), or to specific plants (e.g. Schieb *et al.*, 2015). However, the understanding of more comprehensive bioeconomy features is also developing and shaping the approach to the study of the sector, leading to a growing number of conceptual papers.

Based on the above, Viaggi *et al.* (2012) identified two broad areas of attention: (i) the bulk of specific research fields related to individual issues in the sphere of the bioeconomy: consumer sciences, markets, patenting rights and innovation, as well as the economic and social aspects of bioenergy, biotechnologies and biomaterials; and (ii) the need to address the broad concept of bioeconomy and to approach it in a comprehensive manner from an economic perspective.

The latter is undoubtedly the most challenging one and highlights the need for a more comprehensive view of the bioeconomy that is able to address the challenges of the interconnections among different components of the bioeconomy, different regional needs and different expectations. This demand also involves research and, while this may be common to all disciplines, it is especially true for economic research.

From a 'disciplinary perspective' (Viaggi *et al.*, 2012), a potential disciplinary shift is envisioned from agriculture to bio-based economics and pathways for potential research developments in this direction. The literature has begun to mention an 'economics of the bioeconomy' (Viaggi *et al.*, 2012) or 'bioeconomy economics and policy' (Wesseler *et al.*, 2015).

As several 'bio' concepts have been associated with the study of economics and there is already a proliferation of such concepts, this book does not intend to coin any specific term for the subject nor take the (partial) perspective of one of those already in existence. For this reason, the contents of this book can be roughly identified as being centred on 'bioeconomy economics and policy', with the intention of picking up suggestions from sometimes different and contrasting fields to achieve a better understanding of the bioeconomy.

1.3   Why this Book

Taking a broad view on the existing work, the bioeconomy appears to be one of the most promising yet at the same time potentially controversial fields of economic development. In contrast to the high demand for economic answers and to the wide diffusion of the use of simplified and sometimes naïve economic concepts by stakeholders and scientists in the bioeconomy, socioeconomic research on the transition towards the bioeconomy, remains insufficient and unstructured both at the firm-level (Van Lancker *et al.*, 2016) and the global level. In addition, while there are studies in the field of management or for specific sectors, few really examine the bioeconomy as a whole and get into the specific novelties of this concept. This is the main motivation for this book.

This book is an attempt to bring the sparse literature on the economics of the bioeconomy together, in order to provide an overview and, at the same time, a consistent account of the detailed emerging branches that are developing within it. The objective of the book is to illustrate the advances and perspectives in the development of the bioeconomy, taking both an economic and policy perspective, as well as to identify potential future pathways and issues for this subject. In doing so, the book also seeks to provide a framework for an overall economic interpretation of the bioeconomy and to develop some new bridging concepts in bioeconomy analysis.

The book intends to meet the needs of different user groups, from practitioners to researchers, to get a consistent view of what the bioeconomy is, how it works and what it means in an economic perspective. As this is largely a primer, the book addresses different aspects while trying to provide a systematization of different pieces of literature that are functional to this understanding.

For this reason, the style of the book is somewhere in the middle between a scientific review and a textbook. It attempts to consider most of recent available economic studies and published research directly concerning the bioeconomy and to bring them into a structured economic view of this emerging sector. Furthermore, the book tries to use well-established concepts from economics in order to better support the understanding of bioeconomy issues for those not well-versed in economics.

At the same time, as non-experts in either economics, life sciences or biotechnology may also be interested in the subject matter of this book, several parts are purposely kept very simple and are intended as a true introduction to the bioeconomy for 'newcomers'.

1.4   Scope of the Book

Given the complexity of the concept of bioeconomy, a number of choices had to be made to delimitate the scope of this work.

Different countries/institutions have different definitions of bioeconomy; the book uses a broad definition covering all of the main notions from different country strategies and is ultimately very close to that of the EU. However, the book also includes sections aimed at investigating different delimitations and country approaches and their *raison d'etre*, notably in the direction of consumers and ecosystems.

Health is mostly excluded, as it is a very specific topic with radically 'different' issues compared to other parts of the bioeconomy. However, as this is a major area of interest for addressing the topic at the interface between ethics and economics, as well as between living beings and the economy, the book makes some rather arbitrary references to it.

There is an issue in how to deal with sectors having their own well-established bundle of research (e.g. agriculture). The book tries to accommodate sector views with a comprehensive view of the bioeconomy focusing mainly on linkages and interconnections of these sectors with the whole bioeconomy.

In terms of the disciplinary approach taken, although the focus is economics, the fields of management, policy, planning, sociology and other social sciences are relevant and touched upon in parts of the book. Some interdisciplinary issues are addressed in the various chapters; in addition, a specific chapter is devoted to interdisciplinary topics and perspectives, mostly regarding issues complementary to economics in the field of social sciences.

The book also maintains an emphasis on policy issues, including keeping in mind the need for science to support evidence-based policy making. Policy (especially research and innovation policy) is a key area of attention as the bioeconomy is in its beginning stages and widely promoted through public policies. New markets and entrepreneurship are also very relevant and, in turn, connected to policy. In many countries, policies have yet to be elaborated, but strategy documents are available that will undoubtedly assist in paving the way for future policy interventions. An effort has been made to avoid considering policy in isolation, but rather to emphasize the interplay between the policy mix, the wider institutional setting, market forces and system organization solutions.

The book has an international scope, including a focus on some of the main countries and regions for the bioeconomy; cross-country and cross-continent linkages are also highlighted. However, more insights relate to the geographical areas in which the bioeconomy is more advanced, at least in terms of research.

Both empirical and conceptual chapters are included, with an attempt to bring them together and to yield facts and interpretations for the readers. A large part of the book is based on reviews of existing literature and documents, but some pieces constitute original work developed while putting together the different pieces of the 'bioeconomy puzzle'.

Of course, there are a number of potential topics that have been deliberately excluded or that were simply missed due to the ongoing fragmented literature on the issues addressed in this book. This mainly means that those interested will find a number of ways to delve further into this broad field of investigation.

1.5 Overview

The book is organized in 11 chapters, in addition to this one, that can be broadly grouped in four main parts. The first part includes Chapters 2 and 3. Chapter 2 illustrates the current definitions, strategies, policy and economic information related to the bioeconomy the world over. This part takes a descriptive approach in presenting the available information about the bioeconomy in the economy. Basically, it serves three objectives: (i) to clarify what is meant by the bioeconomy in practice; (ii) to identify the main statistical information and sectors of the bioeconomy; and (iii) to shed light on the main strategy documents and policies on the bioeconomy.

The following treatment is based on an understanding of bioeconomy technologies and their representations, either at the interface with non-bioresource sectors or within the way bioeconomy technology is developing. For this reason, a chapter is dedicated to bioeconomy technologies from an economic perspective (Chapter 3).

The second part (Chapters 4 to 8) describes the current economic analysis and research effort in qualifying and understanding the economics of the bioeconomy. This part tends to systematize the current economic literature analysing the bioeconomy, taking a more disciplinary approach and to connect bioeconomy issues with economic thought and economic instruments. It serves five main purposes: (i) to help describe and understand bioeconomy from an economic perspective; (ii) to provide scholars and practitioners with a set of ideas and approaches to the bioeconomy as an object of economic analysis; (iii) to provide practitioners with concepts and tools that are useful in terms of economic analysis and evaluation of bioeconomy-related issues; (iv) to support policy analysis and design to match societal challenges in the field of bioeconomy; and (v) to attract attention explicitly to the need to further develop an economics of the bioeconomy.

For these reasons, this is the section that most resembles a textbook. The starting chapter is related to approaches to the understanding of the bioeconomy (Chapter 4). The following three chapters investigate the bioeconomy through the three classical perspectives of demand (Chapter 5), supply (Chapter 6) and institutional mechanisms to ensure the functioning of the bioeconomy (Chapter 7), including markets, policies and supply chain organization solutions. Chapter 8 looks more in detail at the political economy of the bioeconomy, a field benefiting from the interpretation of the forces illustrated in the previous chapters.

The third part, Chapters 9 to 11, delves into more detail in a number of issues that are complementary and that help deepen the economic analysis. The first and main chapter of this part brings the discussion towards the understanding of how the bioeconomy can contribute to global sustainability objectives, including issues related to natural resources and sustainable development (Chapter 9). The following two chapters provide a snapshot of two complementary yet stimulating areas linked to the above: (i) impact evaluation and management tools for the bioeconomy (Chapter 10) and (ii) developments in fields of research complementary to economics (such as psychology, sociology and philosophy) (Chapter 11).

The fourth part, Chapter 12, concludes with some general outlooks and perspective views on the future of the bioeconomy, highlighting uncertainties and dependence on political will (and good research). It also discusses once again the economics of the bioeconomy as a potential emerging field of economic research, pointing in particular to the most urgent needs, as well as trying to help researchers to identify priority (and promising) research pathways in this very complex field of investigation.

References

Aguilar, A., Wohlgemuth, R. and Twardowski, T. (2018) Preface to the special issue bioeconomy. *New Biotechnology* 40, 1–4.

Bonaccorso, M. (2016) *The Bioeconomy Revolution*. Milan, Italy, Edizioni Ambiente.

El-Chichakli, B., von Braun, J., Lang, C., Barben, D. and Philp, J. (2016) Five cornerstones of a global bioeconomy. *Nature* 535, 221–223.

European Commission (2012) Innovating for sustainable growth: a Bioeconomy for Europe. Available at https://ec.europa.eu/research/bioeconomy/pdf/official-strategy_en.pdf (accessed on 30 December 2017).

European Commission (2015) Closing the loop – An EU action plan for the Circular Economy. *COM(2015) 614 final*. Available at https://ec.europa.eu/transparency/regdoc/rep/1/2015/EN/1-2015-614-EN-F1-1. PDF (accessed on 30 December 2017).

Foguesatto, C.R., Artuzo, F.D., Oliveira, L. and Souza, Â.R.L. (2017) Research agenda of the bio-economy: A study in the field of agribusiness. *Espacios* 38(4), 15.

German Bioeconomy Council (2015a) Bioeconomy Policy (Part I): Synopsis and analysis of strategies in the G7. Available at biooekonomierat.de/fileadmin/international/Bioeconomy-Policy_Part-I.pdf (accessed on 16 November 2015).

German Bioeconomy Council (2015b) Bioeconomy policy (Part II): Synopsis of national strategies around the world. Available at biooekonomierat.de/fileadmin/Publikationen/berichte/Bioeconomy-Policy_Part-II.pdf (accessed on 16 November 2015).

German Bioeconomy Council (2018) Bioeconomy Policy (Part III): Update report of national strategies around the world. Available at http://gbs2018.com/fileadmin/gbs2018/Downloads/GBS_2018_Bioeconomy-Strategies-around-the_World_Part-III.pdf (accessed 26 May 2018).

Kamal, N. and Che Dir, Z. (2015) Accelerating the growth of bioeconomy in Malaysia. *Journal of Commercial Biotechnology* 21(2), 43–56.

Lewandowski, I. (ed.) (2018) *Bioeconomy: Shaping the Transition to a Sustainable, Biobased Economy*. New York, USA, Springer.

OECD (2017) *The Next Production Revolution. Implications for Governments and Business*. Paris, France, The Organisation for Economic Co-operation and Development.

Schieb, P.-A., Lescieux-Katir, H., Thénot, M. and Clément-Larosière, B. (2015) *Biorefinery 2030: Future Prospects for the Bioeconomy*. New York, USA, Springer.

Van Lancker, J., Wauters, E. and Van Huylenbroeck, G. (2016) Managing innovation in the bioeconomy: An open innovation perspective. *Biomass and Bioenergy* 90, 60–69.

Viaggi, D. (2016) Towards an economics of the bioeconomy: Four years later. *Bio-based and Applied Economics* 5(2), 101–112.

Viaggi, D., Mantino, F., Mazzocchi, M., Moro, D. and Stefani, G. (2012) From Agricultural to bio-based economics? Context, state-of-the-art and challenges. *Bio-based and Applied Economics*, 1(1), 3–11.

Wesseler, J., Banse, M. and Zilberman, D. (2015) Introduction special issue 'The political economy of the bioeconomy'. *German Journal of Agricultural Economics* 64(4), 209–211.

2

What is the Bioeconomy

2.1 Introduction and Overview

This chapter describes the bioeconomy as it is defined today in research and policy documents. As mentioned, there are differing views on these definitions depending on the point of view and needs of individual countries or actors defining the bioeconomy.

To a large extent this is based on the identification of economic sectors to be included in (or excluded from) the bioeconomy. Economic sectors are usually described based on the kind of goods and services they use and produce. Part of the effort here is to distinguish the bioeconomy from non-bio sectors or from non-bio sub-sectors in hybrid sectors such as energy, construction, and so on. In this attempt, the basis in biological raw materials and products is the distinguishing feature of the sectors of the bioeconomy.

Moreover, the bioeconomy is characterized by the interconnections among these sectors. Both sectors and their connections can be qualified based on different aspects, such as their economic value, underlying biomass flows, waste and by-products management and the degree of circularity. These aspects are increasingly important in understanding the functioning of the bioeconomy, its economic potential and its limitations.

This chapter also illustrates the main statistical evidence about the bioeconomy, with particular attention devoted to economically relevant parameters (e.g. contribution to gross domestic product [GDP] and employment). As the bioeconomy is not a traditional aggregate of economic statistics, there are problems with data availability; this issue is also highlighted together with the efforts being made to improve the availability of statistical information.

The annex to this chapter details the configuration of the bioeconomy in selected countries.

2.2 Some History of the Bioeconomy

In a way, the origin of the bioeconomy is rooted in the origins of humankind. The use of biological resources has always been integral to life. Harvesting and hunting living resources have always been part of the activities needed to sustain human life since its origin. The use of biological resources for the production of food, fibre and energy has been a key part of the human economy since its beginning as well. During this time and since domestication and breed selection at the origin of agriculture, humanity has also modified living organisms for its uses and the ecosystems it was living in.

The development of the fossil economy during the industrial revolution, first based on carbon and then on oil, has dramatically modified this picture in the last couple of centuries. Fossil resources have become the most widely used basis

 © D. Viaggi 2018. *The Bioeconomy: Delivering Sustainable Green Growth* (D. Viaggi)

for the production of energy, as well as providing materials for a number of other uses (e.g. plastics). Fossil energy has also supported the development of highly energy-intensive technologies, such as the production of chemical fertilizers or mechanization, as well as very concentrated industrial processes and the globalization of the economy through lower transportation costs.

The awareness of the limited availability of fossil resources as well as climate change concerns linked to CO_2 and other gas emissions, have opened up an agenda for moving away from these sources, a need that was already attracting significant attention in the second half of the 20th century.

In parallel, progress in the biological sciences has created the basis for the development of biotechnology. Key steps in developing the basis of the bioeconomy in the 20th century have included the full understanding of the structure of DNA in 1953, which opened the way for a number of new possibilities in genetically modifying plants and animals. Key steps were the development of new technologies that made it possible to transform and manipulate genes within organisms, including the Cohen–Boyer patent enabling genetic engineering (GE) in medicine, and agro-bacterium that allowed for GE in agriculture (Zilberman *et al.*, 2015).

The first European programme on biotechnology run by the European Commission was in the period 1982 to 1986; since then, biotechnology has consistently been part of the European Union (EU) programmes, with a number of success stories and a major contribution to the field of bioeconomy in recent years (Aguilar *et al.*, 2013).

The opportunities linked to these technologies have been accompanied by the development of a number of other technologies in the broad field of biotechnology during the second half of the 20th century, while policy and public attention, including tensions, around these technologies increased. The first decade of the 21st century has been characterized by intense attention on two issues. First, the role of biotechnologies has become more evident due to raising public concerns, for example, of genetically modified organisms (GMOs). This has brought it to the attention of consumers and has become a major market issue. On the other hand, the capacity for GE is growing, as the cost of sequencing the genome of different species is declining, which opens

up increased possibilities. In addition, new techniques like gene editing are being introduced (Zilberman *et al.*, 2015). This has been accompanied by the growing potential of these technologies in a range of applications.

Second, the development of bioenergy production has become massive, as a means of addressing fossil fuel scarcity and in connection with climate change concerns. Bioenergy has rapidly become a major area of attention in relation to concerns about food-energy competition and world food prices, but also due to its potential implications for water and soil use.

Finally, the potential for new technologies and products using bio-based solutions is now growing in a significant way.

The bioeconomy has meanwhile developed through specific policies, especially in research. In the EU the basis for the bioeconomy was established as of the 5th Framework programme through the 'Cell Factory' scheme and boosted through the knowledge-based bioeconomy (KBBE) section of the 7th Framework programme in preparation for the EU bioeconomy strategy in 2012 (Patermann and Aguilar, 2017). The use of knowledge-based bioeconomy as a policy concept and the growing awareness of the economic contributions of industrial biotechnology at the crossroads of its implementation contributed to launch the current notion of bioeconomy.

At the same time, the movement of agricultural policy towards ecosystem services and ecology issues made it possible to launch the idea of bioeconomy linked to that of more environmental friendly primary sectors (Jordan *et al.*, 2007). More or less in the same period the stream of research related to GMOs started channelling into the wider bioeconomy concept (Chapotin and Wolt, 2007).

During this period the perception started to consolidate that this trend was not just a marginal change, but rather a major transition to something beyond the oil age, leading Dale (2007) to state: 'Strong evidence exists that we are in the early phases of a truly historic transition from an economy based largely on petroleum to a more diversified economy in which renewable plant biomass will become a significant feedstock for both fuel and chemical production.'

Already in 2003, Duchesne and Wetzel (2003), discussing the issue in the Canadian context, highlighted the many potential promises of the bioeconomy:

> The bioeconomy is expected to replace the current information economy and will depend heavily on the manufacturing and trade of bioproducts…The bioeconomy should impact most of Canada's economic sectors…The bioeconomy holds promises to wean the Canadian economy from its dependence on fossil fuels as a primary source of energy as well as platform chemicals in materials and manufacturing, while meeting the Kyoto commitments on greenhouse gas reductions. Finally, the bioeconomy will reduce the environmental impact of economic growth.

Dale (2007) identifies at least six key driving forces and qualifying features of the bioeconomy: (i) increasing yields and the use of the whole biomass; (ii) diversification of bio-based products; (iii) exploiting a diversity of biomass resources; (iv) the need to consider the limits of agricultural productivity and resources; (v) the need to integrate biorefining and agricultural ecosystems in a local social and political context; and (vi) the need to care for the sustainability of the mature bio-based economy and its most important underlying resources, such as biodiversity and soils.

In the decade from 2007 to 2018, a number of country strategies, research and financial programmes and emerging new bio-based solutions have characterized the development of the bioeconomy and a growingly widespread understanding of this notion. These topics will be discussed in the following sections.

Besides the development of different sectors characterizing the bioeconomy and formal definitions, objectives and strategies by different countries and bodies, a major feature of sector evolution is the convergence of industries/technologies and the breaking down of boundaries across sectors and across scientific fields. The elimination of boundaries between the agricultural and energy industries is an example that points to an industry convergence process. Other examples exist in relation to the use of biowaste and the interplay between food and non-food chains at the agriculture level. This is complemented by the notion that the convergence of research-intensive industries follows the convergence of scientific fields and technologies. This convergence of sectors and of research is indeed a key factor supporting the idea of the bioeconomy

as a whole, rather than just a sum of sectors (Golembiewski *et al.*, 2013).

In this context, the concept of bioeconomy has developed in parallel in different areas and with different focuses depending on the country or the institutions involved. The term itself has been a topic of debate, or at least often used in ambiguous ways, especially with reference to the sectors involved (e.g. food, health). The concept of bioeconomy that is emerging (or that is the notion that is deemed most interesting for this book), relies on three mains features. First, it considers all sectors using biological resources in a unified concept, including traditional agriculture and food sectors, and emphasizing the linkages among them. Second, it recognizes the central role of knowledge, research and innovation. Third, it also includes some specification of what the bioeconomy should achieve, aiming at qualifying the bioeconomy as sustainable and circular.

When the bioeconomy, in the meaning adopted in this book, began to consolidate, the term bioeconomy was not new. Works in biological sciences used the term to mean something similar to the economic-like physiology of living organisms. In the 1990s, the terms 'bioeconomy' and 'bioeconomic models' were largely associated with the management of natural resources that grow naturally and are harvested by human activities such as fishing and forestry (see, for example, Christensen (1996) and many textbooks in the field of resource economics). Bioeconomic models have also been used in economic analysis to mean models accounting for both the economic dimension of decision making and for the underlying technical (economic) relationships, connected to the features of living organisms such as cultivated plants. The interconnection between the environmental and social dimensions of sustainability have often been recognized as a weak issue in sustainability assessment, and the term 'bioeconomy model' has been used for models in an attempt to better represent this connection (Lehtonen, 2004).

Notably, since the beginning of the 21st century, the term has been used for biotechnology applications related to the health sector, such as stem cells or organ transplantation (Davies, 2006; Salter *et al.*, 2006). This also highlights the somehow arbitrary exclusion of the medical component of bioeconomy from this book.

Already in 2003 some authors used bioeconomy in a meaning rather close to the current one, focusing on the potential from the use of living organisms through the use of technology and biotechnology (Duchesne and Wetzel, 2003).

Rather than to contrast definitions, it is preferable to highlight here that different uses of the term bioeconomy actually point to different aspects of the same phenomenon, which is the use of biological resources by humans, and that the more recent and consolidating way of using the term is somehow also the most comprehensive one.

However, the need to qualify and to narrow the definition of the bioeconomy purposes remains and this is especially evident for policy strategy purposes, as discussed in the next section.

2.3 Bioeconomy Definitions

The general concept of 'bioeconomy' (or bio-based economy, which is often considered as a synonym) is that of an economy based on the sustainable exploitation of biological resources. However, when putting this concept into practice, a number of qualifications or different focuses can emerge. Let us develop this point based on the experience of bioeconomy strategy documents.

In a review of 45 country strategies concerning the bioeconomy, the German Bioeconomy Council (2015b) found that only about 40% of these strategies use the term bioeconomy or bio-based economy. This is no surprise, because as we saw previously, the bioeconomy comprises several individual sectors that can be identified and promoted even without using the term bioeconomy itself.

Of these strategy documents, only about 30% provided a definition of bioeconomy. The definitions available show different visions of the bioeconomy, as well as different pathways for the development of the sector:

- some countries have a focus on bioscience and biotechnology, mostly also including the health sector (e.g. US, India, South Africa, South Korea);
- others are more focused on primary industries and define bioeconomy as encompassing agriculture, forestry and the marine economy, focusing on new biomass value

chains (e.g. Brazil, Canada, Finland, New Zealand); and
- a third group focuses more on emerging industries and high-tech development (e.g. Netherlands, China, Malaysia, Thailand, Japan, Russia).

The Organisation for Economic Co-operation and Development (OECD, 2009) offers a more narrow definition, centred on biotechnologies:

> A bioeconomy can be thought of as a world where biotechnology contributes to a significant share of economic output. The emerging bioeconomy is likely to involve three elements: the use of advanced knowledge of genes and complex cell processes to develop new processes and products, the use of renewable biomass and efficient bioprocesses to support sustainable production, and the integration of biotechnology knowledge and applications across sectors.

On the contrary, the EU provides one of the more comprehensive definitions of the bioeconomy, based on a mix of sectors and technology identifiers. The EU 'Communication on the bioeconomy' and the accompanying working document (European Commission, 2012) defines the bioeconomy as encompassing

> ...the production of renewable biological resources and their conversion into food, feed, bio-based products and bioenergy. It includes agriculture, forestry, fisheries, food and pulp and paper production, as well as parts of chemical, biotechnological and energy industries. Its sectors have a strong innovation potential due to their use of a wide range of sciences (life sciences, agronomy, ecology, food science and social sciences), enabling and industrial technologies (biotechnology, nanotechnology, information and communication technologies, and engineering), and local and tacit knowledge.

Academic works usually try to identify the bioeconomy using concepts derived from policy strategy definitions connected with disciplinary views that qualify the phenomenon. Two extreme interpretations can be seen in the 'biochemical' view of the bioeconomy and in the 'cultural vision' view of the bioeconomy.

The first extreme, based on the life science identification of the bioeconomy, is linked to the role of photosynthesis (Roy, 2016). Bioeconomy can be identified as the part of the economy linked to the use of the results of photosynthesis in terms of solar energy storage through carbon fixation. In a way this applies to both the fossil economy and bioeconomy. The difference is that the fossil economy is based on energy storage throughout the ages and used at a speed much higher than the fixation and fossilization process. On the contrary, the bioeconomy uses energy fixed by current living organisms and hence assumes (or at least aims for) a rate of consumption of fixed energy keeping pace with solar energy fixation.

At the opposite extreme, the bioeconomy can be seen as a vision of resource use and economic development, often associated with the quest for a transition process. Bugge *et al.* (2016) indeed identify three visions of the bioeconomy:

- the biotechnology vision, emphasizing the importance of biotechnology research, application and commercialization in different sectors of the economy;
- the bioresource vision, focusing on processing and upgrading of biological raw materials, as well as on the establishment of new related value chains; and
- the bioecology vision, highlighting sustainability and focusing attention on ecological processes that allow for an improved use of energy and nutrients and promote biodiversity, including agricultural practices that avoid monocultures and soil degradation.

Similarly, Brunori (2013) identifies different 'clusters' of definitions, based on the keywords biotechnologies, biomass, and biochemical and biophysical processes as key criteria for identification.

This 'visions perspective' approach to the bioeconomy can be partly recognized in the strategy documents developed by countries, in which either one or the other prevail; indeed, they can be viewed as sometimes contrasting when they feed political processes with different orientations towards the bioeconomy. On the other hand, they are not totally competing views but rather approaches highlighting different complementary aspects of the bioeconomy, which, to some extent, can also support each other. In fact, the complex nature of the bioeconomy and its nature of 'master narrative' makes it possible to accommodate these different perspectives in one vision and to achieve political consensus, similar to what happened with the move towards the KBBE

of the EU agriculture research agenda in spite of rather different interpretations of what it meant (Levidow *et al.*, 2013).

The most frequent concept in the literature relates to an economy based on the conversion of biomass, for example, 'The bioeconomy can be defined as an economy based on the sustainable production and conversion of renewable biomass into a range of bio-based products, chemicals, and energy' (De Besi and McCormick, 2015). Or more focused towards objectives: 'In response to growing societal and environmental challenges, the concept of the bioeconomy has emerged in Europe, shifting society away from fossil fuels to utilizing renewable biological resources to meet food, feed, fuel and material needs' (Devaney *et al.*, 2017).

Other definitions highlight even different aspects, pointing to the interwoven nature of the bioeconomy and the need for disciplinary integration. For example, Peyron (2016) states that the bioeconomy can be 'understood as a systemic vision of the activities based on living materials' and that bioeconomy 'innovation is not conceivable without associating economic, commercial and even institutional considerations with biological knowledge and technological know-how'.

Brunori (2013) proposes building a definition of the bioeconomy on the concept of 'biovalue': 'It identifies the goal of the bioeconomy as the capacity to mobilize science to obtain high biovalue returns from low-cost living matter, for example organic waste, and includes the value of non-market goods associated with agriculture and food'. This also links with the process of building an economic definition of the bioeconomy and understanding its economic specificities as discussed more in detail in the next chapter.

Furthermore, other authors propose terminologies to distinguish different fields within the bioeconomy, such as the distinction between bioeconomy and 'biotechonomy' (Blumberga *et al.*, 2016), which is intended as 'the utilization of bio-resources by producing new value-added products that are demanded and therefore are competitive with already existing products in the market. Moreover, these value-added products are produced by using innovative biotechnology methods'.

This discussion clearly emphasizes a tension between the need for systemic concepts and a specification of the object of analysis seeking to capture the emerging functional issues embedded in the bioeconomy. In order to maintain a structured vision of the bioeconomy, this book addresses it first as an aggregate of sectors in this chapter, and then goes on to qualify both the underlying technical relationships and then the economic interpretations in the next chapter.

2.4 Relevance and Building Blocks of the Bioeconomy

This section describes the main characteristics of the various bioeconomy sectors, the relationships across sectors, and the related economic value and actors. The technologies involved and their economic representation are described in Chapter 3.

The bioeconomy can be roughly considered to account for about one-fifth of the turnover in industrialized countries. Precise data are available only for a few non-comparable cases. According to Ronzon *et al.* (2017), the bioeconomy employed approximately 18.4 million people in the EU in 2014 (8.6% of the EU labour force), generating around €2.2 trillion of turnover and 4.2% of EU GDP. Though the same statistics are not available for other areas of the world, bioeconomy sectors are already playing a key role in the economy of several large countries such as the US, India and Brazil. Sparse information and estimates are available for individual countries and partial shares of the bioeconomy, mostly locating it at a few percentage points of GDP (when excluding food and agriculture), but with noteworthy growth rates (Wesseler and Von Braun, 2017).

The share of bio-based products in world trade has increased from 10% in 2007 to 13% in 2014 (El-Chichakli *et al.*, 2016).

The building blocks of the bioeconomy can be identified in the various sectors of the bioeconomy, characterized by the use of biological resources. Some of these sectors are also defined by the type of technology they use (e.g. biotechnology) and others more by the type of resources they use, such as in the wastes sector. Food and agriculture remain the two main subsectors, yet the other components are growing

steadily. For example, in the EU about 20% of bioeconomy employment and 25% of turnover are now generated by non-food and non-agriculture bioeconomy industries (El-Chichakli *et al.*, 2016; Ronzon *et al.*, 2017). The biotechnology industry is estimated to be worth US$300 billion in revenue (Lokko *et al.*, 2017).

The total contribution of the bio-based products industry to the US economy in 2013 was US$369 billion with the employment of four million workers. Moreover, each job in the bio-based industry was responsible for generating 1.64 jobs in other sectors of the economy. Bio-based products accounted for 40% in the textile industry and around 4% in bio-based chemicals and enzymes, while it was still less than 1% in plastics, packaging and bottles (Golden *et al.*, 2015).

A simplified graphical view of the bioeconomy is provided in Fig. 2.1, which depicts the main components of the bioeconomy and their relationships.

A more practical way of representing the bioeconomy, due to its internal and external linkages, is by way of the matrix form used in Table 2.1. The individual blocks are addressed in turn in the text that follows.

2.4.1 Biological and non-biological natural resources

The bioeconomy is embedded in a system connected with other economic sectors and in ecosystems identified by the integration of natural non-biological resources and bio-resources. The distinction between biological resources and non-biological resources is on some level artificial, because in practice both are integrated in the natural environment and in ecosystems working. Yet here they are distinguished as they have a different connection with the bioeconomy. Natural biological resources are living organisms that can be found in nature. Many of them can be of interest to the bioeconomy. They provide biological material (biomass) used as feedstock for bio-based industries, food industries and energy, and are partly used for fishing, forestry and other activities directly using natural biological stocks. The other important contribution of

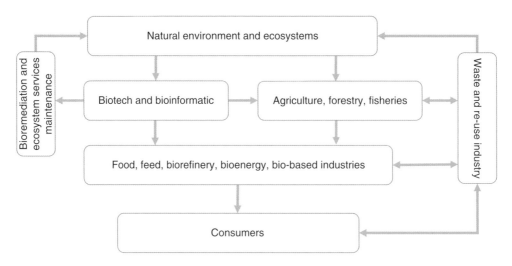

Fig. 2.1. A graphical representation of the bioeconomy (the arrows represent the main flows of biomass). Source: modified from Wesseler and Von Braun (2017), OECD (2009).

biological resources is genetic material that can be used by the biotechnology sector.

Non-biological resources are used as raw materials for a number of bioeconomy activities. Some non-biological resources, such as water, nutrients and solar energy are the basis for life and hence the basic support to the bioeconomy, which can be seen as being based on the transformation of light into chemical energy. This use of natural resources is one the most problematic effects of the bioeconomy on the outside world, and at the same time a potential limitation to its future growth. Some of the most limited resources are now land, water and natural mineral fertilizers.

Both biological and non-biological resources are also the receptors ('sink') of a number of emissions from the economy, including the bioeconomy, providing depollution services to the bioeconomy (as well as to the non-bio sectors).

In a broader sense, these contributions fit better under the concept of 'ecosystem services' (ES) (see next chapters). Ecosystems provide not only biological material, but also a much larger number of services supporting the bio-based sectors or used directly by consumers. Some papers estimate the value of ES in monetary units, allowing for a straightforward understanding of their role. A recent exercise in this direction is provided by Kubiszewski *et al.* (2017), estimating the economic importance of ES as being approximately twice the current world GDP.

2.4.2 Agriculture, forestry and fisheries

Agriculture is the breeding and cultivation of animals and plants for food, fibre, biofuel and medical and other products used for human life. Hunting, forestry, fisheries and aquaculture are usually associated with agriculture. However, there are differences between those activities focused on the harvesting of natural resources (hunting and fisheries) and those activities devoted to cultivating/breeding plants or animals.

Agriculture started about ten thousand years ago. Since its beginning, agriculture has undergone a number of technical changes, of which the most evident took place in the 20th century. Modern agronomy, plant and animal breeding, agrochemicals (pesticides and fertilizers) and mechanization, as well as, more recently, digital technologies, have transformed the sector over time. Technological developments have sharply increased yields from cultivation and reduced the need for inputs (especially labour), but at the same time have caused ecological damage and negative human health effects. More recently, the objectives of research and technology change have also changed with a greater focus on safety, the reduction of environmental effects and the reduction of resource use.

Agriculture is essential for the primary production of biomass for human use. Traditionally, agriculture has produced food directly, but the

Table 2.1. Flow of biological resources. The horizontal sector is a provider of the vertical sector

	Natural biological resources	Non-bio-resources	Biotechnologies	Agriculture, forestry, fisheries	Food industry	Bioenergy industry	Bio-based industry and biomanufacturing	Non-bio sectors	Consumers	Re-use industry	Bioremediation and ecosystem maintenance industry
Natural biological resources			Ecosystems services, genetic material, living organisms	Ecosystems services, living organisms	Ecosystems services, raw materials	Ecosystems services, raw materials	Ecosystems services, raw materials, feedstock	Ecosystems services	Ecosystems services, raw materials, (food, feed, fuel)	–	–
Non-biological resources			Raw materials	Raw materials	Raw materials	Raw materials	Raw materials	Raw materials	–	–	–
Biotechnologies	–	–		Plant variety, animal breeds,	Solutions, modified microorganisms	Modified micro-organisms	Modified living organisms	–	Products	Modified micro-organisms	Modified micro-organisms
Agriculture, forestry, fisheries	–	–	–		Primary materials	Raw materials for energy	Raw materials for manufacturing	–	Direct food and non-food	Waste/ by-products	Modified living organisms
Food industry	–	–	–	Waste directly used		Waste/ feedstock	Waste/feedstock	–	Food	Waste, wastewater	–
Bioenergy industry	–	–	Energy/fuel	Energy/fuel	Energy/fuel		Energy/fuel	Energy/fuel	Energy/fuel	Energy/fuel	Energy/fuel
Bio-based industry and Biomanufacturing	–	–	–	–	Raw materials	Feedstock		Bio-based products	Bio-based products	Waste/ by-products	–
Non-bio sectors	–	–	Non-bio material and services	Non-bio material and services	Non-bio material and services	Non-bio material and services	Non-bio material and services		Non-bio material and services	–	–
Consumers	–	–	–	–	–	Wastes	Wastes	Wastes		Wastes, wastewater	–

Continued

Table 2.1. Continued.

	Natural biological resources	Non-bio-resources	Biotechnologies	Agriculture, forestry, fisheries	Food industry	Bioenergy industry	Bio-based industry and biomanufacturing sectors	Non-bio sectors	Consumers	Re-use industry	Bioremediation and ecosystem maintenance industry
Re-use industry	–	–	Raw materials	Fertilizers, feed, other row materials	Raw materials	Feedstock	Feedstock	–	Food, materials		–
Bioremediation ecosystem maintenance industry	Restored and improved biological components of sites	Restored non-biological components of sites	–	–	–	–	–	–	–	–	

production of raw materials for the food and fibre industries has become increasingly important and is now much more important (in economic terms) than the production of final products. In addition, it is often the only sector providing some sort of livelihood to rural populations, especially in less developed countries. Accordingly, it is also important for social reasons.

Forestry is at the centre of attention in the bioeconomy. Traditionally producing wood and other services, it is now at the core of lignin-based biorefineries. A growing number of papers related to the bioeconomy are dedicated to forests.

The use of aquatic and marine resources is being reshaped by the bioeconomy; new vision terms, such as 'blue growth', and new concepts (biodiscovery, etc.) are emerging, bringing the sea and water resources to the forefront of the bioeconomy, and have even led to the coining of the term 'blue bioeconomy' (Børresen, 2017).

Agriculture and primary productions in general are also increasingly important for the potential availability of biomass from residues and by-products. Ronzon and Piotrowski (2017) estimate that EU agriculture produced 395 Mt of dry matter of primary agricultural residues in 2013. Of this, almost 300 Mt need to be left in the fields for the maintenance of ES and 29 Mt are collected for agricultural use. However, the remaining 66 Mt (about 16.7%) can be collected as feedstock for the bio-based and bioenergy industries.

2.4.3 Biotechnology industry

The biotechnology industry uses a variety of technologies to modify or utilize living systems and organisms for human purposes; there are several definitions of biotechnology, such as:

- the integration of natural sciences and organisms, cells, parts thereof, and molecular analogues for products and services (European Federation of Biotechnology); or
- any technological application that uses biological systems, living organisms or derivatives thereof, to make or modify products or processes for specific uses (United Nations, 1992).

In its wider concept related to the modification of living organisms, biotechnologies harken back to the domestication of animals, the cultivation of plants, and improvements through breeding programmes that employ artificial selection and hybridization. In the late 20th and early 21st centuries, biotechnology expanded to include new and diverse sciences such as genomics, recombinant gene techniques, applied immunology and the development of pharmaceutical therapies and diagnostic tests. Modern usage includes GE, cell and tissue culture technologies and often overlaps with related fields such as bioengineering, biomedical engineering, biomanufacturing, molecular engineering, and so on. Biotechnology techniques are still evolving rapidly (e.g. in the field of new breeding techniques).

Biotechnology is also connected to the biological sciences (animal cell culture, biochemistry, cell biology, embryology, genetics, microbiology, and molecular biology) as well as to knowledge and methods from outside the sphere of biology, including bioinformatics, bioprocess engineering, biorobotics and chemical engineering.

Biotechnology applications can be classified in the following fields:

- green biotechnology: is biotechnology applied to agricultural processes, e.g. the selection and domestication of plants via micropropagation and the designing of genetically modified plants;
- blue biotechnology: includes marine and aquatic applications of biotechnology;
- white biotechnology (industrial biotechnology): is biotechnology applied to industrial processes, e.g. the designing of an organism to produce a useful chemical; using enzymes as industrial catalysts to either produce valuable chemicals or destroy hazardous/polluting chemicals; or
- red biotechnology: is applied to medical processes, e.g. designing of organisms to produce antibiotics; the engineering of genetic cures through genetic manipulation.

In the context of the bioeconomy, the biotechnology industry is an independent sector producing final products for consumers. More importantly, it is a sector supporting other bio-based sectors at large with modified organisms and bio-based solutions. This is the case for agriculture, with the genetic modification of cultivated plants.

It uses living organisms taken from nature or from existing human processes and is hence linked to biodiversity on the one hand, and with the non-bio industries allowing for biotech processes on the other, such as data management in bioinformatics or non-bio materials.

The biotech sector is characterized by a strong structural duality. The commercial development of new products is mainly driven by multinational enterprises, whereas small and medium enterprises contribute primarily to innovation and technological development (Festel, 2015).

2.4.4 Food industry

The food industry is the sector that processes food. Food is any substance consumed to provide nutritional support for the body as it contains nutrients (such as carbohydrates, fats, proteins, vitamins). Food usually originates from plants or animals. It can be obtained directly from hunting/fishing and gathering (from natural biological resources), from agriculture or from the food industry. Today, the bulk of the food energy required by the ever-increasing population of the world is supplied by the food industry, which, in turn procures raw material from agriculture and fisheries. The food industry represents one of the most important manufacturing industries, and its importance is connected to the fact of addressing primary human needs.

Downstream, food is delivered to consumers through the important media of retailers and traders. Waste and by-product streams can be very important and high value, and can be reused in the food industry, used to feed (e.g. for livestock), used in agriculture or just go to waste treatment and disposed (e.g. through landfill).

Growing importance is attributed to the use of food waste/by-products into other bio-based sectors, in particular biorefinery, bioenergy and biowaste sectors.

2.4.5 Bioenergy industry

The bioenergy industry uses biomass either from natural biological resources, from agriculture and forestry or from waste, to generate different types of energy. Biomass is any organic material that has stored sunlight in the form of chemical energy. Some biomass can be used to produce energy directly, such as wood, wood waste, straw, manure, sugarcane, and many other by-products from a variety of agricultural processes.

Bioenergy industry is usually classified according to the different type of energy material it produces, of which the most noteworthy include:

- combustible solids;
- biogas;
- biofuel (ethanol); and
- electricity (potentially derived from one of the above).

The output is used by consumers or industries, including bio-based industries.

The potential of bioenergy is dependent upon the interplay with the energy system in general. Examples include the flow of electricity from diffuse biogas plants into the general electricity system, the blending of biodiesel into carburants and the option of using bio-methane in different forms as a substitute for natural gas.

2.4.6 Bio-based industry

The bio-based industry is the sector processing or producing bio-based products. Bio-based products were defined by the United States Secretary of Agriculture in the Farm Security and Rural Investment Act of 2002 as commercial or industrial products (other than food or feed) that are composed, in whole or in significant part, of biological products or renewable domestic agricultural materials (including plant, animal, and marine materials) or forestry materials or an intermediate feedstock. A broad review of over 200 studies summarizing the current situation and trends in Europe is available in Dammer *et al.* (2017).

Golden *et al.* (2015) identify the following components of the bio-based industry:

- agriculture and forestry (selected products);
- biorefining;
- bio-based chemicals;
- enzymes;

- bioplastic bottles and packaging;
- forest products; and
- textiles.

The bio-based industry is one of the building components of the bioeconomy and as such, it is undergoing profound evolution, but also experiencing ambiguities in the use of the term (sometimes merged with bioenergy for example).

A specific component of bio-based industries is biomanufacturing. Biomanufacturing

> is a type of manufacturing that utilizes biological systems, such as living microorganisms, resting cells, animal cells, plant cells, tissues, enzymes, or *in vitro* synthetic (enzymatic) systems, to produce commercially important biomolecules for use in the agricultural, food, material, energy, and pharmaceutical industries. (Zhang *et al.*, 2016)

In a way it incorporates bio-based features not only in the type of products, but also in the processes and, in particular, in its more advanced forms it is considered to be a very promising means of addressing various challenges currently faced by humankind.

A key component of the bioindustry is biorefinery. Biorefinery can be intended as an industry component, as a type of process or as a facility. In the latter sense, a biorefinery is a facility that allows biomass conversion into fuels (or power or heat) and value-added chemicals. In this way, biorefinery activities are both part of bio-based and bioenergy chains. This is also a key linkage in the future of bio-based processes and bioenergy, as several authors point out that one of the key strategies is the extraction of fine chemicals and high-value-added products connected to energy chains (Vaz, 2015) and that, while biofuels have been the focus of the first wave of the bioeconomy, it is essential that renewable chemicals are developed as well (Kircher, 2015).

Bioplastics, which could be defined as thermoplastic biopolymers that are either biodegradable or at least partly bio-based, are one of the fastest growing markets (Storz and Vorlop, 2013). The largest growth is expected for partly bio-based conventional plastics (especially for bio-PET).

Some downstream sectors are strongly moving towards bioeconomy-based solutions. For example, the construction industry is setting ambitious growth targets for wood-based construction, that is, tripling its market share by 2030 (Hurmekoski *et al.*, 2017).

2.4.7 Bioremediation and ecosystem maintenance industry

Bioremediation uses organisms to remove or neutralize pollutants from contaminated sites. According to the United States Environmental Protection Agency, bioremediation is a 'treatment that uses naturally occurring organisms to break down hazardous substances into less toxic or nontoxic substances'. Examples of bioremediation technologies include phytoremediation, bioventing, bioleaching, and so on. Bioremediation can be generally classified into *in situ* bioremediation, involving the treatment of contaminated material on-site, and *ex situ* bioremediation, involving the removal of contaminated material to be treated elsewhere.

Bioremediation is increasingly being integrated into other bioeconomy sectors, such as bioenergy production and carbon sequestration (Tripathi *et al.*, 2016).

Services for maintaining or improving ecosystems, such as those linked to restoration or regulation (e.g. water flows management, tree re-planting, etc.), is a growing industry.

2.4.8 Waste industry

Waste and by-product use are key issues related to the bioeconomy, especially in connection to its objectives to move towards a circular economy. Wastes are unusable or unwanted, sometimes potentially harmful, materials, while by-products are secondary products derived from a process aimed at the production of some primary material. The waste industry is often considered as a separate industry because legal constraints require waste management activities to be separate from the other processes they are connected to. All sectors of the bioeconomy produce some wastes that can be used as feedstocks in new processes. The waste valorization industry is largely relevant downstream from agriculture and the food industry and produces outputs that are essentially the same as those of the food industry, yet at the same time much more related to bio-based industries and bioenergy.

The waste industry can also be associated with consumer waste sources, such as organic urban wastes or wastewater (see, for example, Puyol *et al.* (2017) for wastewater).

2.4.9   Bio-based and non-bio-based

The bio-based and non-bio-based sectors interact with each other to some degree. In a number of cases, the merging of bio-based products into non-bio-based products is seen as progress towards integration into the bioeconomy and an opportunity to use well-established non-bio-based markets to launch the market of bio-based products, e.g. plastic blending with a share of bio-based plastic or in blended biodiesel. This happens when bio-based sources replace a component of non-bio-based materials. In other cases, such as biowastes, keeping the biological component separate from the non-biological component is at the basis of a suitable utilization of the downstream materials.

2.5   Bioeconomy and Biomass Flows

Since it is based on biological resources, the functioning of the bioeconomy system can be described through biomass flows.

Box 2.1. Quantifying biomass flows

Morrison and Golden (2015) quantify flows of bio-mass by agriculture and forestry and their destination. The majority of the global biomass flow comes from roundwood forestry, followed by cereals and vegetable. The main destination is food, followed by energy and feed. The Sankey bio-mass diagram for the EU is available online at https://ec.europa.eu/jrc/en/publication/biomass-flows-european-union-sankey-biomass-diagram-towards-cross-set-integration-biomass

---

The quantification of biomass flows is not straightforward, given the variety and heterogeneity of biomass sources. Actual and potential biomass availability is a major point of attention regarding its ability to meet industry needs for a growing bioeconomy. It also serves the purpose of better understating the ability of agriculture and the forestry sectors to meet future demand for biomass, as well as connections with resource use (Kalt, 2015).

Biomass needs and availability may have different dimensions, of which the most important can be summarized as follows. The quantity of biomass is the primary dimension. It can be measured in tonnes of biomass, or more rigorously in tonnes of dry matter; one method of measuring quantity for specific purposes involves using amounts of specific contents, such as carbon, energy, protein, sugar, etc.

Quality, in terms of composition and related technological characteristics, is the second issue; stability over time, constant characteristics and the fact that such characteristics are well-known to users are also relevant features for technological processes using biomass. For example, different characteristics of biomass may affect the working of biogas digestors; this topic may be even more important for biomass feedstocks intended for downstream industry in the field of biomaterials.

Location is also relevant for the use of biomass, given that it is usually a product with low value per unit of weight, hence incurring high (relative) transportation costs. For this reason, biomass production (especially for lower value uses) needs to be distributed close to its users, and biorefinery plant models are based on rigorous procurement planning involving short geographical distances.

Flexibility of use is also important. Lack of flexibility in use constrains the actual exploitation use of biomass to a specific use, and hence increases the economic cost due to risk of non-usability. One interesting concept is that of flexible feedstock.

Property rights on biomass, and resources needed for its production, are also key to determine availability and production potential, as well as the effects of its use. A link between biomass distribution and social issues is provided by Temper (2016).

For all these reasons, biomass flows are a key descriptor for the bioeconomy, but, at the same time, they are largely insufficient for a thorough analysis of the economic side of the bioeconomy, which requires a better understanding of the technology connecting different points of the flow and of the related economic values and mechanisms.

2.6 Bioeconomy, Wastes, By-products and Circularity

The degree of circularity is a key issue for the bioeconomy and its sustainability. The circular economy concept relates to the idea that the economy should rely less on external raw materials and more on the re-use of resources that are already in the system.

At present, the circularity of the overall economy (measured by the degree of circularity intended as the share of materials that flow back into the anthropic system) is rather limited. Haas *et al.* (2015) provide an estimation of the degree of circularity of the global and the EU-27 economy for 2005. They show that there is a global flow of roughly 4 Gt/yr (gigatonnes per year) of recycled waste materials; this is of moderate size compared to 62 Gt/yr of processed materials and 41 Gt/yr of outputs. The bioeconomy has a major role in the overall degree of circularity as it accounts for 19 Gt/yr with only a 3% (7% in the EU-27) degree of circularity. One of the reasons for this is that biomass is largely used for energy purposes (including food), which makes it non-recyclable.

In addition, biomass is related to overall circularity indirectly, depending on the production technologies and system considered; if biomass is produced sustainably (i.e. without damaging soil or water resources and without depleting ecological carbon stocks), it can be considered renewable and the emitted CO_2, as well as waste flows, can largely be recycled into new primary biomass within ecological cycles (Jordan *et al.*, 2007), hence making it possible to consider the biomass itself as within the circular flow. Closing the cycle for biomass-related industries hence involves closing the cycle of nutrients (nitrogen and phosphorus), reducing waste from production and consumption, and changing dietary patterns towards less meat-demanding diets (Viaggi, 2015).

While this holds true from a broad view of the sector, in a rather focused understanding of circularity the quantification of waste and by-products and the establishment of circular flows into re-use is one of the key issues in understanding the bioeconomy, in connection with the idea that the minimization of waste and the promotion of re-use is key to increasing the circularity of the bioeconomy (Cardoen *et al.*, 2015). Widely

speaking, there are two issues linked to circularity. The first is waste minimization and the second is waste re-use.

There are innumerable examples of former waste that is now usable as raw materials. Some general types include:

- the use of wastes and by-products, as such, due to their characteristics; this has been historically one of the main ways of closing cycles and avoiding dependency on external materials;
- the use for bioenergy production; this is a particularly important destination of wastes at present, but it is not considered to be the most rational one, as it implies the loss of valuable compounds;
- use of key components of waste feedstock after processing, e.g. banana leaf residue as raw materials for the production of high lignin content micro/nano fibres (Tarrés *et al.*, 2017) or orange peel used to produce compounds for beverages, snacks and fish foods (Vergamini *et al.*, 2015); and
- use of biowaste in biorefinery processes that destructure and use the main compounds of the waste (Venkata Mohan *et al.*, 2016).

Waste from food production and consumption play a very important role in the issue of circularity, in part because of the potential to address not only circularity issues, but also food security and affordability. For this reason, keeping food waste in the cycle of the food industry ensures a higher value added from a societal point of view. However, diversions to non-food production are often used, and sometimes required by law due to food quality and food safety considerations.

The amount of waste use at present is rather differentiated in different sectors due to various technological and organizational issues. Egelyng *et al.* (2016) illustrates several examples for Norway.

2.7 Limitations and Perspectives in Bioeconomy Statistics

Information remains a clear constraint for more focused analyses of the bioeconomy and for policy design, as estimates about the value of bioeconomy sectors remain limited (Ronzon *et al.*, 2017; Wesseler and Von Braun, 2017). The figures on the bioeconomy are very uncertain for three main reasons. First, the sector is changing quickly and updated information at a meaningful level of aggregation is difficult to obtain. Second, definitions of the components of the bioeconomy are still dissimilar in different countries and the basic scope of the bioeconomy is not the same everywhere. Third, current statistical systems do not classify the bioeconomy as a separate sector, but rather bioeconomy components are spread over several sectors coded in national statistics. For this reason, current estimates are based on analytical efforts to disentangle bio-based components from wider sectors (e.g. energy, materials, etc.). Examples of these efforts are available in Golden *et al.* (2015) for the US and Ronzon *et al.* (2017) for the EU.

Therefore, the main challenges that still need to be overcome are data availability and the difficulties that arise due to the fact that economic activities themselves cannot be unambiguously assigned to the bioeconomy or the non-bioeconomy (Efken *et al.*, 2016).

The feeling that current statistical classifications are not suitable for understanding the bioeconomy are corroborated by micro-level studies. For example, Ehrenfeld and Kropfhäußer (2017) provide an inventory of relevant actors in the three Central German states of Saxony, Saxony-Anhalt and Thuringia. They take an in-depth look at the different sectors, outline the industries involved, note the location and age of the enterprises and examine the distribution of important European industrial activity classification (NACE) codes. The results confirm the fact that established industry classifications are insufficient in identifying the plant-based bioeconomy population.

To fill this information gap, initiatives are under way, such as the Bioeconomy Observatory launched by the EU Commission (https://biobs. jrc.ec.europa.eu/). Other initiatives are ongoing in other areas of the world.

Target indicators usually include different aspects of the bioeconomy, focusing on two main areas. The first is the relevance in the economy represented, for example, by turnover, GDP and employment. The second is the innovation effort, represented, for example, by investment in research and patents produced (Wesseler and Von Braun, 2017).

2.8 Overview of Country-level Bioeconomy Strategies and Policies

The future bioeconomy is largely determined by the visions that countries and international bodies have of it and how it is embedded in policy action. The starting point for understanding strategies and policy formulation on the bioeconomy is arguably the publication of the policy agenda on the bioeconomy by the OECD in 2009 (OECD, 2009; Staffas *et al.*, 2013).

Currently, many countries are still in the process of building their bioeconomy strategies, while a number of policies are already in place for individual bioeconomy sectors (e.g. bioenergy). This section provides an overview of the strategies put in place by different countries and international bodies (e.g. OECD, EU). The main strategy types and policy instruments will be further discussed in Chapter 7. Some country strategies and features will be further described (briefly), based on the better documented or paradigmatic cases available, in the annex to this chapter.

Some papers provide overviews of country strategies (McCormick and Kautto, 2013; Staffas *et al.*, 2013; De Besi and McCormick, 2015; Meyer, 2017). Virgin and Morris (2017) provide a broad overview of bioeconomy development in Europe and Africa.

The German Bioeconomy Council has developed two key overviews of bioeconomy policies (German Bioeconomy Council, 2015a, 2015b), one focusing on G7 countries and the other on the situation worldwide. All of the G7 countries have developed a bioeconomy strategy.

The world study identifies 45 countries with bioeconomy strategies or policies having significant impacts on the bioeconomy.

Of the strategies related to bioeconomy development worldwide, only 12% have a mainly holistic bioeconomy approach, notably including four G7 countries (including the US and the EU). Twenty-eight per cent focus on high-tech bioeconomy solutions, followed by 18% with a focus on research and innovation and 16% that have a mostly bioenergy focus. Both bio-based economy and regional bioeconomy have a 7% share, while the blue economy and the green economy have 6% each.

It is worth noting that only 18 out of 45 use the term 'bioeconomy' and only 13 provide a definition of the term. The understanding of country strategies is indeed often affected by the unclear or outright absence of a definition of the terms 'bioeconomy' and 'bio-based economy'; this point is discussed in detail in Staffas *et al.* (2013).

Different areas of the world do not show a clear tendency in one direction or another, nor is there always a clear connection with country features. However, strategies to address bioeconomy issues are very differentiated across countries. This differentiation is justified by the different economic and resource characteristics of each country (Staffas *et al.*, 2013; German Bioeconomy Council, 2015b). Strategies range from the preoccupation with securing access to raw materials to a true transformation of the innovation and ecological system of the country.

In presenting an early overview of bioeconomy strategies in the EU, US, Canada, Sweden, Finland, Germany and Australia, Staffas *et al.* (2013) found that emphasis was often placed on enhancing the economy of the country and providing new employment and business possibilities. On the contrary, the aspects of sustainability and resource availability were only addressed to a limited extent.

Over time the scope of the strategies has become wider. In most countries, motivations rest now in the need to meet grand societal challenges, such as, mainly, climate change mitigation, food security and sustainable resource management (German Bioeconomy Council, 2015b). More recently, the sustainable development goals (SDG) have been used as the starting point of these strategies. Among these, the horizontal issue of ecological balance is widely considered as an objective of the bioeconomy. On the contrary, only China accounts for recent needs brought by mega-urbanization (such as food security).

Meyer's (2017) review selected bioeconomy strategies in the EU, Germany, OECD, Sweden and the US with regard to their context, visions and guiding implementation principles. The author identifies two main visions:

- a biotechnology-centred vision, in which life science and biotechnology are at the core of the bioeconomy as drivers of innovation; and
- a transformation-centred vision, where the focus is on the shift to a bio-based economy.

To some extent, political motivations to promote the bioeconomy and the strategy design depend on countries' resource endowment, specialization and economic development track. Three main pathways have been identified (German Bioeconomy Council, 2015b):

- countries with high biomass availability and high oil imports focus on higher independence and seek to increase value added from their biological resources;
- industrialized countries with a significant share of rural population and primary industry jobs also see bioeconomy development as a way of fostering rural development and social inclusion;
- industrialized countries with fewer primary resources and a smaller share of primary industry focus more on an industrialization of biology and creating value added from bioscience.

Among G7 countries, three main approaches can be found (German Bioeconomy Council, 2015a), the first, pursued by the US and Canada, is based on the large amount of agriculture and forest resources that have been already largely used in the past. In these countries, the focus is on exploitation of these resources through platform chemicals and bioenergy, e.g. through wood pellets, bioethanol and next-generation biofuels. Both countries have invested in biotechnologies, including industrial biotechnologies (conversion technologies). This also includes the health sector, mostly excluded elsewhere.

A second group of countries includes areas with strong industrial structures but poor in natural resources. In these cases, the bioeconomy is viewed more for its innovation potential and as a potential boost for industrial renaissance. These include Germany, Japan, France and the EU as a whole. The focus is mainly on replacing fossil fuels and reducing CO_2 emissions to combat climate change, as well as achieving economic and technological advantages from new ways of processing biomass. Lacking in basic biomass resources, wastes and by-products play an important role here. Moreover, partnerships with other parts of the world, especially emerging countries that are rich in biomass, are important in these strategies.

Another group, basically the United Kingdom (UK) among G7 countries, sees the bioeconomy as a strategy to further develop their excellent services and bioscience research towards strengthening their competitiveness in high-value industries.

The need to increase research, development and demonstration activities in the area of the bioeconomy is a key point in all strategies (Staffas *et al.*, 2013).

2.9 Outlook

The bioeconomy is a complex system that has developed from some of the most basic and ancient human activities to a high-tech and sophisticated sector at the core of the sustainability of the current economic system. The representation of the bioeconomy as a sum of sectors is still somehow artificial as these sectors are still institutionally separated from each other; the same applies for their statistical information. However, the many traditional sectors of the bioeconomy are interconnected through different pathways and the most evident feature of bioeconomy development is in the direction of emphasizing these interconnections. In a way, the bioeconomy is already a consistent and internally linked sector – much more than it appears from research and statistics. In addition, some originally new components such as biotechnology, biorefinery and biomaterial production characterize the sector.

Different parts of the bioeconomy have different trends and statuses. Some are old and well-established mature sectors, such as agriculture; others are emerging by virtue of new technological possibilities, such as those in the bio-based industries. The sector is, overall, growing in importance, especially in its more advanced and innovative sub-sectors (such as the bio-based industry), while the currently technological processes are moving in the direction of making the internal linkages stronger, more varied and complex, as well as changing over time due to changes in markets and technology, as research is allowing for more and more interconnections and exploiting more opportunities. The result of this will be, among others, greater possibilities, differentiated pathways towards final products and hence a higher potential flexibility and hopefully resilience of the system. This will also require higher management capacity, flow monitoring and traceability.

The approach to the bioeconomy is highly differentiated among countries in relation to their resources, but also in terms of their strategies and pathways for the future. The future will depend on a number of factors regarding driving forces and demand scenarios (that are better treated in Chapter 5), but also on the possibilities provided by technology and the mechanisms put in place for its realization.

A basic issue for understanding and describing the bioeconomy is the need for improved data availability. This is frequently an issue for new sectors. Several authors emphasize that limitations in available information is also limiting the certainty of projections and all papers attempting scenario analysis and outlooks basically acknowledge such limitations in their estimates. This points at the general issue of uncertainty, that will be a pervasive topic in the following chapters.

2.10 Annex: Selected Country Examples

2.10.1 Brazil

Brazil is at the forefront of a large process of bioeconomy development in the whole of Latin America (Sasson and Malpica, 2017). In fact, Brazil is one of the largest producers of agro-industrial biomass in the world, which it uses for several purposes, and most notably for bioenergy (Vaz Jr., 2014). Bioenergy is a very important part of Brazil's bioeconomy strategy with renewable bioenergy accounting for about 24% of total country energy production and 5% of the world energy production (Dias and De Carvalho, 2017). The first biotech programmes were developed in Brazil in the 1980s; a survey in 2011 revealed that 40% of the biotech companies worked in the health sector and 10% in agriculture biotech.

Chemical production has a high potential to add value to a vegetable biomass chain, due to the importance of the conventional chemical industry and fine chemical industry in the economy of the country (Vaz Jr., 2014). Derived compounds can be used as building blocks, intermediaries of synthesis and specialties. Given the current bioenergy strategy of Brazil, the connection between residual biomass from the biodiesel and bioethanol industries and chemical production is of strategic interest (Vaz, 2015).

2.10.2 China

A recent overview of bioeconomy development in China is provided by Wang *et al.* (2017). China, being the largest world market due to its 1.4 billion population and the fastest modernizing economy in the world, as well as a country with the most evident environmental and climate challenges, is of key interest for the bioeconomy. Indeed, China was also the first Asian country to establish its own national bioeconomy initiative (Kamal and Dir, 2015) and the development of the bioeconomy is presently supported by both government and large industry.

China is also a key country in the fight against climate change, which can be attained and balanced by a clear bioeconomy strategy. As in many Asian countries, it is expected that the bioeconomy will contribute to make it easier, achieving climate change reduction targets even if this will have a very high cost for these countries (Lee, 2016).

2.10.3 European Union

The Communication on the bioeconomy in the EU was released in 2012. The bioeconomy is seen as offering significant opportunities for the realization of a competitive, circular and sustainable economy with a sound industrial base that is less dependent on fossil carbon and at the same time contributes to climate change mitigation. The EU has invested a good deal in research and innovation in this field and the European Commission is committed to maintaining a leading role in European bioeconomy strategies (Bell *et al.*, 2017).

Scarlat *et al.* (2015) estimate the current bioeconomy market at about €2.4 billion (including agriculture, food and beverage, agro-industrial products, fisheries and aquaculture, forestry, wood-based industry, biochemicals, enzymes, biopharmaceuticals, biofuels and bioenergy), employing 22 million people and using about 2 Gt of biomass. Similar, though updated, estimates are given by Ronzon *et al.* (2017). The European bioeconomy relies on a number of well-established traditional bio-based industries, including agriculture, food, feed, fibre and forest-based industries. New sectors, such as biomaterials and green chemistry, are being developed. Given the EU

environment (internally rather heterogeneous), the transition toward the bioeconomy will depend not only on advancements in technology and increased cost-effectiveness, but also on the sustainable availability of biomass. In Europe, the demand for biomass for the whole bioeconomy is increasing year by year. The EU is already a net importer of biomass for bioenergy, and imports could be even more relevant in the near future (Sánchez *et al.*, 2016).

Within Europe, several strategies have been produced by Member Countries, reflecting different perspectives and visions for the transition to the bioeconomy. Twelve country strategies in Europe are analysed by De Besi and McCormick (2015), showing that a common direction for the bioeconomy is developing in Europe. The strategies highlight the important role of the regional level in facilitating collaboration between industries and research institutions. The analysis also highlights the need for the development of a European bio-based product market to boost bioeconomy expansion.

McCormick and Kautto (2013) provided an overview of the bioeconomy in Europe noting that different actors tend to use different definitions of the bioeconomy, but that there are also similarities, such as the emphasis on economic output and a broad, cross-sectoral focus. The paper also finds that, together with benefits and opportunities, significant risks and trade-offs are also identified.

Several works provide analyses and projections related to bioeconomy sub-sectors in Europe. Among others, Schipfer *et al.* (2017) provide projections on the use of advanced biomaterials in Europe. According to this study, advanced biomaterials could reach a share of between 4% and 11%.

2.10.4 Finland

Due to its natural endowment of forest resources, forest bioeconomy is key for Finland. Forest bioeconomy has already shown a remarkable potential for economic growth, in the direction of both industrial uses of wood resources and energy production (Karttunen *et al.*, 2016). This focus has pushed research toward an analysis and simulation of forest growth and management, as well as to investigate their connection with industry and energy development through prices and costs. An analysis of impacts on the overall economy is also an important target for the future, including through modelling exercises (Karttunen *et al.*, 2016).

2.10.5 Germany

Germany is at the forefront of the bioeconomy in Europe. The German Federal Government was one of the first worldwide to put the bioeconomy on its research policy agenda by adopting the 'National Research Strategy BioEconomy 2030' in 2010, which was followed by a national policy strategy two years later (Schütte, 2017).

The German approach recognizes that 'biologization' as the guiding principle of the bioeconomy can achieve a fundamental change in industry. In addition, it emphasizes that the targeted use of biological resources by industry can help to successfully reconcile ecology and economy (Schütte, 2017).

The bioeconomy in Germany in 2010 was estimated to account altogether for about five million employees (10% of all employees) and €140 billion (6% of gross national product), a strong increase compared to the year 2002 (Efken *et al.*, 2016).

Since the initial strategy was launched, Germany has successfully implemented several measures to establish the bioeconomy and is continuing on this path. The current aim is to strengthen the process towards a sustainable economy, with special importance attributed to technological innovation. It is emphasized that the bioeconomy research policy will need to be better targeted towards the achievement of the SDGs (Schütte, 2017). International cooperation is also important and indeed Germany is one of the foremost promoters of worldwide initiatives in the field of bioeconomics.

Meyer (2017) studied selected bioeconomy strategies in Germany and compared them with the EU, OECD, Sweden and US. The guiding principle of strategy implementation and biomass use is sustainability, supplemented by priority for food, prevention of land use conflicts, priority for residual and waste biomass, cascading and coupled use and consideration of ecological and socioeconomic impacts.

Projections for biomass prices in Germany emphasize the projected increase in biomass costs due to competition for biomass (including competition from the food industry) and how this potentially inhibits the growth of the bioeconomy (Millinger and Thrän, 2016). This also hinders investment in research and development in bioenergy as compared to, for example, solar energy.

A multiregion input–output model to monitor the advances of the bioeconomy in Germany is presented in Budzinski *et al.* (2017).

2.10.6 India

Due to the amount and variety of biomass, India is expected to be of particular importance for bioeconomy development. Shrivastava *et al.* (2016) estimated that the main biomass sources were from sugar crops (117.4 Mt), followed by oil crops (97.3 Mt) and starch crops (29.7 Mt). The biodegradable fraction of waste was estimated at 25.5 Mt per year.

The budget allotted to bioeconomy projects was in the range of €350 million (Shrivastava *et al.*, 2016).

Residues and waste are an important component of the Indian bioeconomy. Cardoen *et al.*, (2015) carried out an assessment of: (i) the residues generated in the field or on the farm; (ii) waste generated by post-harvest losses; and (iii) by-products from the processing of agricultural produce (e.g. sugarcane biogases). The relevance of waste for the country and the technologies needed to allow wastes to contribute to a circular bioeconomy in India is also discussed in Venkata Mohan *et al.* (2017).

India is expected to be one of the countries with the largest pharmaceutical industry development (Lee, 2016). Furthermore, India is expected to be the largest and most reactive market for genetically modified products in Asia (Lee, 2016).

Ahn *et al.*, (2012) assessed the priorities, capabilities and competitiveness of the emerging bioeconomy in India. The authors identified access to talent and access to funding as the main capability needs and noted that several actions are needed, such as building and coordinating infrastructure, accelerating technology and capital flows, public–private sector collaboration to enable biotechnology start-ups, partnering between academia and government to accelerate technology transfer, and seeking international investment and alliances by companies.

2.10.7 Malaysia

Malaysia is considered to have one of the most competitive biotechnology sector in the Asia–Pacific region (Kamal and Dir, 2015). The country has taken steps in promoting the bioeconomy, of which the most significant is the Bioeconomy Transformation Programme, launched in October 2012 (the second country in Asia, after China). The programme aims to bring Malaysia into a high-income country status by 2020 by focusing on bio-based industries. The sector has been identified as having enormous potential to further develop the country due to the abundance and variety of natural resources available (Sadhukhan et al., 2016). Policies are in place to support the bioeconomy and especially the development of the biotech industry. See Arujanan and Singaram (2017) for a critical analysis.

The country has also set the ambitious goal of voluntarily reducing emission intensity by 40% with a major role to be played by agriculture and forest resources.

According to Sadhukhan et al. (2016), the palm oil industry will continue to play a major role in contributing to gross national income in Malaysia. Businesses tend to focus on products that are low-risk and enjoy subsidies (e.g. bioenergy and biogas). The authors report expectations that the bioeconomy will give a major contribution to the replacement of fossil resources especially through the development of bio-based products, food and pharmaceutical ingredients, fine, specialty and platform chemicals, polymers, biofuels and bioenergy. The process integration through innovative biorefinery configurations is also under way.

2.10.8 Russian Federation

In Russia, attention to a bioeconomy based on renewable biomass resources and pushed by sustainable technologies is driven by environmental and resource issues, such as the depletion of mineral resources, climate change, population growth and pollution.

A study carried out in 2011–2013 found that the biotech research and development (R&D) sector, although having high potential, was lagging behind the EU and US. Promising sectors in which efforts should be focused include high-performance genomics and post-genomics research platforms, systems and structural biology, microbial metabolic engineering, plant biotechnology and microbial strains and consortia for the development of symbiotic plant–microbial communities (Grebenyuk and Ravin, 2017). A review of recent initiatives at the federal and local level in the field of biotechnology is provided in Osmakova et al. (2017).

A critical issue is the provision of renewable resources for a number of key sectors, such as medical, food, and chemical industries, agriculture, ecology, the maritime industry and forestry. A critical point is the commercialization of new biotech technologies, which in turn needs strong governmental support through a range of instruments, including intellectual property rights management. In addition, innovative infrastructures and training are needed (Kasatovaa et al., 2016).

2.10.9 Spain

Spain launched its own strategy on bioeconomy in January 2016 aimed at boosting a bioeconomy based on the sustainable and efficient production and use of biological resources within a circular economy approach (Lainez et al., 2017). The targeted sectors are food, agriculture and forestry. A key issue is that these sectors are conditioned by water availability. As with other countries with biomass availability limitations, it also includes a focus on industrial use and the valorization of wastes and residues. The strategy also puts a focus on rural and coastal development, and, in this direction, it is supported by commitments by both research and innovation, policy making and regional structures (Lainez et al., 2017).

As a pilot application for the EU, Cardenete et al. (2014) developed a social accounting matrix for Spain, with a highly disaggregated agricultural account for the year 2000 and measured linkages, key sectors and employment multipliers of the Spanish agri-food and other bio-based accounts. The production of bioenergy appears to be the main key sector related to agri-food and other bio-based accounts, whereas livestock and related bio-based products have the most significant effects on the whole economy.

2.10.10   United States

The US bioeconomy is largely focused on conversion of biomass into energy, fuels and products. Scenarios for the US are given by Rogers *et al.* (2017), reporting the assessment of the size and benefits of the Billion Ton Bioeconomy, a strategy to enable a sustainable market for producing and converting a billion tonnes of US biomass to bio-based energy, fuels and products by 2030. In 2014, the utilized biomass was estimated in 365 Mt, which would displace approximately 2.4% of fossil energy consumption and avoid 116 Mt of CO_2-equivalent emissions. The effect could increase up to three times these figures by 2030. Again in 2014, bio-based activities were estimated to have directly generated more than US$48 billion in revenue and 285,000 jobs. The authors estimate that achieving a Billion Ton Bioeconomy could expand direct bioeconomy revenue by a factor of five by 2030.

This strategy would require developing integrated systems, supply chains and infrastructure to efficiently grow, harvest, transport and convert large quantities of biomass in a sustainable way (Rogers *et al.*, 2017).

Another study shows the size and significance of logistical efforts in the development of the bioeconomy (Ebadian *et al.*, 2017). With reference to the only corn stover bioeconomy linked to biorefinery, the authors estimate a required workforce to run the logistics operations of 50 000 units in the US.

References

Aguilar, A., Magnien, E. and Thomas, D. (2013) Thirty years of European biotechnology programmes: From biomolecular engineering to the bioeconomy. *New Biotechnology* 30(5), 410–425. doi: 10.1016/j.nbt.2012.11.014.

Ahn, M.J., Hajela, A. and Akbar, M. (2012) High technology in emerging markets. Building biotechnology clusters, capabilities and competitiveness in India. *Asia-Pacific Journal of Business Administration* 4(1), 23–41. doi: 10.1108/17574321211207953.

Arujanan, M. and Singaram, M. (2017) The biotechnology and bioeconomy landscape in Malaysia. *New Biotechnology* 40(Pt A), 52–59. doi: 10.1016/j.nbt.2017.06.004.

Bell, J., Paula, L., Dodd, T., Németh, S., Nanou, C., Mega, V. and Campos, P. (2017) EU ambition to build the world's leading bioeconomy – Uncertain times demand innovative and sustainable solutions. *New Biotechnology* 40(Pt A), 25–30. doi: 10.1016/j.nbt.2017.06.010.

De Besi, M. and McCormick, K. (2015) Towards a bioeconomy in Europe: National, regional and industrial strategies. *Sustainability* 7(8), 10461–10478. doi: 10.3390/su70810461.

Blumberga, D., Muizniece, I., Blumberga, A. and Baranenko, D. (2016) Biotechonomy framework for bioenergy use. *Energy Procedia* 95, 76–80. doi: 10.1016/j.egypro.2016.09.025.

Børresen, T. (2017) Blue bioeconomy. *Journal of Aquatic Food Product Technology* 26(2), 139. doi: 10.1080/10498850.2017.1287477.

Brunori, G. (2013) Biomass, biovalue and sustainability: Some thoughts on the definition of the bioeconomy. *EuroChoices* 12(1), 48–52. doi: 10.1111/1746-692X.12020.

Budzinski, M., Bezama, A. and Thrän, D. (2017) Monitoring the progress towards bioeconomy using multiregional input-output analysis: The example of wood use in Germany. *Journal of Cleaner Production* 161, 1–11. doi: 10.1016/j.jclepro.2017.05.090.

Bugge, M.M., Hansen, T. and Klitkou, A. (2016) What is the bioeconomy? A review of the literature. *Sustainability* 8(7), 691. doi: 10.3390/su8070691.

Burns, C., Higson, A. and Hodgson, E. (2016) Five recommendations to kick-start bioeconomy innovation in the UK. *Biofuels, Bioproducts and Biorefining* 10(1), 12–16. doi: 10.1002/bbb.1633.

Cardenete, M.A., Boulanger, P., Del Carmen Delgado, M., Ferrari, E. and M'Barek, R. (2014) Agri-food and biobased analysis in the Spanish economy using a key sector approach. *Review of Urban and Regional Development Studies* 26(2), 112–134. doi: 10.1111/rurd.12022.

Cardoen, D., Joshi, P., Diels, L., Sarma, P. M. and Pant, D. (2015) Agriculture biomass in India: Part 2. Post-harvest losses, cost and environmental impacts. *Resources, Conservation and Recycling* 101, 143–153. doi: 10.1016/j.resconrec.2015.06.002.

Chapotin, S.M. and Wolt, J.D. (2007) Genetically modified crops for the bioeconomy: Meeting public and regulatory expectations. *Transgenic Research* 16(6), 675–688. doi: 10.1007/s11248-007-9122-y.

Christensen, S. (1996) Optimal management of the Iceland–Greenland transboundary cod stock. *Journal of Northwest Atlantic Fishery Science* 19, 21–29.

Dale, B.E. (2007) Designing a new industry for sustainability: Life cycle analysis for the emerging bioeconomy. In *ACS National Meeting Book of Abstracts*. Presented at the 233rd ACS National Meeting, 25–29 March, Chicago, Illinois, USA.

Dammer, L., Carus, M., Iffland, K., Piotrowski, S., Sarmento, L., Chinthapalli, R. and Raschka, A. (2017) Study on current situation and trends of the bio-based industries in Europe. Final report. Available at https://www.bbi-europe.eu/sites/default/files/bbiju-pilotstudy.pdf (accessed 7 July 2017).

Davies, G. (2006) Patterning the geographies of organ transplantation: Corporeality, generosity and justice. *Transactions of the Institute of British Geographers* 31(3), 257–271. doi: 10.1111/j.1475-5661.2006.00222.x.

Devaney, L., Henchion, M. and Regan, A. (2017) Good governance in the bioeconomy. *EuroChoices* 16(2), 41–46. doi: 10.1111/1746-692X.12141.

Dias, R.F. and De Carvalho, C.A.A. (2017) Bioeconomia no Brasil e no mundo: panorama atual e perspectivas. *Revista Virtual de Quimica* 9(1), 410–430. doi: 10.21577/1984-6835.20170023.

Duchesne, L.C. and Wetzel, S. (2003) The bioeconomy and the forestry sector: Changing markets and new opportunities. *Forestry Chronicle* 79(5), 860–864.

Ebadian, M., Sokhansanj, S. and Webb, E. (2017) Estimating the required logistical resources to support the development of a sustainable corn stover bioeconomy in the USA. *Biofuels, Bioproducts and Biorefining* 11(1), 129–149. doi: 10.1002/bbb.1736.

Efken, J., Dirksmeyer, W., Kreins, P. and Knecht, M. (2016) Measuring the importance of the bioeconomy in Germany: Concept and illustration. *NJAS - Wageningen Journal of Life Sciences* 77, 9–17. doi: 10.1016/j.njas.2016.03.008.

Egelyng, H., Romsdal, A., Hansen, H.O., Slizyte, R., Carvajal, A.K., Jouvenot, L., Hebrok, M., Honkapää, K., Wold, J.P., Seljåsen, R. and Aursand, M. (2016) Cascading Norwegian co-streams for bioeconomic transition. *Journal of Cleaner Production* 172, 3864–3873. doi: 10.1016/j.jclepro.2017.05.099.

Ehrenfeld, W. and Kropfhäußer, F. (2017) Plant-based bioeconomy in central Germany: A mapping of actors, industries and places. *Technology Analysis and Strategic Management* 29(5), 1–14. doi: 10.1080/09537325.2016.1140135.

El-Chichakli, B., von Braun, J., Lang, C., Barben, D. and Philp, J. (2016) Five cornerstones of a global bioeconomy. *Nature* 535, 221–223.

European Commission (2012) Innovating for Sustainable Growth: a Bioeconomy for Europe. Available at https://ec.europa.eu/research/bioeconomy/pdf/official-strategy_en.pdf (accessed on 30 December 2017).

Festel, G. (2015) Technology transfer models based on academic spin-offs within the industrial biotechnology sector. *International Journal of Innovation Management* 19(4), 1550031. doi: 10.1142/S1363919615500310.

German Bioeconomy Council (2015a) Bioeconomy policy (Part I) Synopsis and analysis of strategies in the G7. Available at biooekonomierat.de/fileadmin/international/Bioeconomy-Policy_Part-I.pdf (accessed on 16 November 2015).

German Bioeconomy Council (2015b) Bioeconomy policy (Part II) Synopsis of National strategies around the world. Available at biooekonomierat.de/fileadmin/Publikationen/berichte/Bioeconomy-Policy_Part-II.pdf (accessed on 16 November 2015).

Golden, J.S., Handfield, R.B., Daystar, J. and McConnell, T. E. (2015) An economic impact analysis of the U.S. biobased products industry: A report to the Congress of the United States of America. Available at https://www.biopreferred.gov/BPResources/files/EconomicReport_6_12_2015.pdf (accessed 12 March 2017).

Golembiewski, B., Sick, N. and Leker, J. (2013) Agriculture and energy industry in the setting of an emerging bioeconomy: Are there any signs of convergence on the horizon in 2013? In *Proceedings of PICMET 2013: Technology Management in the IT-Driven Services*. Presented at PICMET '13 Conference 'Technology Management in the IT-Driven Services. 28 July – 1 August, San Jose, California, USA.

Grebenyuk, A. and Ravin, N. (2017) The long-term development of Russian biotech sector. *Foresight* 19(5), 491–500. doi: 10.1108/FS-06-2016-0024.

Haas, W., Krausmann, F., Wiedenhofer, D. and Heinz, M. (2015) How circular is the global economy?: An assessment of material flows, waste production, and recycling in the European Union and the world in 2005. *Journal of Industrial Ecology* 19(5), 765–777. doi: 10.1111/jiec.12244.

Hurmekoski, E., Pykäläinen, J. and Hetemäki, L. (2017) Long-term targets for green building: Explorative Delphi backcasting study on wood-frame multi-story construction in Finland. *Journal of Cleaner Production* 172, 3644–3654. doi: 10.1016/j.jclepro.2017.08.031.

Jordan, N., Boody, G., Broussard, W., Glover, J.D., Keeney, D., McCown, B.H., McIsaac, G., Muller, M., Murray, H., Neal, J., Pansing, C., Turner, R.E., Warner, K. and Wyse, D. (2007) Sustainable development of the agricultural bio-economy. *Science* 316(5831), 1570–1571. doi: 10.1126/science.1141700.

Kalt, G. (2015) Biomass streams in Austria: Drawing a complete picture of biogenic material flows within the national economy. *Resources, Conservation and Recycling* 95, 100–111. doi: 10.1016/j.resconrec.2014.12.006.

Kamal, N. and Dir, Z. C. (2015) Accelerating the growth of the bioeconomy in Malaysia. *Journal of Commercial Biotechnology* 21(2), 101–112. doi: 10.5912/jcb686.

Karttunen, K., Ahtikoski, A., Hynynen, J., Salminen, H. and Ranta, T. (2016) Impact of forest management decision making on forest biomass supply in regional level of Finland. In *European Biomass Conference and Exhibition Proceedings*, pp. 194–199. doi: 10.5071/24thEUBCE2016-1BV.4.17.

Kasatovaa, A.A., Vagizovaa, V.I. and Tufetulova, A.M. (2016) Bioeconomy's potential for development and commercialization opportunities for biosphere projects in Russia. *Academy of Strategic Management Journal*, 15(Special Issue 1), 210–217.

Kircher, M. (2015) Sustainability of biofuels and renewable chemicals production from biomass. *Current Opinion in Chemical Biology* 29, 26–31. doi: 10.1016/j.cbpa.2015.07.010.

Kubiszewski, I., Costanza, R., Anderson, S. and Sutton, P. (2017) The future value of ecosystem services: Global scenarios and national implications. *Ecosystem Services* 26, 289–301. doi: 10.1016/j.ecoser.2017.05.004.

Lainez, M., González, J. M., Aguilar, A. and Vela, C. (2017) Spanish strategy on bioeconomy: Towards a knowledge based sustainable innovation. *New Biotechnology* 40(Pt A):87–95. doi: 10.1016/j.nbt.2017.05.006.

Lee, D.-H. (2016) Bio-based economies in Asia: Economic analysis of development of bio-based industry in China, India, Japan, Korea, Malaysia and Taiwan. *International Journal of Hydrogen Energy* 41(7), 4333–4346. doi: 10.1016/j.ijhydene.2015.10.048.

Lehtonen, M. (2004) The environmental–social interface of sustainable development: Capabilities, social capital, institutions. *Ecological Economics* 49(2), 199–214. doi: 10.1016/j.ecolecon.2004.03.019.

Levidow, L., Birch, K. and Papaioannou, T. (2013) Divergent paradigms of European agro-food innovation: The knowledge-based bio-economy (KBBE) as an R&D agenda. *Science Technology and Human Values* 38(1), 94–125. doi: 10.1177/0162243912438143.

Lokko, Y., Heijde, M., Schebesta, K., Scholtès, P., Van Montagu, M. and Giacca, M. (2017) Biotechnology and the bioeconomy: Towards inclusive and sustainable industrial development. *New Biotechnology* 40(Pt A), 5–10. doi: 10.1016/j.nbt.2017.06.005.

McCormick, K. and Kautto, N. (2013) The bioeconomy in Europe: An overview. *Sustainability* 5(6), 2589–2608. doi: 10.3390/su5062589.

Meyer, R. (2017) Bioeconomy strategies: Contexts, visions, guiding implementation principles and resulting debates. *Sustainability* 9(6), 1031. doi: 10.3390/su9061031.

Millinger, M. and Thrän, D. (2016) Biomass price developments inhibit biofuel investments and research in Germany: The crucial future role of high yields. *Journal of Cleaner Production* 172, 1654–1663. doi: 10.1016/j.jclepro.2016.11.175.

Morrison, B. and Golden, J.S. (2015) An empirical analysis of the industrial bioeconomy: Implications for renewable resources and the environment. *BioResources* 10(3), 4411–4440. doi: 10.15376/biores.10.3.4411-4440.

OECD (2009) The Bioeconomy to 2030: Designing a Policy Agenda. Main Findings and Policy Conclusions. Organisation for Economic Co-operation and Development Paris.

Osmakova, A., Kirpichnikov, M. and Popov, V. (2017) Recent biotechnology developments and trends in the Russian Federation. *New Biotechnology* 40(Pt A), 76–81. doi: 10.1016/j.nbt.2017.06.001.

Patermann, C. and Aguilar, A. (2017) The origins of the bioeconomy in the European Union. *New Biotechnology* 40(Pt A), 20–24. doi: 10.1016/j.nbt.2017.04.002.

Peyron, J.-L. (2016) The bioeconomy and innovations in the forestry and wood sector. Bioéconomie et innovations dans la filière forêt-bois. *Revue Forestiere Francaise* 68(2), 107–114. doi: 10.4267/2042/61858.

Puyol, D., Batstone, D. J., Hülsen, T., Astals, S., Peces, M. and Krömer, J. O. (2017) Resource recovery from wastewater by biological technologies: Opportunities, challenges, and prospects. *Frontiers in Microbiology* 7, 2106. doi: 10.3389/fmicb.2016.02106.

Rogers, J.N., Stokes, B., Dunn, J., Cai, H., Wu, M., Haq, Z. and Baumes, H. (2017) An assessment of the potential products and economic and environmental impacts resulting from a billion ton bioeconomy. *Biofuels, Bioproducts and Biorefining* 11(1), 110–128. doi: 10.1002/bbb.1728.

Ronzon, T. and Piotrowski, S. (2017) Are primary agricultural residues promising feedstock for the European bioeconomy? *Industrial Biotechnology* 13(3), 113–127 doi: 10.1089/ind.2017.29078.tro.

Ronzon, T., Piotrowski, S., M'Barek, R. and Carus, M. (2017) A systematic approach to understanding and quantifying the EU's bioeconomy. *Bio-based and Applied Economics* 6(1), 1–17. doi: 10.13128/BAE-20567.

Roy, C. (2016) Les potentiels de la bioéconomie: De la photosynthèse à l'industrie, de l'innovation aux Marchés. *Futuribles: Analyse et Prospective* 410, 69–80.

Sadhukhan, J., Martinez-Hernandez, E., Murphy, R.J., Ng, D.K.S., Hassim, M.H., Siew Ng, K., Yoke Kin, W., Jaye, I.F.M., Leung Pah Hang, M.Y. and Andiappan, V. (2016) Role of bioenergy, biorefinery and bioeconomy in sustainable development: Strategic pathways for Malaysia. *Renewable and Sustainable Energy Reviews* 81(2), 1966–1987. doi: 10.1016/j.rser.2017.06.007.

Salter, B., Cooper, M. and Dickins, A. (2006) China and the global stem cell bioeconomy: An emerging political strategy? *Regenerative Medicine* 1(5), 671–683. doi: 10.2217/17460751.1.5.671.

Sánchez, D., Del Campo, I., Janssen, R., Rutz, D., Fritsche, U., Iriarte, L., Fingerman, K., Diaz-Chávez, R., Junginger, M., Mai-Moulin, T., Visser, L., Elbersen, B., Nabuurs, G.J., Elbersen, W., Staritsky, I. and Pelkmans, L. (2016) Towards the development of a European bioenergy trade strategy for 2020 and beyond (Biotrade2020plus project). In *European Biomass Conference and Exhibition Proceedings*, pp. 1356-1363. doi: 10.5071/24thEUBCE2016-4CO.6.6.

Sasson, A. and Malpica, C. (2017) Bioeconomy in Latin America. *New Biotechnology* 40(Pt A), 40–45. doi: 10.1016/j.nbt.2017.07.007.

Scarlat, N., Dallemand, J.-F., Monforti-Ferrario, F. and Nita, V. (2015) The role of biomass and bioenergy in a future bioeconomy: Policies and facts. *Environmental Development* 15, 3–34. doi: 10.1016/j.envdev.2015.03.006.

Schipfer, F., Kranzl, L., Leclère, D., Sylvain, L., Forsell, N. and Valin, H. (2017) Advanced biomaterials scenarios for the EU28 up to 2050 and their respective biomass demand. *Biomass and Bioenergy* 96, 19–27. doi: 10.1016/j.biombioe.2016.11.002.

Schütte, G. (2017) What kind of innovation policy does the bioeconomy need? *New Biotechnology* 40(Pt A), 82–86. doi: 10.1016/j.nbt.2017.04.003.

Shrivastava, D., Joshi, P., Sarin, N. B., Claps, D. and Sharma, N. (2016) An insight to India's biomass production and biowaste management: Scope and challenges. In *European Biomass Conference and Exhibition Proceedings*, pp. 176–180. doi: 10.5071/24thEUBCE2016-1BV.4.4.

Staffas, L., Gustavsson, M. and McCormick, K. (2013) Strategies and policies for the bioeconomy and bio-based economy: An analysis of official national approaches. *Sustainability* 5(6), 2751–2769. doi: 10.3390/su5062751.

Storz, H. and Vorlop, K.-D. (2013) Bio-based plastics: Status, challenges and trends. *Landbauforschung Volkenrode* 63(4), 321–332. doi: 10.3220/LBF-2013-321-332.

Tarrés, Q., Espinosa, E., Domínguez-Robles, J., Rodríguez, A., Mutjé, P. and Delgado-Aguilar, M. (2017) The suitability of banana leaf residue as raw material for the production of high lignin content micro/nano fibers: From residue to value-added products. *Industrial Crops and Products* 99, 27–33. doi: 10.1016/j.indcrop.2017.01.021.

Temper, L. (2016) Who gets the HANPP (human appropriation of net primary production)? Biomass distribution and the bio-economy in the Tana Delta, Kenya. *Journal of Political Ecology* 23(1), 410–433.

Tripathi, V., Edrisi, S.A., O'Donovan, A., Gupta, V.K. and Abhilash, P.C. (2016) Bioremediation for fueling the biobased economy. Trends in Biotechnology, 34(10), 775–777. doi: 10.1016/j.tibtech.2016.06.010.

United Nations (1992). Convention on Biological Diversity. Available at https://www.cbd.int/doc/legal/cbd-en.pdf (accessed 24 May 2018).

Vaz, S. (2015) A renewable chemistry linked to the Brazilian biofuel production. *Applied Adhesion Science* 1, 13–18. doi: 10.1186/s40538-014-0013-1.

Vaz Jr., S. (2014) Perspectives for the Brazilian residual biomass in renewable chemistry. *Pure and Applied Chemistry* 86(5), 833–842. doi: 10.1515/pac-2013-0917.

Venkata Mohan, S., Modestra, J.A., Amulya, K., Butti, S.K. and Velvizhi, G. (2016) A Circular Bioeconomy with Biobased Products from CO_2 Sequestration. *Trends in Biotechnology* 34(6), 506–519. doi: 10.1016/j.tibtech.2016.02.012.

Venkata Mohan, S., Chiranjeevi, P., Dahiya, S. and Naresh Kumar, A. (2017) Waste derived bioeconomy in India: A perspective. *New Biotechnology* 40(Pt A), 60–69. doi: 10.1016/j.nbt.2017.06.006.

Vergamini, D., Cuming, D. and Viaggi, D. (2015) The integrated management of food processing waste: The use of the full cost method for planning and pricing Mediterranean citrus by-products. International Food and Agribusiness Management Review. *International Food and Agribusiness Management Association* 18(2), 153–172. Available at: http://www.scopus.com/inward/record.url?eid=2-s2.0-84928946054& partnerID=tZOtx3y1.

Viaggi, D. (2015) Research and innovation in agriculture: beyond productivity? *Bio-based and Applied Economics* 4(3), 279–300.

Virgin, I. and Morris, E. J. (2017) *Creating Sustainable Bioeconomies: The Bioscience Revolution in Europe and Africa*. London, UK, Routledge.

Wang, R., Cao, Q., Zhao, Q. and Li, Y. (2017) Bioindustry in China: An overview and perspective. *New Biotechnology* 40(Pt A), 46-51 doi: 10.1016/j.nbt.2017.08.002.

Wesseler, J. and Von Braun, J. (2017) Measuring the Bioeconomy: Economics and Policies. *Annual Review of Resource Economics* 9, 275–298. doi: 10.1146/annurev-resource-100516-053701.

Zhang, Y.-H. P., Sun, J. and Ma, Y. (2016) Biomanufacturing: history and perspective. *Journal of Industrial Microbiology and Biotechnology* 44(4-5), 773–784. doi: 10.1007/s10295-016-1863-2.

Zilberman, D., Graff, G., Hochman, G. and Kaplan, S. (2015) The political economy of biotechnology. *German Journal of Agricultural Economics* 64(4), 212–223.

3

Technology and Innovation in the Bioeconomy

3.1 Introduction and Overview

Innovation, and hence technology, is one of the main focuses of the bioeconomy. From the perspective of technology, the bioeconomy can be viewed as a continuing evolutionary process of transition from systems of mining non-renewable resources to farming and processing renewable bio-based ones (Zilberman *et al.*, 2013). However, in addition to this, a number of features are unique to the bioeconomy or at least highly relevant to it. Some of these include: biotechnologies that affect genomes, biomass breaking down to building blocks, biorefinery concepts and the high use of information technologies, technologies for closing cycles in the economy (waste recovery) and technologies that affect the provision of ecosystem services by affecting the environment. Knowledge and information are also key issues throughout the bioeconomy.

A key issue for the scope of this book is the way a meaningful representation of bioeconomy technologies can be developed for the purposes of economic studies (Viaggi, 2016). Another key topic is the uptake mechanisms of specific bioeconomy technologies that may, in fact, be connected to an increasingly intricate network of intersectoral relationships. It should be emphasized that the combination of technologies is of particular importance. The same applies from a disciplinary perspective, as interdisciplinarity, multidisciplinarity and transdisciplinarity are

characterizing features of the bioeconomy, involving microbiology, molecular biology, chemistry, genetics, chemical engineering, agronomy, economics, logistics and digitalization studies (Krüger *et al.*, 2017) embedded in complex institutional and market processes.

This chapter starts with a description of the main bioeconomy-relevant technologies, together with a review of some of the main non-bioeconomy-specific technologies with important applications in connection with the bioeconomy. In the second part, we provide a simplified representation of these technologies able to better understand some of their key features in an economic perspective and to support the economic analysis of the bioeconomy in the chapters that follow.

3.2 Key Bioeconomy Technologies

In this section we investigate the features of the main bioeconomy technologies. In light of its extremely broad scope, the bioeconomy encompasses a very wide range of technologies that are continually evolving over time. Accordingly, it is not the ambition of this section to give a detailed account of the features of each technology, nor is it possible for the book to delve into detail on each of the potentially relevant technologies for the future. The idea is rather to aggregate them by their functional role and features with an eye to a better understanding of the economic issues behind each type. In this respect, this section aims to be comprehensive, but by no means exhaustive. A broad review of bioeconomy technologies is available in Sillanpää and Ncbi (2017).

3.2.1 Environmental and ecosystems management and bioremediation

The bioeconomy at large starts with the monitoring, management and control of the environment in which organisms live. There are several aspects to this, with each relying on completely different solutions. First, there is a knowledge issue about the state of ecosystems and the prediction of their evolution. This includes technologies from sensors, to bio-indicators, remote sensing and satellite information at large. Detection and prediction of climate trends and atmosphere compositions is clearly a major issue for the bioeconomy as a whole, as this interacts and determines organisms' life in natural and cultivated environments. A second broad area of intervention concerns the management of artificial ecosystems, for example, whole territories through the provision of basic resources (such as water), infrastructure, the regulation of the population of wild animals, and so on. An area of noteworthy recent development is that of urban ecosystems. Finally, technologies can be used to eliminate specific localized problems, such as bioremediation technologies, which remedy pollution problems through the action of biological agents. As a result, in most cases there are ways of modifying the environment in which human beings carry out their activities.

Though this is a very heterogeneous set of technologies, the main message is that humans are more and more proactively influential in affecting the overall biosphere, and this actually provides a major background for all other bioeconomy technologies.

3.2.2 Harvesting and cultivating biological organisms: primary production in farming, forestry and fisheries

Hunting and fishing, followed by farming and forestry, may well be the oldest activities practised in the bioeconomy. Traditonally, cultivation and rearing are distinguished from harvesting. From a technological point of view cultivation combines artificial (capital) resources, labour and natural resources to produce goods. In harvesting, the prevailing activity is the withdrawal of biomass from the natural stock (population). Both require effort (capital, labour), but in harvesting the population of living organisms stay in their environment and are not as artificially controlled as in the case of cultivation. Agriculture and acquaculture are mostly in the range of cultivation, whereas fisheries and forestry are mainly based on harvesting techniques.

Technologies used for cultivation have changed significantly over time. The characterizing features of the 20th century have been genetics, chemistry and mechanization. Presently major changes are occurring, especially in the field of conservation agriculture and information and communication technology (ICT) connected to precision farming.

Farming uses land, water and biological organisms in an extensive manner. For this reason, it has a very important role in regulating the production of ecosystem services, or in affecting (often negatively) the environment, depending on the circumstances. Being exposed to weather and climate variables, as well as the dynamics of pests and the variability of living beings, agriculture is characterized by the high variability of outcomes and uncertainty. Using highly heterogeneous land and local resources, agricultural systems are very diversified. An essential aspect is agrobiodiversity and the structure and ecological features of agricultural systems (Jordan et al., 2016).

Forestry is also attracting attention from the bioeconomy. Forest management has traditionally been the basis for the timber economy and has come to be attached to a number of additional uses, such as recreation. It has become a focus of attention once again with regard to the production of biomass for wood-based chains, including the innovative options coming from biorefinery for energy and bio-based materials. Forestry is now less a matter of harvesting natural resources and increasingly an innovation-based management/cultivation of an anthropized resource, especially in order to meet the challenges of the bioeconomy (see for example, Deleuze et al., 2016).

Fisheries is traditionally a key area of production. The use of the seas is one of the major focuses of the bioeconomy strategies in some countries. Aquaculture is also a major area of expansion. New forms of farming are emerging, with high expectations, such as seaweed aquaculture (Stévant *et al.*, 2017) or microalgae cultivation discussed by Hingsamer *et al.* (2016). Insects, as a source of protein, are also attracting growing attention.

Agriculture and primary production also produce large quantities of residues and by-products. Their actual exploitation, however, needs a wider adoption of *ad hoc* harvesting machinery, the optimization of logistical processes and the maturation of the new bio-based value chains.

3.2.3 Food preservation and processing technologies

The food industry uses a number of technologies, many of which are related to biological processes or to non-biological processes applied to biological raw materials. Many of these are linked to conservation or seasoning, which was the focus of earlier technologies and related research. This includes salting, fermentation, refrigeration, thermal treatments, drying, smoking, and so on, with a focus on avoiding spoilage and destroying organisms that are harmful to human health. Others (more recently) are more connected to obtaining the right ingredient composition or final desired food attributes. Canning and packaging are important steps in food preservation and facilitate transport. Cooking is also an important step in food preservation and preparation for final consumption. There have been significant improvements in the ability of industry to break down and recombine ingredients. Some of these technologies are nowadays very similar to the same technologies that can be generically applied in the preservation or breaking down of building blocks used in non-food industries. However, typical food technologies, compared to other bio-based technologies, tend more often to maintain at least part of the original biomass without dismantling it.

Security and safety are primary issues in food production, as, more generally, are issues related to health, which implies much greater attention to processes and the characteristics of final products.

Since not all biomass can ultimately be consumed, and that quality and health requirements prevent the consumption of food that does not meet safety standards, the food industry is a large producer of waste from

processing and the same applies to consumers in the final consumption stage.

3.2.4 Genetic modification technologies

One of the most characterizing groups of technologies related to the bioeconomy is that of biotechnologies focused on modifying the genetic features of useful organisms. The main feature of these technologies is that they modify the properties and ability of living organisms, especially in terms of the transformation of inputs into productive outputs (level and stability of yields). Moreover, the reduction of resource needs (e.g. water), resistance to stress and pathogens and other adaptations are sought through such technologies.

The modification of the genetic properties of organisms usually involves the creation of variability and selection of organisms with appropriate characteristics.

A degree of modification is now pervasive in many cultivated plants and animals. Plant breeding is still viewed as a cornerstone of the bioeconomy, especially in view of ensuring a sufficient supply of biomass while reducing resource use (Małyska and Jacobi, 2017), for example, improving the output/input potential, especially in connection to critical resources and towards more environmentally friendly solutions than those provided by traditional industrial agriculture. An example of this is the engineering of a plant to avoid the need for the external application of pesticide (e.g. Bt corn). In general, biotechnologies are expected not only to increase competitiveness and productivity, but also to contribute to a wide range of sustainable development goals, that is, in relation to food safety and sustainable industrial development (Lokko *et al.*, 2017).

Though the range of possibilities of this kind has widened enormously over time, biotechnologies are still constrained to the potential and variability of living organisms and their genetic resources, which implies that modifying the genetic features of organisms still involves trade-offs in terms of performance. Also, the modification of organisms takes time not only to bring about changes, but also to select suitable lines and to patent them so that they are ready for commercialization. Finally, the modification of living organisms is often seen as potentially harmful for the environment and society and often entails very careful permission procedures and potential public opposition or market segmentation. See the example of genetically modified organisms discussed in detail in Chapter 5 (Bennett *et al.*, 2013).

The number of technologies usable to achieve some kind of genetic modifications is already very high and ever growing. More recent technologies focus a lot on information-intensive techniques and tend to be less disruptive in terms of consistency among individual characters used. Genome editing is considered to be a new promising technology in this direction, using technology such as CRISPR-Cas. Gene drivers also use CRISPR/Cas9 gene editing for innovative pest control methods (Brown, 2017).

Given that information on genetic features is key for new genetic modification practices, genomics is expected to have a major role in shaping the future of the bioeconomy (Jiménez-Sánchez and Philp, 2016). This applies to the sequencing of individual species of cultivated crops and animals. Improvements in sequencing have been one of the cornerstones of biotechnology and a number of new upcoming solutions (e.g. next-generation sequencing) are seen as strategic for future developments. In addition, metagenomics allows for the study of genetic material recovered directly from the environment (Montella *et al.*, 2016). In the light of this information-intensive approach, facilities collecting 'genetic' information (biodiversity and gene bank facilities) are key for future developments of the bioeconomy (Overmann, 2015).

Genetic technologies are not focused on plants alone, but are also devoted to microbial strains and enzyme development (Bott and Eggeling, 2017; Krüger *et al.*, 2017). Another case of non-vascular plants concerns fungi. For example, Meyer *et al* (2015) discuss the role and potential improvement of *Aspergillus* used in some of the top-performing microbial cell factories.

3.2.5 Breaking down (and recomposing) biomass

Several technologies are related to breaking down biomass into building blocks. The simplest

include drying and physical separation (e.g. filtration); more sophisticated versions include biocatalysts such as enzymes and microorganisms.

Enzymatic breakdown is a key aspect of bioeconomy technology. In a recent paper Heux *et al.* (2015) illustrate the role of biocatalysts (enzymes and microorganisms) as key tools of white biotechnology and one of the key technological drivers for the growing bioeconomy. Biocatalyst applications already include the chemical and agro-food industries, as well as product manufacturing (e.g. antibiotics, paper pulp, advanced polymers and food).

Box 3.1. Perspectives of biocatalysts

Heux *et al.* (2015) provide a broad review of state of the art approaches to biocatalysts, with a particular focus on the biorefinery sector. In particular, it highlights the importance of three areas of innovation: (i) technologies underpinning the development of industrial biocatalysts, in particular the discovery of new enzymes and enzyme improvement using directed evolution techniques; (ii) cell engineering to shape the metabolism of microorganisms; and (iii) 'omic' technologies for understanding and guiding microbial engineering toward more efficient microbial biocatalysts. With regard to biocatalysts Krüger *et al.* (2017) argue that sustainability of existing biotechnological processes can be improved with a better study and use of extremophiles.

Basic building blocks can be combined and used to produce new products; one case of paramount importance is that of plastic. This sector, thanks to a number of innovation in the fields of chemical catalysis, biotechnology, and plastics engineering, now increasingly focuses on the production of durable, bio-based plastics (Storz and Vorlop, 2013).

3.2.6 Production of bioenergy

An important component of bioeconomy technology is the production of bioenergy. The basic principle is the breaking down of biomass that makes fixed energy available. Bioenergy production also includes the production of suitable compounds that can be used for energy production (e.g. production of biogas from fermentation).

A common feature is that at the end of the process (when energy is produced) biomass is generally completely destroyed, except for the non-organic components of the biomass. This may entail the destruction of valuable compounds and is the main reason for advocating for the connection of bioenergy production with biorefinery processes (see below).

Bioenergy technologies are differentiated according to the type of biomass transformed into energy as well as according to the process involved, which also implies different properties and processes. The following is a rough distinction:

- direct combustion of biomass, such as wood or wood pellets;
- production and combustion of biogas for the production of electricity; and
- production and combustion of biofuel, such as biodiesel.

The different processes have different properties in terms of plants and infrastructures, transportability and suitability of the process for different energy use.

Besides the intermediate form of combustible substances, a very important difference is the source of biomass, which is related in various ways to the different technologies. Roughly speaking, biomass can derive from harvesting, cultivation or waste. An important point concerns the use of sources not competing with food.

Beyond the technologies themselves, there are issues related to impacts on land use and landscapes, especially when raw materials are massively procured from agriculture and the production of bioenergy entails a strong concentration of single species in some areas or the introduction of new species into an agroecosystem.

3.2.7 Biorefinery

A component of bioindustry is biorefinery, which is a central concept in the development of the bioeconomy. Biorefinery can be intended both as a type of process, as a facility or a form of industrial organization. The International Energy Agency (IEA) has defined biorefining as the sustainable processing of biomass into a spectrum of bio-based products (food, feed, chemicals and other materials) and bioenergy (biofuels, power and/or heat). In general, it is hence intended as a process whereby different types of compounds (renewable fuels and a wide spectrum of intermediate materials or end-products) are obtained from bio-based resources. In a narrower definition, a biorefinery is a facility that allows biomass conversion into bioenergy and value-added compounds. More specifically, a biorefinery is a combination of functionally integrated production processes based on symbiotic connections and synergies that allow an increase in the overall efficiency in the use of energy and resources. A biorefinery may produce several low-volume, high-value chemical products, together with low-value, high-volume fuel, heat and power (Francavilla *et al.*, 2016) (Fig. 3.1).

The term comes from the analogy with the petroleum refinery, which produces multiple fuels and products from petroleum. Another view of biorefinery is that of a whole change towards an 'industrial methabolism', allowing for the full use of plants as compounds, materials and energy (Octave and Thomas, 2009). Biorefinery in Europe and worldwide is now promoted as one of the key economic organizational concepts in the bioeconomy and already accounts for a huge body of literature.

Biorefinery is based on the idea of a stepwise treatment of biomass starting from the extraction of the highest added-value components down to the production of energy, in such a way as to make the most out of the limited biomass resource availability (cascading approach). The range of products obtained is hence determined by the order of actions. By producing multiple products, a biorefinery takes advantage of the various components of biomass and their intermediates, therefore maximizing the value derived from the biomass feedstock. In this sense, the biorefinery concept is often seen as a way of improving the economic performance of the bioeconomy, especially through the substitution of standalone systems for the conversion of biomass into bioenergy (Sawatdeenarunat *et al.*, 2016; Budzianowski, 2017).

The cascading approach is key to the concept of biorefinery. Cascading can be intended as

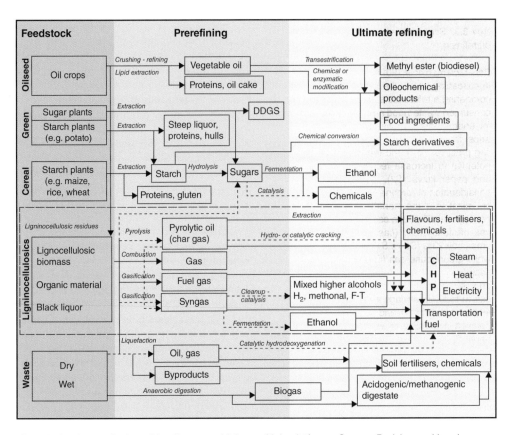

Fig. 3.1. A scheme of a large biorefinery combining multiple platforms. Source: Budzianowski and Postawa (2016).

Box 3.2. Optimizing biorefineries

The production of conventional bioenergy (e.g. biofuels, biopower and bioheat) is, for the most part, not profitable due to competition with low price fossil fuels (Budzianowski, 2017). This is also due to the fact that complete biomass disintegration through combustion, gasification or fermentation for energy production is highly inefficient compared to cascading approaches in which high-value, low-volume bioproducts are coupled with bioenergy production. The integrated production of bio energies and bioproducts may be achieved by: (i) coupling existing biofuel plants with new bio-based processes; (ii) retrofitting existing bioindustries with new bioenergy facilities; or (iii) constructing new integrated facilities. It is a common view of several authors that the move from traditional anaerobic digestion solutions to a new biorefinery concept in which digestion for energy is the final step is the most promising pathway. See Sawatdeenarunat et al. (2016) for a discussion of the opportunities offered by this strategy.

prioritizing the use of biomass for socially preferable products (notably materials) over its use for energy (Keegan et al., 2013). Cascading potentially increases cost efficiency and decreases emissions and could be a key strategy for systems with high bioeconomy potential, especially when bioenergy is already well developed. However, the cascading approach also entails trade-offs, especially when considering multiple effects at the system level. In addition, the suitability of the cascading approach depends on a number of circumstances in different product markets, including alternative (fossil fuel) sources. Barriers to the more widespread use of the cascading approach are discussed by Keegan et al. (2013), and are identified mainly in underdeveloped supply chains to facilitate the reuse of biomass resources and a regulatory framework that focuses support on the energy use of biomass only.

A very relevant distinction can be drawn between first- and second-generation biorefinery,

---

Box 3.3. Scenarios and cost-benefits of biorefining

Bais-Moleman *et al.* (2017) analyse the trade-offs in cascading use in the European wood sector comparing a reference scenario in which (post-consumer) waste wood and paper are re-utilized for energy only (*S0*) with two alternative scenarios: the current waste wood and paper recycling practices (*S1*) and the maximum technical potential to increase recycling of waste wood and paper flows (*S2*). The analysis includes consideration of forgone fossil-fuel substitution, optimization at the manufacturing level and forest regrowth. Through cascading use, the wood use efficiency ratio (cascade factor) would increase by 23% (*S0* vs *S1*) and 31% (*S0* vs *S2*), while greenhouse gas (GHG) emissions would decrease by 42% in scenario S1 and 52% in scenario *S2*. However, in the short term, this will result in reduced savings in the energy sector by 49% in scenario S1 and 48% in *S2* due to the delayed availability of waste wood and waste paper fibres.

---

depending on the sources of biomass (Scoma *et al.*, 2016). First-generation biorefineries make use of dedicated crops or primary cultivations; as a result, they cause competition with food and other bio-based industries. Second-generation biorefineries use agricultural, industrial, zootechnical, fishery and forestry biowastes as the main feedstock. The shift from first- to second-generation biorefinery leaves aside ethical and social issues generated by first-generation approaches, and also helps environmental and economic issues associated with waste disposal.

A very relevant aspect of biorefinery is its potential to valorize waste as a renewable feedstock to recover bio-based materials and energy through sustainable biotechnology. This approach holistically integrates remediation and resource recovery, including potential food waste and can be implemented using different technologies (Mohan *et al.*, 2016).

3.2.8 Use of biowaste

A set of processes relevant for the bioeconomy is represented by technologies allowing the valorization of biomass from wastes. This also highlights the circular economy approach of the bioeconomy. Biowastes can come from several sources, including agriculture and food wastes, biowastes from other bio-based processes and biowastes from non-bio-based products, such as urban organic wastes and wastewater.

Different technologies are available for the use of biowastes, sometimes overlapping with those already described. Studies highlight a noteworthy pattern of investments in the past decade that sees movement from chemical to biological technical bases (Poz *et al.*, 2017) and from bulk low-value uses of biomass to high-value processes supported by a range of new technologies (Peinemann and Pleissner, 2017).

Approaches valorizing food wastes into food and feed are clearly of paramount interest. This may imply the use of several of the technologies already discussed above for food production itself, though a good deal of innovation and new proposals are emerging in this field. Use in a biorefinery concept with extraction of value-added chemicals and ending in energy production is also a valuable solution. These are especially interesting when leading to the extraction of chemical platforms that can be used for multiple purposes (Scoma *et al.*, 2016; Dahiya *et al.*, 2017).

Box 3.4. Examples of the use of food streams

According to Matharu *et al.* (2016), global food supply chain wastes (FSCWs) amount to 1.3 Gt. This potential is highly relevant, in particular, for certain tropical FSCW for three major countries: Brazil, India and China. A very wide set of processes and products can build on food wastes, including platform molecules and functional materials (Matharu *et al.*, 2016). Considering the increasing global demand for vegetable proteins, proteins from FSCW may become a dominant area. Poltronieri and D'Urso (2016) provide a thorough discussion of advances in technology and plant designs for the valorization of food by-products through fermentation (feed-batch liquid fermentation, solid-state fermentation) into bio-based bio-chemical/biofuel production. They also illustrate pathways in the biosynthesis of fibres, sugars and metabolites. Egelyng *et al.* (2016) illustrate several examples of technologies that can be used on selected streams of food industry residues in Norway, allowing for the use of co-streams. Examples include automation and scanning technologies, a collection system for fish waste, use of second-grade vegetables for smoothies and potato peels for biodegradable plastics.

Composting is a classic example of (well-known) use of biowastes, which can be very useful as it allows for the use of wastes that are otherwise not of commercial interest for more sophisticated uses (e.g. food or feed), including as a potential fertilizer.

Box 3.5. Potential of composting

Composting can valorize biomass from other biomass processing and/or that is unsuitable for other processes and, in doing so, has the potential to provide multiple beneficial effects to soil quality (Viaene *et al.*, 2016). However, compost is rarely used in intensive agriculture. The paper identifies market and financial, policy and institutional, scientific and technological, and informational and behavioural barriers to this. The most important barriers to sector development are: the shortage of woody biomass, strict regulation, considerable financial and time investment and a lack of experience and knowledge on the part of processors. The main barriers to apply compost are complex regulations, manure surpluses, variable availability, transport costs and variability of compost quality and composition.

Among options for conversion of waste to biomass as a promising approach to improve the bioeconomy co-cultures are also identified as efficient strategies. For example, Tossavainen *et al.* (2017) cultivate algae and bacteria in diluted wastes in composting leachate liquid treatment.

Distributed biowaste composting from rural households may have a specific role in closing the cycles in rural areas, change the overall performance of the waste management system in terms of GHG emissions and contribute to rural development objectives (Mihai and Ingrao, 2016).

Box 3.6. Wastes and wastewater as a bioresource

Scoma *et al.* (2016) gathered data on some of the main high-impact biowastes found in Europe (in particular in southern Europe) and discussed the bio-based chemicals, materials and fuels that can be produced from such residues, focusing in particular on chemical platforms.

Waste utilization can also involve waste streams from sectors outside the bioeconomy, such as waste water (Puyol *et al.*, 2017).

The use of biowastes is still hindered by a lack of information about biowaste availability and a lack of integrated modular approaches to technology development, as well by legislative constraints (Scoma *et al.*, 2016).

Some of the main areas of interest to ensure a circular economy include: (i) the cascading approach to using bio-based resources, including food waste; (ii) the potential for innovation in new bio-based materials, chemicals and processes contributing to the circular economy; and (iii) the recycling of wood packaging and separate collection of biowaste (European Commission, 2015).

3.2.9 Synthetic biology and cell-free systems

One of the forefront technologies of biotechnology is 'synthetic biology'. Synthetic biology is an interdisciplinary branch of biology and engineering, which combines various disciplines such as biotechnology, evolutionary biology, genetic engineering, molecular biology, molecular engineering, systems biology, biophysics and computer engineering. It is now claimed to have a huge economic potential due to its approach to research and development, and the unique nature of the carefully designed, stakeholder-inclusive, community-directed evolution of the field (Flores Bueso and Tangney, 2017).

Cueva (2017) provides a recent update on the perspectives of synthetic biology. The author presents numerous methods for the development of novel synthetic strains, synthetic biological tools and synthetic biology applications. In particular, while there are still limitations in synthetic biology research, innovation is propelling the development of technology, the standardization of synthetic biological tools and the use of suitable host organisms.

Another stream of technology development that goes beyond living organisms while staying bio-based are cell-free biosystems. Cell-free biosystems, previously used mainly for research, are now an emerging manufacturing platform for the production of high-value protein and carbohydrate drugs, low-value biocommodities (e.g. H_2, ethanol and isobutanol), as well as fine chemicals (You and Percival Zhang, 2013). The authors argue that, due to potential high yields and hence low production costs, cell-free biosystems could become a disruptive technology in the production of high-impact, low-value bio-based commodities.

3.2.10   Characterization and tracing

As biomass is extremely varied, a number of technologies related to physical, chemical and biological analyses are aimed at characterizing the features of biomass or bio-based products at large. This concerns forms of biomass qualification before processing, status of biomass during processing or the quality of final products aimed at consumers. Accordingly, these technologies can actually be useful at any stage of the product chain.

In addition, a growing field relates to technologies for detecting origin that allow for the traceability of goods and products. This is accomplished by recording procedures during production, processing and transport, but also through chemical analysis and the identification of genomic characteristics of products.

Innovation in this field is very relevant (e.g. sensors, bio-indicators, bio-based essays, etc.) and increasingly connected to the digitalization of processes and packaging features.

3.3   Complementary Fields and Technologies

As mentioned, bio-based technologies are connected to a bundle of innovative areas and fields of science, some of which are especially relevant and often mentioned in documents and definitions.

3.3.1   Downstream technologies

A number of economic sectors use bio-based materials and their scope may well increase in the future. This may result in new, or rather the expansion/adaptation of existing technologies. Classic examples are the fuel sectors and the energy industry. A less obvious, but increasingly relevant example, is wood multi-storey construction (Toppinen *et al.*, 2016).

The technologies related to these sectors are, of course, too wide and varied for the scope of this book. What needs to be highlighted here is that their changes affect the profitability of bio-based products and hence the development of the bioeconomy. On the other hand, the diffusion of these technologies is affected by the costs of raw materials connected to bio-based technologies (see Chapter 6 for more details).

3.3.2   Digital information and communication technologies

Information and ICT is probably the most relevant area of research supporting the bioeconomy, together with biology in general. ICTs are a rapidly evolving bundle of technologies used to collect, treat and use data as well as for communication. The fact that the bioeconomy is a knowledge-based business, together with the high degree of complexity of the processes considered, gives a major role to ICT. The interplay with other bioeconomy technologies is now manifold, including:

- process monitoring and control;
- decision-making support;
- communication among firms, consumers and other stakeholders; and
- traceability.

A key sector at the border with the bioeconomy is bioinformatics. Bioinformatics is an interdisciplinary field that develops methods and software tools for analysing and interpreting biological data. It is a largely interdisciplinary field, combining computer science, statistics, mathematics and engineering.

Several aspects of this technology are now at the forefront of innovation in the bioeconomy, such as the internet of things and blockchains. The topic of 'big data' is a central issue for the bioeconomy, as the storage and processing of huge amount of data is key to support better diagnostics, choices and innovation in genetic, traceability, precision farming, remote sensing, and so on.

The economic role of these tools is sometimes difficult to capture, yet at the same time very relevant. It can be interpreted in terms of the cost and value of information depending on its potential consequences for action.

Another peculiarity is that it interacts with the coordination among actors, leading to new mechanisms of collective interaction and decision making, as is experienced everyday through social media. This is, in turn, another hot topic related to the aim of sustainable and participatory bioeconomy development.

The nature of ICT technologies is to connect the different processes discussed above, so that technical data processing and decision making are increasingly part of the same societal decision process.

3.3.3   Nanotechnologies

Nanotechnology is a field of science and technology dealing with manipulation of matter at the 'nano' scale, usually between 1 and 100 nanometres (one billionth of a metre). Nanotechnologies are another field of innovation widely recognized as complementary to the bioeconomy and identified as one of key enabling technologies (KET) by the European Commission (Parisi *et al.*, 2017). The same authors note, however, that concrete contributions are still uncertain, as agricultural applications are not used on the market yet, in spite of a growing number of publications and patents. Not only do they still show high costs of investment compared with expected profitability in agriculture, but there is still uncertainty about public acceptability, especially in applications linked to food.

3.3.4   Engineering

Knowledge needs to be engineered into practical innovation solutions to be usable. Engineering has a strong role in making innovation happen. Moreover, the interplay between engineering and bio-concepts, as shown by terms such as 'genetic engineering' or 'synthetic biology' as a branch of engineering, are demonstrating an interplay between engineering and the bioeconomy, linked to an increasing control over biological systems and a greater ability to build artificial biology-like complex systems.

Many current trends are moving towards the concept of the engineering of biological systems, which can help bridge the current gap

between the exponential growth of knowledge about biology and the still somehow slow development of sustainable bio-based value chains (Lopes, 2015). A major and required pathway of life sciences in the bioeconomy is the transition of contemporary, gene-based biotechnology from being a trial-and-error endeavour to becoming an authentic branch of engineering (de Lorenzo and Schmidt, 2018).

3.3.5   Local and tacit knowledge

While this is not a technology, the role of local and tacit knowledge is emphasized in definitions of the bioeconomy by some authors, in particular to connect the bioeconomy to territorial development in an ecosystem services and social sustainability perspective. Within the general role of knowledge, this is the component accounting for:

- distributed experience and past knowledge existing among players and institutions; and
- allowing adaptation to local conditions made possible by this experience.

This is especially relevant in adapting technologies and their use to very specific combinations of local conditions, in which a lot of highly relevant information is already available to local actors, as it is common in agriculture, fisheries or natural resources use in general.

3.4   Some Specific Features of Bioeconomy Technologies

3.4.1   Renewable and non-renewable, natural and artificial

Much emphasis in the promotion of the bioeconomy is linked to the need to overcome dependency on fossil fuels and to achieve an economy based on renewable resources. Zilberman *et al.* (2013) provides a key reference for introducing this issue, starting with the classification of resources.

The first distinction in resource economics is between renewable and non-renewable resources. Non-renewable resources, e.g. minerals and oil, have finite stocks. However, for some of them, such as fossil fuels, availability may increase with discoveries in the short run or in the very long run due to, for example, CO_2 fixation. This is usually beyond the relevant time horizon for human society but it is important to keep in mind the whole picture, as the basic problem with fossil fuels is that the time needed for regeneration is so much longer than the speed of consumption that the two are essentially incomparable. Renewable resources can be produced at a certain level infinitely over time. Yet renewable resources can be depleted or even exhausted if the rate of use is faster than the rate of regeneration. However, at a certain level of use, renewable resources can potentially be sustained forever, contrary to non-renewable resources.

A large body of literature analyses the economics of non-renewable resources, with a key topic being that of the optimal rate of use over time, taking into account scarcity price, interest rates and innovation (e.g. backstop technology, i.e. a technology that can substitute the use of non-renewable resources when their price becomes so high as to render them no longer profitable). The literature also identifies conditions for the optimal use of renewable resources, especially considering stock-harvest relationships linked to resource management arrangements (Bergstrom and Randall, 2016).

Renewable resources can be divided into physical (e.g. water, wind and sunlight) and living systems (e.g. forests and fish) (Zilberman *et al.*, 2013).

Renewable resources based on living systems are divided in the literature into systems that are harvested and systems that are farmed (Zilberman *et al.*, 2013). This distinction is becoming less clear over time for different reasons: (i) on the one hand, cultivation is carried out in a way that is more aware of relationships with ecosystems and more careful in preserving the stocks of resources that are behind the cultivated harvest; (ii) the management of natural resources is more active and aware, at least in terms of monitoring and harvesting control; and (iii) activities are being developed, such as aquaculture, that are substituting classic harvesting with the harvesting of a more controlled population, largely based on artificial reproduction and seeding (hence, in fact, becoming cultivation).

Many cases of transition exist from harvesting to cultivation. For some of the main agricultural commodities, it happened thousands of years ago when agriculture emerged. The most evident cases of contemporary change from harvesting to farming relate to fish production, which is shifting rapidly from fishing to fish farming, and now also increasingly for biofuels. An interesting case for the bioeconomy is that of biological processes being harnessed to produce fine chemicals, representing another transition from non-renewable to renewable resource use, as well as from harvesting to cultivation. Hence, a key element of the bioeconomy is the extension of cultivation rather than harvesting processes (Zilberman *et al.*, 2013). Yet at the same time, interventions into natural living systems are more and more frequent, hence rendering them less and less 'natural' and more and more 'anthropized', with at least some effect, or more and more often a controlled stock.

As part of this increasing human role in affecting the evolution of biological capital, a key element for the bioeconomy is the extension of breeding processes and genetic modification. Using advances in biotechnology and synthetic biology, the process of modifying and designing new organisms to develop valuable products will become even more pronounced in the future (Zilberman *et al.*, 2013).

The productivity of renewable resources is dependent on climatic and biophysical conditions, which, for living systems, affects the dynamics of living populations. As a result, the bioeconomy must adapt to climate diversity and to climate change. On the other hand, by controlling the flows of carbon and other key compounds linked to climate, the bioeconomy can play a role in reducing climate change. As a result, the development of the bioeconomy is intrinsically linked in multiple ways to the control of the earth's environment.

These trends altogether go in the direction of weakening the distinction between cultivation and harvesting and moving towards different layers of control of biological processes with different levels of intensity that can be classified as: (i) control over the environment and ecosystems; (ii) control over genetic features; (iii) direct control over populations; (iv) provision of input to biological systems; and (v) control of biomass transformation.

Moreover, the contemporary influence of human beings on the environment, and the use of biological processes in industrial production (e.g. biomanufacturing), is progressively blurring the difference between harvesting, cultivation and industrial processing, towards a hybrid cultivation/engineering process of worldwide biomass (somehow cultivating the world). For this reason, in this book we use the term 'exploitation' to mean this mixed concept of harvesting, cultivation and industrial processing.

At the same time, the trends illustrated above reduce the scope for truly natural capital as opposed to 'artificial' capital (capital in economics means goods produced by people used again in a production process). They expand the number of ecological components that are affected, and hence to some extent anthropized, by human action. In a way, the development of the bioeconomy and the evident ability of humankind to modify the global environment is breaking the distinction between artificial and natural capital and moving towards a notion of anthropized natural capital characterized by a continuum of human influences and on a decreased separability between different components. In this way, natural biological resources are increasingly interwoven not only to artificial physical capital and to biological humans, but also to knowledge stock and newly developed artificial biological components. While several concepts that have been developed relate to 'bio-features', none of them has been pragmatically used in economics (but see section 3.5 for an improved discussion). We propose the use of the concept of anthropized biological capital (ABC), intended as the biological component of natural capital (to distinguish it from non-biological resources), which is now affected by humans in terms of properties, amount and status (e.g. modified genetic pool). This capital changes over time by the work of humankind and requires investment and maintenance expenditure in order to keep its properties consistent with human needs. In this light we can then distinguish two areas of human activity, that is, exploitation and maintenance/investment. This view is also consistent with the growing recognition of the role of humanity in affecting the status of the earth, like, in particular, in the Anthropocene theory (Balter, 2013).

While much of the focus of the bioeconomy and its tendency to alter technology emphasizes its role in replacing products that are derived from non-renewable resources (e.g. fossil fuels) with products derived from renewable biological resources (e.g. biofuel), more complex links should be emphasized. First, the bioeconomy is also related to non-biological resources and processes, either renewable or non-renewable. Second, the production of biomass is largely connected to non-renewable resources, such as fossil fuels and fertilizers. The consequence of the former point is that an economy based on renewable resources is not exclusively dependent on the bioeconomy. The consequence of the latter is that actual renewability need not only ensure biomass production, but also the renewability of resources used for this production.

3.4.2 Technology links and flexibility

Based on the ability of breaking down biomass into single compounds and on the ability to recombine them, as well as modifying the performance of living organisms, the number of points in the production chain in which the process is separable is growing exponentially. This potentially makes it possible to increase (exponentially as well) the number of compounds/products, the number of uses of each individual compound and the number of potential pathways of compounds into the bioeconomy system. As a result, a feature that is emerging as key for the bioeconomy is that of flexibility.

Flexibility may concern compounds, processes, plants and location. This reflects into different concepts and declinations, notably flexible feedstock, flexible transformation processes (or plants) and flexible biomass or platforms. It is also the basis for the flexibility of agriculture, food and bioeconomy systems, hence being part of the determinants of dynamics, adaptation and resilience.

Focusing on individual products, according to the IEA Bioenergy Task 42 'platforms' can be defined as 'intermediate products from biomass feedstocks towards products or linkages between different biorefinery concepts or final products' (IEA Bioenergy, 2014). Platforms are currently at the heart of the biorefinery concept and are one of the most important features in the classification of biorefinery activities. Platform products include oils, syngas, hydrogen, pulp, lignin and C6 sugars (Taylor *et al.*, 2015).

An example is provided in Fig. 3.2, where C5 and C6 sugars represent the intermediate platform products. These products may derive from several original feedstocks and produce several different downstream products.

Bomtempo *et al.* (2017) provide a framework for evaluating the potential and the current stage of bio-based platform chemicals. The framework identifies the following features: (i) being an intermediate molecule; (ii) having a structure able to generate a number of derivatives; (iii) being produced at a competitive cost; (iv) allowing for exploitation of scale and scope economies; (v) being inserted into a complete innovation ecosystem, including suitable governance systems and able to create value; and (vi) facilitation by a focal company leading the platform (mostly the producer of the platform molecule).

The concept of a flexible biorefinery facility is also a growing concept in the literature, making it possible to obtain different products in the same facility but also to adapt to feedstock availability and product demand. An example is given in Wendisch *et al.* (2016), and these concepts are discussed in more detail in Chapter 6.

A relevant issue is that an ongoing increase in the flexibility of technology may modify its connection with location and increase the role of institutional factors in attracting business and allowing for the development of the bioeconomy (Schrager and Suryanata, 2018).

3.4.3   Uncertainty

Uncertainty is a major issue in the bioeconomy. Some causes of uncertainty arise for technological reasons, that is, output or attributes of processes can be unknown in advance due to incomplete control over the process or simply a lack of information. These add to uncertainties from the economic and social environment, such as prices, consumer reactions and changes in legislation. It is worth noting that these are not disconnected from each other. Uncertainty surrounding technology contributes to uncertainty-related behaviour; also, other features of bioeconomy technologies, such as complexity of biological organisms or the increase in separability points in processes have an impact on the (increased) number of potential impacts and increase the type and range of effects, hence adding further uncertainty.

Economists distinguish at least three types of uncertainty:

- risk: in which there are some known states of nature and decision makers can attach some probability to them;
- uncertainty (Knightian uncertainty): is when states are naturally defined or can be constructed, but probabilities are unknown;
- fundamental uncertainty: occurs in the presence of 'structural ignorance' in which the states can neither be naturally defined or constructed.

It is recognized that in fields involving technical change, Knightian uncertainty is pervasive,

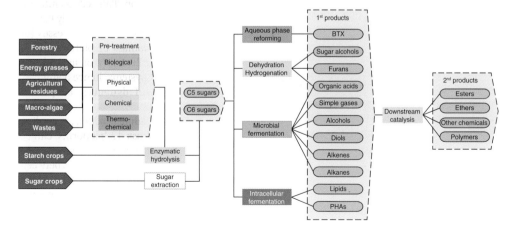

Fig. 3.2. Mapping of pathways through the sugar platform. Source: Taylor *et al.* (2015).

making risk–benefit balancing difficult for individuals and societies (Herring and Paarlberg, 2016). Herring and Paarlberg (2016) argue that this situation is especially prominent in the field of genetic engineering. In this field, risk is indeterminate because there is to date no established hazard from crop biotechnology from which a probability distribution of risk can be constructed and there is no way to predict unknown future hazards. On the other hand, proving the absence of risk is impossible for science. In many cases, especially in long and diversified impact pathways, the issue tends towards fundamental uncertainty. As a result, risk in this sphere is largely socially constructed and its management is largely to be understood in the context of the politics of science rather than economics.

The notion of total risk or non-probabilistic risk has been used in sustainability analysis to account for the very complex and unpredictable sources of uncertainty in global socioecological systems. Recent literature promotes a change in paradigm of risk conceptualization and measurement, towards the notion of 'tychastic risk measure' that make explicit the time component of risk measurement and which aims not to predict the future, but rather to make sure socioecological systems are ready to deal with future risky events (Aubin *et al.*, 2014; Bates and Saint-Pierre, 2018).

3.4.4   Steps in technology development

In the flow from research to innovation, a major distinction is between basic and applied research. However, this distinction is progressively becoming less pertinent in the description of the complex development process from basic ideas to marketing. More suitable for a detailed description of the research and innovation process and the different states of a technology is the classification of different technology readiness levels (TRL) as identified, for example by the European Commission:

- TRL 1: basic principles observed;
- TRL 2: technology concept formulated;
- TRL 3: experimental proof of concept;
- TRL 4: technology validated in laboratory;
- TRL 5: technology validated in relevant environment (industrially relevant environment in the case of KET);

- TRL 6: technology demonstrated in relevant environment (industrially relevant environment in the case of KET);
- TRL 7: system prototype demonstration in operational environment;
- TRL 8: system complete and qualified; and
- TRL 9: actual system proven in operational environment (competitive manufacturing in the case of KET; or in space).

The bioeconomy uses a number of low TRL levels of technology development, but in most cases it is related to relatively high levels of TRL. A growing distinction is that of identified KET, such as biotechnologies or ICT, that are in a way a horizontal basis for more applied technology developments. In some cases, this also falls under the name of 'cross-cutting technologies'. As mentioned above, bioeconomy development is largely driven by opportunities found within these cross-cutting technologies.

3.5   Economic Representation of Bioeconomy Technologies

3.5.1   Approach

In this section we investigate bioeconomy technology and its economic representation with a twofold objective. First, this section aims to elicit and formalize some of the main characteristics of bioeconomy technology as seen in the previous sections. Second, it seeks to provide simplified representations of the technology suitable for supporting considerations related to supply and matching of demand and supply in the following chapters.

The illustration proceeds in four steps:

1. First, we specify the concept of good and the related output features that determine human appreciation; in order to do so, we represent technology through the set of possible combinations of attributes that define a good. This is a way of approaching the increased openness of both biomass de-structuring and re-composition as well as of multiple characteristics that consumers attach to goods.

2. Second, we move one step backwards and investigate the relationship between input and output. This is a classic issue in economics, where technology can be investigated in its usual representation as a series of points on a production function, that is, as the shape, identified by its boundaries, of the production possibilities. Taking into account multiple factors, this takes the form of the shape of iso-yield curves in the factor space, while, considering multiple outputs, this takes the form of production possibility frontiers in the output space. Technology change is represented through changes in this shape and related boundaries.

3. Third, we move another step backwards and investigate harvesting/exploitation of natural resources and its relationship with natural capital. Technological change here can modify harvesting yields, costs and capital management, and, in turn, the amount of natural capital.

4. Finally, we close the loop and re-connect final output and capital dimensions both in their annual exploitation decision problem and in the long-term vision of sustainable development.

3.5.2 Products as bundles of attributes: decomposition, re-composition of biomass and product design

This section looks at the products themselves and attempts to characterize, from an economic point of view, the technologies that allow for a higher level of disaggregation of biomass in more simple and basic components that can later be used as the building blocks of a variety of products. Many more types of products can be identified both now and in the future thanks to this higher number of processes and flexibility (though this framework only accounts for a partial and somehow disappointing way of flexibility).

In order to deal with this, we represent a good as a combination of attributes. This concept is widely used in consumer economics (see Chapter 5) but is usually not used at all in production economics. There are a number of practical cases, however, in which the idea is used. Prices, for example, are linked to quantity levels of specific relevant compounds, which may be considered as attributes (e.g. sugar in sugar beet, proteins in some cereals, etc.). Policy objectives are stated in terms of some specific characteristics, i.e. higher need for proteins. In practice, the growing use of assessment methods (e.g. life

cycle assessment (LCA) which qualifies a product based on a list of performance attributes) goes in the same direction and shows the practical relevance of this approach (see Chapter 10).

A special case of this logic is when the degree of pureness of the obtained materials has an added value; this may be thought of as a case in which the potential value added of a good comes from an extreme (high) value of an attribute and low values of others. We can consider, widely speaking, the attributes of the process as part of the characteristics of the goods itself. However, very often, it is useful to distinguish the features of the process that do not modify the 'intrinsic quality' of the final product, and in this case we can distinguish attributes of the process from the attributes of the product. Environmental and ethical implications can be also considered. Note here the relevant distinction between product innovation and process innovation on the one hand and between different types of goods concerning the ability of consumers to identify the good's characteristics before purchase, on the other hand (see also in the next chapters).

An important feature of this representation is that different stakeholders can attribute different values to individual attributes, which allows for a better understanding of differences in perceived values of a good. This makes it possible to manage notions such as that of externality, in which private and public values attributed to a good's production (or private values given by different individuals) may differ.

We take up this idea of attributes as the relevant point of attention in order to represent technologies as possibilities in the space represented by combinations of attributes. Figure 3.3 represents goods in a two-attribute space (for simplification purposes). In the figure, a different good is defined as a combination of the level of two attributes. Thus, P_1, P_2, P_3 are different goods as identified by the attribute levels 1, 2 and 3, respectively.

In fact, there could be uncertainty about the actual attribute levels due, for example, to variability in composition, which means that the product is actually a cloud around the most likely central point (as in the case of the cloud around P_1).

Both of these characteristics (good defined by a combination of attributes and uncertainty about the actual level of attributes of a good) are not exclusive of biological products but are very

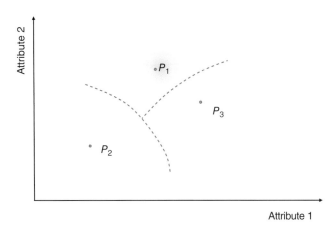

Fig. 3.3. Goods representation in a two-attribute space. P_1, P_2, P_3 are different goods as identified by the attribute levels 1, 2 and 3, respectively.

important in this field because products coming from biological resources tend to be very complex and with variable features due to the number of potential determinants of these features. This also implies issues in good definition and explains practices widely used in the bioeconomy, such as the definition of standardization quality levels, as well as traceability and origin detection. Standards could be thought of as drawing lines that provide a boundary among different areas of the figure with each area identified as a different product.

Innovation can have an impact in different ways, such as:

- making intermediate combinations possible that were not available before;
- making more extreme combinations possible, either in the direction of isolating one single attribute or of having a higher amount of both attributes, either in the sense of total value of attributes or relative values of attributes (for example, changing the relative concentration of compounds); and/or
- reducing uncertainty about the characteristics of the product (concentrating the cloud around one point or well-identified individual points).

These technical changes can be the result of higher flexibility in the process, for example, the ability to break down biomass in individual components in the upstream processes, or to combine them with different alternative solutions.

Enlarging the decision space in the field of potential sets of attributes may help finding

potentially better solutions as the process becomes more flexible and combinations of attributes can better fit utility, as well as finding types of goods that respond to combinations of attributes with higher overall benefits.

In addition, technology can modify the amount of resources needed to obtain the same or different combinations of attributes (i.e. reduce costs), which leads us to the next point.

3.5.3 Input–output relationships

The general specification of a technology can be illustrated through a relationship between multiple inputs and outputs. This general form is usually illustrated in economics through three different partial representations: one input–one output relationship, one factor–one factor relationship and one output–one output relationship (Debertin, 2012).

Cases of input–output relationships are illustrated in Fig. 3.4.

X is the amount of a variable input, whereas Y is a product obtained with that input. Classic examples of production functions pertain precisely to the production of biological resources (e.g. agricultural processes). For example, X is a fertilizer and Y is wheat. Curve A illustrates a typical example of such a relationship. Increasing the amount of fertilizer used in a process, the amount of wheat produced increases. There is, however, a limitation to this and above certain levels of fertilizer use the amount of wheat produced decreases. This illustrates limitations due to the availability of

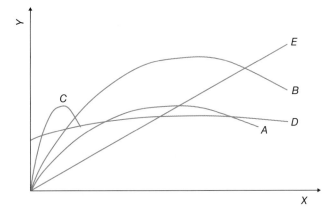

Fig. 3.4. Input–output relationships with different technologies. *A–E*, different examples of input versus output (see the text for more details); *X*, amount of a variable input; *Y*, product obtained with that input.

other factors (assumed to be fixed in this representation) (e.g. land) and the inherent limitation of living organisms to react to outside inputs due to their physiological features. The other curves represent technology alternatives to obtain the same product, such as, for example, different varieties of the same crop. Case *B* is that of a variety with a higher yield for the same amount of input for every level of input. The availability of variety *B* would represent a technology improvement compared to *A*, by expanding production possibilities. Curve *C* provides an example of different properties. Curve *C* barely reaches the same level of production of *A* and certainly not that of *B*, but achieves its highest production level at a much lower level of inputs; hence, the crop may be more suitable when the price of input *X* is high or when strong restrictions in input availability may be expected, for example, in areas at risk of drought if input is water. Case *D* pushes this in another direction. It never reaches the production level of the other cases, but allows a high yield even without factor *X*; in addition, it is very stable across different amounts of factor *X* and hence can be suitable for situations of uncertainty in input availability. These relationships can apply to a number of other cases of bio-based processes (e.g. biogas production plants) in which living organisms transform an input and other input different from *X* remain stable. Case *E*, on the contrary, represents a linear relationship, that is, the amount of output is proportional to the amount of input and is more frequent in non-biological processes or when other inputs are allowed to change at the same time.

Uncertainty considerations can be attached to this model in the form of potential variability of outcome around the production line; that is, given a certain level of input, the production can be located on the vertical axis around the function with a certain distribution of probability. This occurs, for example, in the case of uncertain yields for the same amount of fertilizers or with uncertain biogas production for the same amount of biomass input.

The input–input relationship and the output–output relationship with different technologies are illustrated in Fig. 3.5.

In the input–input representation, the axes measure the amount of two variable factors (X_1 and X_2); curve A illustrates the combinations of X_1 and X_2 that make it possible to obtain the same amount of a good (say again, Y). The curve hence illustrates the substitution possibilities between the two factors. Again, curves B and C present alternatives that can be implemented due to technological change. For example, curve B allows for the achievement of the same amount of good with lower levels of factors; this roughly applies also to C, but what is most evident is that the degree of substitution also changes and, in particular, the same level of production can be achieved with much lower levels of X_2 than before, provided a higher level of X_1 is used. This representation illustrates technology change both in terms of reduction of input per unit of output on multiple input dimensions and, what is more important, as inducing changes in the combinations of input most suitable from an economic point of view (non-neutral technology change). As an example of this reasoning, assume X_1 is

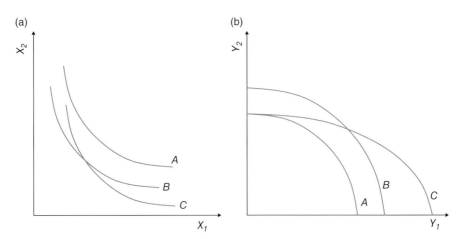

Fig. 3.5. (a) Input–input and **(b)** output–output relationship with different technologies. *A–C*, different examples of input–input or output–output relationships (see the text for more details) X_1, X_2, variable production factors; Y_1, Y_2, products.

biomass and X_2 a fossil input in a blended fuel. Moving from *A* to *B*, the amount of both input is reduced, while moving from *A* to *C* allows to achieve the same yield with a much lower proportion of the fossil input. Other relevant examples may include the trade-off between different biomass sources, the trade-off between labour and other input, the trade-off between knowledge intensity and biomass use, and so on.

Figure 3.5b illustrates the complementary view of trade-offs among the production of two outputs (Y_1 and Y_2) given a certain amount of production factors. This representation illustrates the multi-output facility such as a biorefinery (producing both energy and bio-based compounds) or a farm (e.g. producing multiple crops, or a crop and bioenergy). Under certain conditions, the relationship may be linear, but the interesting case is when it is a curve, such as in the case *A*, as this brings to the economic problem of the optimal combination of output. Technology change here is represented as an expansion of production possibilities. Compared to curve *A*, both curve *B* and *C* represent effects of technology changes. However, while *B* illustrates an improvement of the production possibilities for both of the products in roughly the same way, *C* illustrates the case of a much higher increase of production possibilities of Y_1 compared with Y_2.

A key technological parameter to measure performance is yield, intended as the units of output per unit of input.

3.5.4 Natural capital–harvesting relationships

Producing by using biological resources implies working with living organisms, that is, entities that are able to reproduce, grow or reduce in number over time. Living organisms are also in ecological relationships with the surrounding environment and with other organisms.

Taking a single resource, the basic theory of renewable resources, widely used to deal with resources such as fish or forests, explains the behaviour based on a stock-flow representation, where the maximum potential harvest is a function of the stock. The stock itself (*S*), in a given ecosystem, assuming it starts from a few individuals, tends to grow over time (*t*) and to reach an equilibrium, as shown in Fig. 3.6(a) (curve *A*).

This equilibrium, assuming stability of external conditions, is determined by external limits to grow due to limits of resources, predators or congestion of individuals of the same species.

ABC is not in ecological equilibrium but the stock can change depending on human action, and eventually reach a different equilibrium or follow some non-equilibrium pathway over time depending on the interactions with human activities.

Different human activities can affect the stock:

1. First, through harvesting, the stock may be decreased. The stock, however, tends to

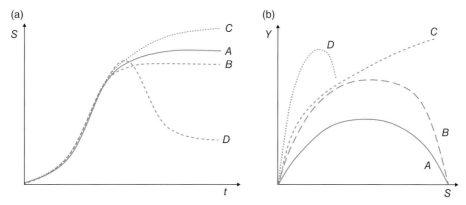

Fig. 3.6. Trends in the relationship between resource stock and harvest. *A–D*, different scenarios (see the text for details); *S*, stock; *t*, time; *Y*, yield.

compensate for this thanks to growth and reproduction. Assuming a constant harvesting level, the stock could find an equilibrium with harvesting at a different level of capital stock (e.g. function *B*).

2. On the other hand, capital can keep growing because human activities keep investing in it (function *C*). This can actually involve different types of actions, such as: (i) removing ecological limits to growth, e.g. through fertilization, fighting natural enemies or threats (e.g. fires); (ii) investing resources to reproduce individuals (e.g. reforestation); and (iii) increasing quality, for example, as a result of research in terms of, for example, genetic diversity and potential for production. This can primarily affect yields and hence harvesting potential, but could also, indirectly, affect the stock.

3. Human action can also play a role in keeping the stock very low by excluding parts of the stock that are not considered useful and reproducing through a well identified but small set of 'reproducers', as happens for cultivated varieties. Starting from an initial (supposedly) wide stock of a biological resource, the selection can also mean significantly reducing the stock, as in the case of the function *D* in Fig. 3.6.

As mentioned, the notions of harvesting (such as fisheries, forestry) and cultivation are increasingly less distinguishable so we use it interchangeably with generic exploitation that can include some degree of cultivation activities. Curve *A* in Fig. 3.6(b) depicts the relationship between stock and maximum yield in a population in natural conditions.

The shape of the figure is determined by the fact that low stock produces low yields (e.g. due to low ability to reproduce), but also high stock produces low yields due to high congestion by the population and overexploitation of supporting resources (e.g. food sources). Examples of this are usually available for natural stocks, such as forestry (see recent examples published in Heinonen *et al.*, 2017) or fisheries.

Investment in natural capital by humans, or human activities, can modify this relationship, for example, increasing yield per unit of stock in a homogeneous way throughout different stock levels (curve *B* in Fig. 3.6(b)). Other actions/investments can modify different aspects in a less neutral way, for example, avoiding the decrease in production due to congestion (e.g. modifying the environment or providing nutrients) as in curve *C* in Fig. 3.6(b) (though it cannot expect to grow forever with stock at least for earth limits). Finally, investment can move maximum yields at different levels of stock, as represented by curve *D* in Fig. 3.6(b), for example, by stimulating selective reproduction of the population.

As long as investment in *ABC* modifies the stock–yield relationship, this can also affect the effort–harvest relationships, which have implications for the optimal level of effort and harvesting. For example:

- moving from *A* to *B* in Fig. 3.6(b) increases the optimal harvesting effort (by increasing the marginal revenue with additional harvesting efforts) and hence decrease the optimal stock level;

- moving from *A* to *C* in Fig. 3.6(b) allows higher harvesting at high stocks and most likely would provide incentives for an increase in stocks; and
- moving from *A* to *D* would allow higher harvest only at low level of capital stock (i.e. towards the left-hand side of Fig. 3.6b). This would be a pathway of investment that increases efficiency of the stock, but makes it optimal to keep a lower capital (in some cases this can be associated with type *D* effects of human action in Fig. 3.6(a)).

3.5.5 Trade-offs between natural/anthropized capital and yearly goods production

The nature of the development of the bioeconomy can be seen as an evolution of trade-offs between annual production possibilities and the size of the *ABC*, assuming, in each case, long-term equilibrium, as depicted in Fig. 3.7, where the horizontal axis represents the capital stock, while the vertical axis represents the yearly production.

Curve *A* represents a technology situation where there is a trade-off between yearly production and *ABC*, making it possible to revisit (qualify) both the concept of decoupling between environment and production and the paradigms connecting growth and the environment. This is the case, for example, in an economy based on fossil resources where biological capital is reduced to leave more non-biological resources available (e.g. space/land, fertilizers,

water) for fossil-based production of biomass. This is roughly, though exaggerated, the situation of the current (partly past) fossil economy. Curve *B* represents a first move towards a bioeconomy in which production is partly synergic (top part) and partly competitive (right/bottom part of the curve) with the stock of anthropized capital. This is a situation in which a management of the *ABC* contributes to better production, but capital stock above a certain level can only be maintained with a reduction of production (and hence consumption) by society. This is somehow close to the current technology. A stronger improvement in technology could go in the direction of a better, more direct relationship between capital stock and production as in curve *C*. Production is more bio-dependent (so lower for lower levels of *ABC*) but keeps growing with growing amount of capital, though less than proportionally (which means that higher levels of capital need to give up proportionally more production). Curve *D* represents a major shift in technology towards more extreme bio-dependency, but also more than proportional benefits from the use of *ABC*. It may be identified with a high-tech ecosystem services-based technology, in which there are synergies from higher *ABC*, rather than obstacles to production. Curve *E* represents the total decoupling from *ABC*, in which any level of production can be achieved independently from the *ABC*, which is in turn determined by other factors. These could be seen as foreseeable futuristic technologies in which production is fully achieved by solar-based technologies producing organic artificial matters independently from *ABC*.

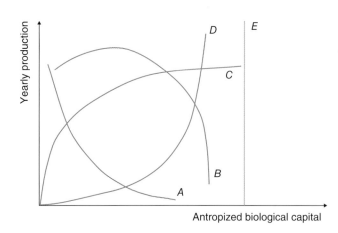

Fig. 3.7. Trade-offs between natural/anthropized capital and yearly goods production.

In all cases, except curve E, the possibility to modify the characteristics of the ABC and, at the same time, to exploit it for production, contributes to the yearly trade-off between production and investment in ABC. For each unit in time, given a certain amount of resources, society has the question of how much to improve the ABC and how much effort to put in its short-term exploitation. This issue will be further developed in Chapter 4, section 4.4.3.

3.5.6 Towards a unified view of bioeconomy technology

The integrative nature of the bioeconomy goes in the direction of exploiting the suggestions coming from the simple models above, towards an integrated view in which the production function can be seen as a relationship between input and output linking anthropized capital (intended as a mix of natural and artificial capital) with output defined as a set of attributes. This direction is partly taken in practice by models connecting resources and output qualified by their characteristics. An example is provided by Vale *et al.* (2017), coupling an input–output model with LCA techniques and a participatory approach. In a way this goes in the same direction of LCA methods applied to product and multi-attribute approaches used in environmental analysis.

However, to the best knowledge of the author there is neither a full theoretical development of this integrated approach, nor a complete empirical implementation, including also resource stocks.

This approach can be envisaged as going in the direction of a revision of the concept of production processes. Production processes can be seen as processes modifying the amount of stock and flows of attributes in a system, in such a way that the total value of such attributes is increased. In this concept, changes in attributes due to the activation of a process can be positive or negative in value, so the result will depend on the balance among changes in these opposite directions. Research and innovation in this context act by increasing the range of possible attribute sets that are feasible in a system, which also includes the productive/reproductive ability of biological resources.

Goods and products in this view reduce their roles as relevant objects of attention, as the value for society depends more on substitutes, design choices, upstream and downstream connections, rather than from their characteristic of 'contingent' combination of attributes.

3.6 Outlook

Bioeconomy technologies are characterized by specificities due the use of biological resources, by the shift towards sustainable bio-based solutions and by the trend towards an increasingly greater ability to break down and recompose biomass. Technology is changing rapidly and will bring continuous change in the near future, which will be key to contribute to the ability of the bioeconomy to answer societal challenges (Wesseler and Von Braun, 2017).

A Delphi study identified seven priority project areas for developing a sustainable bioeconomy, which are also a good proxy of the way forward in the development of the bioeconomy (El-Chichakli *et al.*, 2016): (i) new food and sustainable agri-food systems; (ii) biosmart and integrated urban areas; (iii) the next-generation biorefineries; (iv) artificial photosynthesis; (v) marine bioeconomy; (vi) the development of consumer markets; and (vii) international regulation and governance of the bioeconomy.

The OECD Secretariat (OECD, 2017) lists a number of key technologies to be further developed and to improve the transition from laboratory to the market: biomass pre-treatment and consolidated bioprocessing; growth on carbon compounds; computational enzyme design; minimal cells for bio-contained microbial factories; small-scale fermentation models; and gene and genome editing in production strains.

Some individual species may have a central role. For example, *Saccharomyces cerevisiae*, being the major bioethanol producer, has a central position among biofuel-producing organisms and researchers are assessing a number of pathways to improve its use and performance (Petrovič, 2015). The properties of derived biofuels should be more similar to those of oil-derived fuels than those of ethanol, as the current trend in the development of biofuels is to synthesize molecules that can be used as drop-in fuels for existing engines.

Some of the biggest scientific breakthroughs are expected from synthetic biology. Aro (2016) see the possibility of a major breakthrough and highlights the example of direct conversion of solar energy to a fuel from sunlight, water and CO_2, using engineered photosynthetic microorganisms or in completely synthetic living factories.

Technology features are also important, with productivity, robustness and flexibility being perceived as key for the future performance of the sector. At the same time, the problems of managing the inherent variability of biological matter pushes more and more for standardization as well as deconstruction and reconstruction (Mackenzie *et al.*, 2013; de Lorenzo and Schmidt, 2018). Indeed, uncertainties, as well as, low performance compared with expectations have also been an obstacle to acceptance of innovation in the sector.

The bioeconomy is embedded in a wide industrial revolution based on a combination of different fields, so bioeconomy technologies cannot be thought of in isolation, but rather need to exploit synergies and opportunities related to all other fields of change (OECD, 2017). While the number of potential synergies and interconnections is growing exponentially, the most suitable combinations may be largely context-specific (Sarkar *et al.*, 2017).

The need to conceptualize multiple knowledge interaction as a way of understanding and developing the bioeconomy has also been highlighted as a key subject for future investigation (Wield *et al.*, 2017). A major area for evolution is that of knowledge and technology linked to soft technologies, such as big data and information technologies.

As a result of the above, the understanding of technology and its representation are among the most important and challenging issues of the bioeconomy studies. One key aspect concerning economic studies is that technology should no longer be considered in isolation from the institutional context. Some of the future solutions could bring major changes in bioeconomy markets not only affecting input–output relationships, but also modifying consumer/societal perceptions, hence qualifying them as less 'invasive' and more acceptable. The space for decision making is increasing and the view of technology as an objective issue, a technical possibility, on which decision makers take decisions based on their simple interest, is increasingly far from current technology management systems. On the contrary, final targets and objectives are increasingly embedded in the governance systems that affect research and technology development. Works scoping future technology pathways emphasize the link with society and the need to accelerate the building of a collaborative and engaged community between government agencies, industry, academia and the public (Friedman and Ellington, 2015). Issues such as responsible innovation are at the forefront of these approaches (Shortall *et al.*, 2015). A huge issue currently attracting attention is openness in science. This is not a unique topic for biological science and is connected to the value of research output. The issue is discussed by Levin and Leonelli (2017), arguing against the 'one-size-fits-all' view of openness and rather seeing it as a rather contextualized approach. At the same time, preferences are not independent from technology use and regulatory approaches. These interdependences will be better understood in the next six chapters. New ways of analysing technologies and supporting choices are also needed. This will be better discussed in Chapter 10.

References

Aro, E.-M. (2016) From first generation biofuels to advanced solar biofuels. *Ambio* 45(Suppl 1), S24–31. doi: 10.1007/s13280-015-0730-0.

Aubin, J.-P., Chen, L. and Dordan, O. (2014) *Tychastic Measure of Viability Risk*. Cham, Switzerland Springer International Publishing. doi: 10.1007/978-3-319-08129-8.

Bais-Moleman, A.L., Sikkema, R., Vis, M., Reumerman, P., Theurl, M.C. and Erb, K.-H. (2017) Assessing wood use efficiency and greenhouse gas emissions of wood product cascading in the European Union. *Journal of Cleaner Production* 172, 3942–3954. doi: 10.1016/j.jclepro.2017.04.153.

Balter, M. (2013) Archaeologists say the 'anthropocene' is here: But it began long ago. *Science* 340(6130), 261–262. doi: 10.1126/science.340.6130.261.

Bates, S. and Saint-Pierre, P. (2018) Adaptive policy framework through the Lens of the Viability Theory: A theoretical contribution to sustainability in the anthropocene era. *Ecological Economics* 145, 244–262. doi: 10.1016/j.ecolecon.2017.09.007.

Bennett, A.B., Chi-Ham, C., Barrows, G., Sexton, S. and Zilberman, D. (2013) Agricultural biotechnology: Economics, environment, ethics, and the future. *Annual Review of Environment and Resources* 38, 249–279. doi: 10.1146/annurev-environ-050912-124612.

Bergstrom, J.C. and Randall, A. (2016) *Resource Economics: An Economic Approach to Natural Resource and Environmental Policy*, Fourth Edition. Northampton, Massachusetts, USA, Edward Elgar Publishing.

Bomtempo, J.-V., Chaves Alves, F. and De Almeida Oroski, F. (2017) Developing new platform chemicals: What is required for a new bio-based molecule to become a platform chemical in the bioeconomy? *Faraday Discussions* 202, 213–225. doi: 10.1039/c7fd00052a.

Bott, M. and Eggeling, L. (2017) Novel technologies for optimal strain breeding. *Advances in Biochemical Engineering/Biotechnology* 159, 227–254. doi: 10.1007/10_2016_33.

Brown, Z. (2017) Economic, Regulatory and International Implications of Gene Drives in Agriculture. *Choices*, 2nd Quarter 32(2), 1–8.

Budzianowski, W.M. (2017) High-value low-volume bioproducts coupled to bioenergies with potential to enhance business development of sustainable biorefineries. *Renewable and Sustainable Energy Reviews* 70, 793–804. doi: 10.1016/j.rser.2016.11.260.

Budzianowski, W.M. and Postawa, K. (2016) Total chain integration of sustainable biorefinery systems. *Applied Energy* 184, 1432–1446. doi: 10.1016/j.apenergy.2016.06.050.

Cueva, M. (2017) 3rd Congress on Applied Synthetic Biology in Europe (Costa da Caparica, Portugal, February 2016). *New Biotechnology* 35, 62–68. doi: 10.1016/j.nbt.2016.12.001.

Dahiya, S., Kumar, A.N., Shanthi Sravan, J., Chatterjee, S., Sarkar, O. and Mohan, S.V. (2017) Food waste biorefinery: Sustainable strategy for circular bioeconomy. *Bioresource Technology* 248(Pt A), 2–12. doi: 10.1016/j.biortech.2017.07.176.

Debertin, D.L. (2012) *Agricultural Production Economics*. Available at www.uky.edu/~deberti/ap/ap.pdf (accessed 25 May 2018).

Deleuze, C., Richter, C., Ulrich, E., Musch, B., Descroix, L., Pousse, N., Dreyfus, P., Bock, J., Riond, C. and Legay, M. (2016) Gestion des peuplements en forêt publique: Nouvelles pistes de recherche, développement et innovation. *Revue Forestiere Francaise* 68(6), 547–558. doi: 10.4267/2042/62401.

de Lorenzo, V. and Schmidt, M. (2018) Biological standards for the knowledge-based bioeconomy: What is at stake. *New Biotechnology* 40(Pt A):170–180. doi: 10.1016/j.nbt.2017.05.001.

Egelyng, H., Romsdal, A., Hansen, H.O., Slizyte, R., Carvajal, A.K., Jouvenot, L., Hebrok, M., Honkapää, K., Wold, J. P., Seljåsen, R. and Aursand, M. (2016) Cascading Norwegian co-streams for bioeconomic transition. *Journal of Cleaner Production* 172, 3864–3873. doi: 10.1016/j.jclepro.2017.05.099.

El-Chichakli, B., von Braun, J., Lang, C., Barben, D. and Philp, J. (2016) Five cornerstones of a global bioeconomy. *Nature* 535, 221–223.

European Commission (2015) Closing the loop – An EU action plan for the Circular Economy. COM(2015) 614 final. Available at https://ec.europa.eu/transparency/regdoc/rep/1/2015/EN/1-2015-614-EN-F1-1.PDF (accessed essed on 30 December 2017).

Flores Bueso, Y. and Tangney, M. (2017) Synthetic biology in the driving seat of the bioeconomy. *Trends in Biotechnology* 35(5), 373–378. doi: 10.1016/j.tibtech.2017.02.002.

Francavilla, M., Intini, S. and Monteleone, M. (2016) Designing an integrated technological platform centered on microalgae to recover organic waste and obtain multiple bioproducts. *European Biomass Conference and Exhibition Proceedings*, pp. 294–299, doi: 10.5071/24thEUBCE2016-1CV.4.16.

Friedman, D.C. and Ellington, A.D. (2015) Industrialization of biology. *ACS Synthetic Biology* 4(10), 1053–1055. doi: 10.1021/acssynbio.5b00190.

Heinonen, T., Pukkala, T., Mehtätalo, L., Asikainen, A., Kangas, J. and Peltola, H. (2017) Scenario analyses for the effects of harvesting intensity on development of forest resources, timber supply, carbon balance and biodiversity of Finnish forestry. *Forest Policy and Economics* 80, 80–98. doi: 10.1016/j.forpol.2017.03.011.

Herring, R. and Paarlberg, R. (2016) The political economy of biotechnology. *Annual Review of Resource Economics* 8(1), 397–416. doi: 10.1146/annurev-resource-100815-095506.

Heux, S., Meynial-Salles, I., O'Donohue, M.J. and Dumon, C. (2015) White biotechnology: State of the art strategies for the development of biocatalysts for biorefining. *Biotechnology Advances* 33(8), 1653–1670. doi: 10.1016/j.biotechadv.2015.08.004.

Hingsamer, M., Jungmeier, G., Kleinegris, D. and Barbosa, M. (2016) Modelling and assessment of algae cultivation for large scale biofuel production – sustainability and aspects of up-scaling of algae biorefineries. *European Biomass Conference and Exhibition Proceedings*, pp. 1457–1459. doi: 10.5071/24thEUBCE2016-4AV.1.31.

IEA Bioenergy (2014) IEA Bioenergy Task42 Biorefining. Available at http://www.ieabioenergy.com/wp-content/uploads/2014/09/IEA-Bioenergy-Task42-Biorefining-Brochure-SEP2014_LR.pdf (accessed 20 May 2018).

Jiménez-Sánchez, G. and Philp, J. (2016) Genomics and the bioeconomy: Opportunities to meet global challenges. In Kumar, D. and Chadwick, R. (eds) *Genomics and Society: Ethical, Legal, Cultural and Socioeconomic Implications*. London, Elsevier, pp. 207–238. doi: 10.1016/B978-0-12-420195-8.00011-2.

Jordan, N.R., Dorn, K., Runck, B., Ewing, P., Williams, A., Anderson, K.A., Felice, L., Haralson, K., Goplen, J., Altendorf, K., Fernandez, A., Phippen, W., Sedbrook, J., Marks, M., Wolf, K., Wyse, D. and Johnson, G. (2016) Sustainable commercialization of new crops for the agricultural bioeconomy. *Elementa* 4, 000081. doi: 10.12952/journal.elementa.000081.

Keegan, D., Kretschmer, B., Elbersen, B. and Panoutsou, C. (2013) Cascading use: A systematic approach to biomass beyond the energy sector. *Biofuels, Bioproducts and Biorefining* 7(2), 193–206. doi: 10.1002/bbb.1351.

Krüger, A., Schäfers, C., Schröder, C. and Antranikian, G. (2017) Towards a sustainable biobased industry: Highlighting the impact of extremophiles. *New Biotechnology* 40(Pt A), 144–153. doi: 10.1016/j.nbt.2017.05.002.

Levin, N. and Leonelli, S. (2017) How does one 'open' science? Questions of value in biological research. *Science Technology and Human Values* 42(2), 313–323. doi: 10.1177/0162243916672071.

Lokko, Y., Heijde, M., Schebesta, K., Scholtès, P., Van Montagu, M. and Giacca, M. (2017) Biotechnology and the bioeconomy – Towards inclusive and sustainable industrial development. *New Biotechnology* 40(Pt A), 5–10. doi: 10.1016/j.nbt.2017.06.005.

Lopes, M.S.G. (2015) Engineering biological systems toward a sustainable bioeconomy. *Journal of Industrial Microbiology and Biotechnology* 42(6), 813–838. doi: 10.1007/s10295-015-1606-9.

Mackenzie, A., Waterton, C., Ellis, R., Frow, E. K., McNally, R., Busch, L. and Wynne, B. (2013) Classifying, constructing, and identifying life: Standards as transformations of 'the biological'. *Science Technology and Human Values* 38(5), 701-722. doi: 10.1177/0162243912474324.

Małyska, A. and Jacobi, J. (2017) Plant breeding as the cornerstone of a sustainable bioeconomy. *New Biotechnology* 40(Pt A), 129–132. doi: 10.1016/j.nbt.2017.06.011.

Matharu, A. S., de Melo, E. M. and Houghton, J. A. (2016) Opportunity for high value-added chemicals from food supply chain wastes. *Bioresource Technology* 215, 123–130. doi: 10.1016/j.biortech.2016.03.039.

Meyer, V., Fiedler, M., Nitsche, B. and King, R. (2015) The cell factory *Aspergillus* enters the big data era: Opportunities and challenges for optimising product formation. *Advances in Biochemical Engineering/Biotechnology*. doi: 10.1007/10_2014_297.

Mihai, F.-C. and Ingrao, C. (2016) Assessment of biowaste losses through unsound waste management practices in rural areas and the role of home composting. *Journal of Cleaner Production* 149, 91–132. doi: 10.1016/j.jclepro.2016.10.163.

Mohan, S.V., Butti, S.K., Amulya, K., Dahiya, S. and Modestra, J.A. (2016) Waste biorefinery: A new paradigm for a sustainable bioelectro economy. *Trends in Biotechnology* 34(11), 852–855. doi: 10.1016/j.tibtech.2016.06.006.

Montella, S., Amore, A. and Faraco, V. (2016) Metagenomics for the development of new biocatalysts to advance lignocellulose saccharification for bioeconomic development. *Critical Reviews in Biotechnology* 36(6), 998–1009. doi: 10.3109/07388551.2015.1083939.

Octave, S. and Thomas, D. (2009) Biorefinery: Toward an industrial metabolism. *Biochimie* 91(6), 659–664. doi: 10.1016/j.biochi.2009.03.015.

OECD (2017) *The Next Production Revolution. Implications for Governments and Business*. Paris, France, The Organisation for Economic Co-operation and Development.

Overmann, J. (2015) Significance and future role of microbial resource centers. *Systematic and Applied Microbiology* 38(4), 258–265. doi: 10.1016/j.syapm.2015.02.008.

Parisi, C., Vigani, M. and Rodríguez-Cerezo, E. (2017) Agricultural nanotechnologies: What are the current possibilities?. *Nano Today* 10(2), 124–127. doi: 10.1016/j.nantod.2014.09.009.

Peinemann, J.C. and Pleissner, D. (2017) Material utilization of organic residues. *Applied Biochemistry and Biotechnology* 184(2):733–745. doi: 10.1007/s12010-017-2586-1.

Petrovič, U. (2015) Next-generation biofuels: A new challenge for yeast. *Yeast* 32(9), 583–593. doi: 10.1002/yea.3082.

Poltronieri, P. and D'Urso, O.F. (2016) *Biotransformation of Agricultural Waste and By-products: The Food, Feed, Fibre, Fuel (4F) Economy*. London, Elsevier.

Poz, M.E.D., da Silveira, J.M.F.J., Bueno, C.S. and Rocha, L.A. (2017) Bio-based energy scenarios: Looking for waste. *Procedia Manufacturing* 7, 478–489. doi: 10.1016/j.promfg.2016.12.048.

Puyol, D., Batstone, D.J., Hülsen, T., Astals, S., Peces, M. and Krömer, J.O. (2017) Resource recovery from wastewater by biological technologies: Opportunities, challenges, and prospects. *Frontiers in Microbiology* 7, 2106. doi: 10.3389/fmicb.2016.02106.

Sarkar, S.F., Poon, J.S., Lepage, E., Bilecki, L. and Girard, B. (2017) Enabling a sustainable and prosperous future through science and innovation in the bioeconomy at Agriculture and Agri-Food Canada. *New Biotechnology* 40(Pt A), 70–75. doi: 10.1016/j.nbt.2017.04.001.

Sawatdeenarunat, C., Nguyen, D., Surendra, K.C., Shrestha, S., Rajendran, K., Oechsner, H., Xie, L. and Khanal, S.K. (2016) Anaerobic biorefinery: Current status, challenges, and opportunities. *Bioresource Technology* 215, 304–313. doi: 10.1016/j.biortech.2016.03.074.

Schrager, B. and Suryanata, K. (2018) Seeds of accumulation: Molecular breeding and the seed corn industry in Hawai'i. *Journal of Agrarian Change* 18, 370–384. doi: 10.1111/joac.12207.

Scoma, A., Rebecchi, S., Bertin, L. and Fava, F. (2016) High impact biowastes from South European agro-industries as feedstock for second-generation biorefineries. *Critical Reviews in Biotechnology* 36(1), 175–189. doi: 10.3109/07388551.2014.947238.

Shortall, O.K., Raman, S. and Millar, K. (2015) Are plants the new oil? Responsible innovation, biorefining and multipurpose agriculture. *Energy Policy* 86, 360–368. doi: 10.1016/j.enpol.2015.07.011.

Sillanpää, M. and Ncbi, C. (2017) *A Sustainable Bioeconomy. The Green Industry Revolution*. New York, USA, Springer.

Stévant, P., Rebours, C. and Chapman, A. (2017) Seaweed aquaculture in Norway: recent industrial developments and future perspectives. *Aquaculture International* 25(4), 1373–1390. doi: 10.1007/s10499-017-0120-7.

Storz, H. and Vorlop, K.-D. (2013) Bio-based plastics: Status, challenges and trends. *Landbauforschung Volkenrode* 63(4), 321–332. doi: 10.3220/LBF-2013-321-332.

Taylor, R., Nattrass, L., Alberts, G., Robson, P., Chudziak, C., Bauen, A., Marsili Libelli, I., Lotti, G., Prussi, M., Nistri, R., Chiaramonti, D., López Contreras, A., Bos, H., Eggink, G., Springer, J., Bakker, R. and van Ree, R. (2015) *From the Sugar Platform to Biofuels and Biochemical: Final Report for the European Commission Directorate-General Energy*, N° ENER/C2/423-2012/SI2.673791. E4Tech, RE-CORD and Wageningen University.

Toppinen, A., Röhr, A., Pätäri, S., Lähtinen, K. and Toivonen, R. (2016) The future of wooden multistory construction in the forest bioeconomy: A Delphi study from Finland and Sweden. *Journal of Forest Economics* 31, 3–10. doi: 10.1016/j.jfe.2017.05.001.

Tossavainen, M., Nykänen, A., Valkonen, K., Ojala, A., Kostia, S. and Romantschuk, M. (2017) Culturing of Selenastrum on diluted composting fluids; conversion of waste to valuable algal biomass in presence of bacteria. *Bioresource Technology* 238, 205–213. doi: 10.1016/j.biortech.2017.04.013.

Vale, M., Pantalone, M. and Bragagnolo, M. (2017) Collaborative perspective in bio-economy development: A mixed method approach. *Collaboration in a Data-Rich World: 18th IFIP WG 5.5 Working Conference on Virtual Enterprises, PRO-VE 2017, Vicenza, Italy, September 18-20, 2017, Proceedings*. New York, USA, Springer, pp. 553–563. doi: 10.1007/978-3-319-65151-4_49.

Viaene, J., Van Lancker, J., Vandecasteele, B., Willekens, K., Bijttebier, J., Ruysschaert, G., De Neve, S. and Reubens, B. (2016) Opportunities and barriers to on-farm composting and compost application: A case study from northwestern Europe. *Waste Management* 48, 181–192. doi: 10.1016/j.wasman.2015.09.021.

Viaggi, D. (2016) Towards an economics of the bioeconomy: four years later. *Bio-based and Applied Economics* 5(2), 101–112.

Wendisch, V. F., Brito, L. F., Gil Lopez, M., Hennig, G., Pfeifenschneider, J., Sgobba, E. and Veldmann, K. H. (2016) The flexible feedstock concept in Industrial Biotechnology: Metabolic engineering of Escherichia coli, Corynebacterium glutamicum, Pseudomonas, Bacillus and yeast strains for access to alternative carbon sources. *Journal of Biotechnology* 234, 139–157. doi: 10.1016/j.jbiotec.2016.07.022.

Wesseler, J. and Von Braun, J. (2017) Measuring the bioeconomy: economics and policies. *Annual Review of Resource Economics* 9, 275–298. doi: 10.1146/annurev-resource-100516-053701.

Wield, D., Tait, J., Chataway, J., Mittra, J. and Mastroeni, M. (2017) Conceptualising and practising multiple knowledge interactions in the life sciences. *Technological Forecasting and Social Change* 116, 308–315. doi: 10.1016/j.techfore.2016.09.025.

You, C. and Percival Zhang, Y.-H. (2013) Cell-free biosystems for biomanufacturing. *Advances in Biochemical Engineering/Biotechnology* 131, 89–119. doi: 10.1007/10_2012_159.

Zilberman, D., Kim, E., Kirschner, S., Kaplan, S. and Reeves, J. (2013) Technology and the future bioeconomy. *Agricultural Economics* 44(S1), 95–102. doi: 10.1111/agec.12054.

4

Approaches to (the Economics of) the Bioeconomy

4.1 Introduction and Overview

Defining the bioeconomy and its boundaries in economics is a particularly difficult task given the number of interconnections that it embodies, both as a concept and as a sector. The definition of the bioeconomy in itself is largely driven by policy action and the contents of bioeconomy strategies worldwide. The term bioeconomy, on the other hand, is used to identify different 'types of objects', notably ranging from a list of sectors to, more ambitiously, a new development model. In addition, the bioeconomy's features are largely driven by the connection between the bioeconomy and surrounding concepts: sustainability, the circular economy, climate change, ecosystem services, the green economy and agroecology. In this respect, the bioeconomy as a political vision is increasingly referred to as a 'sustainable and circular bioeconomy'. From a purely conceptual (but also cultural, political and economic) point of view, one stimulating as well as confusing aspect is that the concept itself is developing in a context characterized by the emergence of a number of other 'bio-concepts' that, in a way, makes it even more difficult to identify a common understanding of the bioeconomy (Birch and Tyfield, 2012).

On the other hand, economics faces a number of challenges that arise due to the interconnected issues brought about by bioeconomy technologies, including the specificities of living organisms and biomass, the growing decomposition

and re-combinations of raw materials and the separability of production processes.

The evolution of technology is closely linked to funding research and innovation to develop new and improved technologies. Research and its link to technology production and transfer is a key object of research in bioeconomy studies (Festel and Rittershaus, 2014; Golembiewski *et al.*, 2015), including those related to agriculture (and primary production in general) knowledge systems. Technology is also relevant in terms of the connections between different goods on the supply side and related costs. Knowledge and information are key issues running through the bioeconomy, not only as a stock of knowledge-guiding innovation, but more generally as governance issues, as well as management and technology use components, and in interfacing with stakeholders and supporting the decision-making process. It is no surprise that a large branch of the literature is devoted to disentangling the economic and governance issues related to innovation processes.

In this context, economics-related disciplines are working on definitions and on the qualification of the bioeconomy from an economic perspective, as well as identifying a core set of ideas about its economic representation. This chapter summarizes this work.

After reviewing some of the relevant trends in the economy as a whole, the chapter first takes a rather standard approach for describing the bioeconomy based on past literature and rather standard economic notions. It then moves to the opposite perspective, namely addressing new economic concepts arising specifically from the recent bioeconomy investigation. It then addresses some of the qualifying features of the bioeconomy and the related literature, that is, the territorial versus the industrial view of the sector, the sustainability and circular economy features, the innovation dimension and the unique role of human beings in the bioeconomy. Thereafter it attempts to identify a general economic representation of the bioeconomy, the purpose of which is to provide a framework for the subsequent chapters, involving the three main features of the bioeconomy: (i) its overall organization as a 'value web'; (ii) its transition from fossil to renewable resources; and (iii) its fundamental trade-offs between exploitation and investment in anthropized biological capital.

4.2   The Bioeconomy in the Economic System

4.2.1   Contextual trends in economy and economics

A number of contextual trends affect the bioeconomy and interact with bioeconomy specificities, though they are not necessarily specific to the bioeconomy. One is the globalization of the economy, thanks to new technologies, transportation and market opportunities flowing from the liberalization policies of the past half a century.

Another connected issue is the financialization of the economy and the increasing roles of derivatives and financial products in allocating capital and defining values based on expectations.

The complexification/specialization of activities and the development of a variety of hybrid forms of economic organization is also a long-term trend, taken up by legislation and aimed at improving the performance of the economic system. In this context, vertical and horizontal integration have also been at the forefront of the adaptation of agriculture and food systems to a more and more competitive environment.

The fact of living in an information-rich context and the increase of awareness by consumers is also a common trend.

Environmental, social and ethical concerns are also a growing issue that has become paramount over the last 50 years and is now marking both technology development and marketing actions in every sector.

Innovation and technology, their roles as values and drivers of individual behaviour and quality of life, are also major features of our world in its different forms.

Finally, in spite of a wave of liberalism in the economy that started in the 1980s, the role of policy in relation to food and the environment has become generally more important. The same also applies to some extent to public research, while at the same time private research by industry has taken on an increasingly important role.

Most of these trends, and the related conceptual background, are now largely embedded in the bioeconomy.

4.2.2   The bioeconomy as (part of) an economic system

The basic representation of the bioeconomy is that of a system in which enterprises process raw materials represented by biological resources in order to meet consumer and societal needs. We can then describe the two main components of the bioeconomy as represented by enterprises supplying goods for their own profit and consumers consuming such goods for their own utility.

Both sides require some relevant qualifications. First, the processing of raw materials is obtained through a complex system of steps that imply not only the interplay of several enterprises, but also appropriate connections among them. This occurs through markets, but also via forms of integration and chain organization. As long as the bioeconomy tends to promote an increasingly pronounced breaking down and recombination of biological materials, the number of potential transactions (i.e. points when the process is technically separable) is increasing. This implies that transactions are more important and concern increasingly sophisticated contractual relationships, business models and institutional connections on the supply side.

On the other hand, considering consumers as simple users of bioeconomy products is a rather narrow view. Many bioeconomy technologies also have environmental or ethical implications and sometimes consumers take these into account in their consumption decisions. Yet individuals come into contact with bioeconomy production processes more frequently in their role as citizens than as consumers. The results of this contact are at least threefold:

- Consumers may have very different points of view about bioeconomy products, and this makes consumers and market segmentation a key issue.
- There is a huge role for information, not only in qualifying and making these segments understandable and recognizable, but also in affecting actual behaviour.
- There is a role for collective action and interaction with the supply side and with other consumers and public institutions that is much greater than in other fields.

Many bioeconomy issues can be cast in the same approaches used for the traditional sectors producing and processing biomass, such as agriculture, food and forestry. However, the qualifying elements of the bioeconomy can rather be found in four additional aspects:

- First, the specificities of bioeconomy technologies, with a focus on dynamics, technology change and innovations in biotechnology and biomass processing technology, as well as their integration with information and community technology.
- Second, the system perspective, in which, due to technology and institutional linkages, the qualifying factor is the interconnection and interplay between these sectors, the chains and sub-chains within the sectors and their actors.
- Third, the bioeconomy declination of general societal aims such as circularity and sustainability (and perhaps more).
- Finally, and partly due to the above, considering the novelty in the interaction between territory (land, or more generally ecosystems) and industrial processes, as well as between ecological and social components.

Given the above, bioeconomy research is largely covered by standard neoclassic economics related to consumer and firm behaviour, as well as technology change, and complemented by attention to coordination mechanisms among actors, which include not only coordination among different elements of supply, but also the role of public policies and coordination between supply and demand. Due to the features of the bioeconomy, a number of works enlarge this field of analysis using well-established economics concepts, while at the same time considering the unique features of the bioeconomy, for example, using instruments from transaction costs economics, economics of regulation or consumer behaviour and the role of information.

In addition, economics offers already well-established pieces of literature to treat various parts of the bioeconomy, such as agriculture and resource economics. Past works on the representation of the economic system (Quesnay) are mentioned as references to represent the aggregate. The connection with energy flows and ecological processes can find a basis in authors such as Georgescu–Roegen and Martinez–Allier, respectively, and in the subsequent streams of literature. The nature of renewable resource typical of living organisms has been widely addressed in the study of resource economics, while the environmental and future generation thinking of many bioeconomy issues can be covered by notions already developed by externality and public good economics. However, some of these fields have in some way been renewed by recent challenges. For example, issues such as the circularity of the economy tend to break the unidirectional flow of materials from nature to consumers and actually highlight feedback flows and the role of consumers as suppliers.

The treatment in this book does not take one specific theoretical approach, but rather uses insights from different strands of economic literature together when relevant. In addition, it focuses more on the most recent and likely novel approaches and insights, largely based on the assumption that there is some scope (and need) for new developments in economics to address the specificities of the bioeconomy seen above.

4.3 Bioeconomy-specific and Complementary Approaches

4.3.1 Emerging 'bio-concepts'

Together with applying existing economic concepts to interpret the bioeconomy, the current literature has also investigated the impact of bio-economy technologies, in particular biotechnologies, on the concepts used in economics, with implications in terms of economic theorization/conceptualization of bioeconomy, especially in political–economic terms. Birch and Tyfield (2012), Birch (2017) and Pavone and Goven (2017) provide a thorough review of these studies. Birch (2017) criticizes the oversaturation of bio-concepts and lists several examples, notably: biovalue, bioeconomics, biocapital (different versions), biocitizenship, knowledge-based bio-economy, bioeconomy, biopower, biowealth, bio-subjectivities, biolabour, biosubject, bio-object, bioknowledge and biocommodification. Some of these terms are indeed interesting attempts at making sense of various aspects of the bioeconomy in economic terms.

The terms bioeconomics and bioeconomic were already used to identify different subjects of research, including the theories of Georgescu–Roegen, the study of the relationships between biological resources and economic systems and the economic side of biopolitics (Rose, 2001). Georgescu–Roegen (1975) first introduced the term 'bioeconomic' as a joint consideration of the biological aspects of human life and bodies (bio), similar to other species, and the particular importance taken by exosomatic instruments ('instruments produced by man but not belonging to his body') in human history and growth (economic). Though this concept in itself is quite far away from the current meaning attributed to 'bio' in bioeconomy, indeed the reasoning of Georgescu–Roegen (1975) goes in the direction of highlighting the need to rely more or exclusively on the flow of solar energy to ensure long-lasting prosperity for humankind. This goes together with ethical aspects linked to the reduction in consumption and regards for future generations.

The concept of biovalue has been developed in human applications, in particular with regard to stem cells (Waldby, 2002), with the meaning of 'the surplus of in vitro vitality produced by the biotechnical reformulation of living processes'. The concept of biovalue is core to the identification of bioeconomy as a result of a mainly political process according to Rose (2001): 'Biopolitics becomes bioeconomics, driven by the search for what Catherine Waldby has termed "biovalue": the production of a surplus out of vitality itself'.

Biovalue has been advocated as a central concept for the bioeconomy (Brunori, 2013). Brunori (2013) goes even further in making a distinction between 'natural biovalue', which is produced by quality agriculture, and 'biotechnological biovalue'.

Several related concepts make the links between economic aspects of the 'bio' features and social–political aspects, such as the notions of 'biocapital' (Rajan, 2006) or of 'life as surplus' (Cooper, 2008).

Birch and Tyfield (2012) constructively argue against these approaches, especially when they seek to identify a clear-cut specificity and uniqueness via the 'bio-' nature of the processes studied, while indeed some of the processes observed come from the unique interactions between the biological nature of bioeconomy issues and contextual trends. Central to this debate is the view of the bioeconomy as a socio-techno-economic restructuring in which technological change is interwoven with a number of transitions in society and economic institutions. Three lines of observation from Birch and Tyfield (2012) are illustrated and developed below.

First, labour remains central to any political–economic analysis, which is consistent with the traditional Marxist approach to value. However, the relevant forms of labour are changing, and attention may now refer to new forms of labour (e.g. immaterial, informatics, relational, etc.) and new labour processes (e.g. in use, consumption, etc.). This is especially true in terms of the knowledge labour that is turned into an asset through intellectual property rights and is still necessary to turn biological matter into commercial products. Yet entrepreneurship, risk taking, creativity and leadership are also central to the bioeconomy. Finally, the connection with the human body in the medical bioeconomy makes it possible to consider tissue donation, for example, as a special sort of labour.

The second point is that the bioeconomy is putting value on assets rather than commodities, largely through the use of financial mechanisms that are not bioeconomy-specific, that is, financial asset values are more important than revenues from the sale of biotechnological commodities. 'Speculation does not relate to biological promises but to more mundane political–economic ones (e.g., rising value of shares), although these two are not entirely separable'

(Birch and Tyfield, 2012), and cannot be totally separated from (or better are strictly linked to) socially constructed values related to 'bio-'.

Finally, as stated by the authors,

> the 'actually existing' bioeconomy is not a set of social relations based on production (e.g. value, capital) or speculation (e.g. surplus) in life science-related industries, but is embedded in particular market relations and institutions of asset value realization. That is, the political economy of the life sciences depends on the realization of value from financial and knowledge assets through exchange on markets that are not only characterized by social order and social structures... but also by social expectations. (Birch and Tyfield, 2012).

For the bioeconomy, as for other fields, these expectations are performative, self-fulfilling and self-reinforcing. However, trends derived from these phenomena have proven to have limitations linked to the 'real economy' as it is shown by periodic financial crises. In addition, they do not necessarily lead to homogeneous or convergent political economies. Rather, markets can take different pathways and geographies that help understand different bioeconomies (see also the section on country strategies) in contrast to universal theories or concepts. As Birch and Tyfield (2012) put it,

> thinking about the bioeconomy in this way helps to avoid making general theoretical claims that have limited empirical or analytical support. Furthermore... it also enables a rethink of the transformation of modern capitalism by analyzing the transformation of economic processes and not just technoscientific ones.

4.3.2 Sustainability and circularity

Sustainability and circularity are currently perceived as two key aspects of the bioeconomy. They are related to each other in as much as circularity is a key strategy to reduce impact on resources, making the bioeconomy more sustainable.

There are several definitions of sustainability. In very broad terms, sustainability is related to the endurance of systems and processes. In economic terms, sustainability entails ensuring a satisfactory level of economic welfare today, without compromising the welfare of future generations. While widely used, the concept of

sustainability has proven to be difficult to define in practice and, even more so, to measure.

The focus on sustainability does not come from a need of the bioeconomy itself; rather, the wider context in which the bioeconomy is developing has been characterized by the widespread use of the concept of sustainability as the increasingly important aim of agriculture and food systems and, related to that, of sustainability-oriented technology change (Viaggi, 2015). In this sense, sustainability for the bioeconomy means not only answering the need to substitute non-renewable resources with renewable ones, but rather complying with a much wider range of societal and ecosystem needs.

In recent years, concepts such as resilience and vulnerability have increasingly accompanied or replaced that of sustainability. Notably, all of these issues have found noteworthy parallel use in ecology, environmental economics and development economics, highlighting the importance of dynamics and the relevance of 'potential' changes. Another feature of these concepts is their attempt to be comprehensive, which is at times pursued at the cost of difficulties with accurate definitions and measurement. With respect to the incorporation of 'green' concerns into private business, a key concept is that of eco-efficiency.

The links between bioeconomy and sustainability can also be interpreted through new ecological approaches. A very relevant framework in this direction is that of the socio-ecological systems (SES), largely focusing on suitable representations of the interaction between society and ecosystems to ensure sustainability. An SES can be defined as:

> 1. a coherent system of biophysical and social factors that regularly interact in a resilient, sustained manner; 2. a system that is defined at several spatial, temporal, and organizational scales, which may be hierarchically linked; 3. a set of critical resources (natural, socioeconomic, and cultural) whose flow and use is regulated by a combination of ecological and social systems; and 4. a perpetually dynamic, complex system with continuous adaptation. (Redman et al., 2004)

SESs emphasize the integration between social and ecological systems and highlight their linkages through feedback mechanisms that display resilience and complexity features. One of the features of this approach is to assume (based on observation) that resource users invest their time and energy to achieve sustainability, in contrast to the prevailing economic theory, in which self-interest damages the environment and government intervention is needed to ensure sustainability. In this regard, the approach promotes looking beyond markets and states for the governance of complex economic systems. In addition, it recognizes that no one size fits all, and that different solutions may well fit different needs and contexts. The approach considers as key for the diagnostics of the sustainability of a SES the description of its sub-systems and of their relationships (Ostrom, 2009).

While the SES approach looks at the interplay between social systems and ecological systems at the local level, other approaches, guided by sustainability concerns, look at the issue from a more global perspective. An interesting concept in this direction is the planetary boundary framework developed by Rockström et al. (2009). A recent work by Dearing et al. (2014) merges the planetary boundary framework and the social 'doughnut' framework into a new framework for defining safe and just operating spaces for sustainable development at regional scales.

These concepts, while very promising in terms of the conceptualization of sustainability at the interface of social and ecological components, are generally poor in terms of accounting for actual incentive systems and precise consideration of technological change potential.

Very relevant insights for the bioeconomy are provided by the concepts of the green economy and the circular economy. These concepts, together with the bioeconomy, are compared and discussed by D'Amato et al. (2017) based on a review of more than 2000 papers. According to the authors, while the three concepts have commonalities they are most often classed as different concepts, though their definitions and differences are far from being clear. Besides having a different geographical focus (the bioeconomy is more polarized in the EU), the bioeconomy focuses more on biological resource-based innovation and is more connected to land use practices, and hence with rural development concerns.

The green economy is the more comprehensive concept. The green economy has been put forward as a more positive and environmentally focused way of seeing the economy. It is intended as an economy seeking to reduce environmental and ecological impacts and that fosters sustainable development without degrading the environment. It also incorporates the idea of fairness.

The circular economy concept is also highly relevant for the bioeconomy and circularity is indeed one of the environmental objectives that is most clearly incorporated in the current vision of the bioeconomy. Circularity has been a well-known issue in environmental and resource economics since early works in the field (e.g. the famous paper by Boulding (1966)). Despite its wide use, the concept of circular economy is itself debated. Though already considered in classical environmental economics textbooks, its definition is being renewed over time, especially considering the wider view of circularity brought forward by the multiple dimensions of the sustainability concept (Korhonen et al., 2018). In a simplified way, circularity is measured by the degree to which resources are taken from the economy itself rather than from the environment through, for example, recycling. As such, this can be considered a key feature contributing to sustainability by reducing the pressure on resources. Korhonen et al. (2018) propose an alternative definition of circularity:

> Circular economy is an economy constructed from societal production-consumption systems that maximizes the service produced from the linear nature-society-nature material and energy throughput flow. This is done by using cyclical material flows, renewable energy sources and cascading-type energy flows... Circular economy limits the throughput flow to a level that nature tolerates and utilizes ecosystem cycles in economic cycles by respecting their natural reproduction rates.

From an economic point of view, this implies understanding the economic values attached to external versus re-used resources (Viaggi, 2015) (see Chapter 6).

4.3.3 Territorial versus industrial views: Ecosystem services and the bioeconomy

The literature emphasizes two contrasting views of the bioeconomy: one more focused on its territorial and ecological dimension and the other taking the form of an industrial process, potentially detached from the regional and eco-systemic dimension in which it is carried out. The distinction between these contrasting views applies, in particular, to one of the most qualifying concepts linked to the bioeconomy, namely that of biorefinery (Ceapraz et al., 2016). This is because most biorefinery plants are also linked to bioenergy and tend to use a large amount of biomass, hence highlighting the connection (or disconnection) with surrounding areas in which this biomass is produced. It seems reasonable to expect that a sustainable bioeconomy needs the integration of these two views. Meanwhile, these perspectives have stimulated the study of the bioeconomy from different theoretical standpoints.

First, the bioeconomy can be viewed as a whole in an ecosystem services (ES) perspective or at least in a context characterized by attention to ES as a structuring way of analysing the interaction between humans and ecosystems (Schmidt et al., 2012; Viaggi, 2015). According to TEEB, (2010), ES are the direct and indirect contributions of ecosystems to human well-being and are most often categorized into four types: (i) provisioning, such as the production of food and water; (ii) regulating, such as the control of climate and disease; (iii) supporting, such as nutrient cycles and crop pollination; and (iv) cultural, such as spiritual and recreational benefits. In contrast to other approaches, the ES approach takes ecosystems directly into account and links them to the uses that human beings can make of the services they provide.

The approach has gained a broad consensus and has increasingly been adopted in policymaking. The use of ES presents several advantages, among which the inclusion in the same framework of those services directly linked to 'traditional' productivity measures (provisioning) and those that relate to other ecosystem roles in human life, thus making explicit the various trade-offs, synergies and relative weights. ES make it possible to link issues related to the economics of sustainability with the ecosystem context and to cast sustainability issues in a territorial and landscape dimension (van Zanten et al., 2013). Of significant importance is the fact that the concept seems to be particularly suitable for policy communication.

The ES framework makes it possible to evaluate the bioeconomy in a broad ecosystem perspective, emphasizing opportunities and trade-offs at the landscape level. These are caused by the whole evolution of bioeconomy chains, but are often expressed at the level of primary activities such as agriculture, forestry and fisheries, which adapt to the needs and evolution of the downstream sectors, but, at the same time, have a primary role in ecosystem management.

The interplay between bioeconomy development and ecosystem services, and the trade-offs this implies, may change in a relevant way depending on the resources involved. For example, in forest management, carbon fixation and recreation can be the main issues, while in semi-arid agricultural systems, agro-biodiversity and pressures on water abstractions may be the most relevant focus.

Box 4.1. Bioeconomy and ecosystem services trade-offs in forestry

Häyrinen et al. (2016) investigate sources of competitive advantage for the Finnish forest sector, together with forest owner views on the future use of forests in Finland, their perceptions on evolving sectorial interlinkages and the position of the forest sector now and in the future bioeconomy. The study showed that forest owners deem the greatest potential for strengthening the sector toward the bio-economy to come from collaboration with energy and construction businesses. In addition, new possi-bilities linked to forest-based recreational services, cooperation with nature-based tourism and in increasing value-added wood products were identified. The study shows a case in which future value creation could be based more on forest ecosystem services and in diversifying the utilization of forests than on the dominant mindset driven by raw material supply.

Heinonen et al. (2017), also studying the case of Finnish forestry, found that lower harvested volumes would increase the total carbon balance of forestry (higher carbon fixation) and increase the values of biodiversity indicators, namely volumes of deciduous trees, amounts of deadwood and areas of old forest.

More from an industrial perspective, value chains, value webs and network concepts are increasingly being used in bioeconomy-related literature, especially to represent the supply side (industry) and product drivers and are being integrated into wide-ranging web concepts (Virchow *et al.*, 2016). The value chain concept originated from the Porter's classical, firm-based value chain. The concept evolved first in terms of accounting for the operation of multiple firms on the same product value chain and then in the direction of accounting for the geographical distribution of economic components of the same product in the globalized economy. Three main concepts to account for these latest developments are the global commodity chain, the global value chain and the global production networks. These are steps forward in explaining how global industries are organized and how this affects the development and opportunities of the regions and companies involved (Coe *et al.*, 2008).

However, this is not deemed to be sufficient and scholars advocate the use of perspective complex pathways of biomass and the integration of social, economic and environmental perspectives (Mangoyana *et al.*, 2013), notably to go beyond the single chain approach and to account for the functioning of global production networks with a number of interchain relationships.

One related proposal is the holistic concept of biomass-based value webs, a concept already being used for describing mainly the supply side of the bioeconomy in some papers (Scheiterle *et al.*, 2016; Virchow *et al.*, 2016).

Following Virchow *et al.* (2016), the biomass-based value web approach utilizes the 'web perspective' to understand the linkages between several value chains and the way they are governed. In the web approach, the activities resulting in a final product are viewed as a system in which the flows of 'values' (materials, semi-finished products, design, production, financial and marketing services) are connected vertically, horizontally and diagonally and organized in complex and dynamic configurations (Henderson *et al.*, 2002). The web approach allows to account for the manifold products that are derived from one biomass raw product. It also looks at the whole mix of products produced by firms

(and especially by farms) as well as to different value chains in which the firms and households participate. The web perspective can support the search for higher efficiency in resource use, by exploitation of synergies and cascading organization, identification of opportunities to close cycles, reduction of competition, better use of innovation potential, and a more comprehensive assessment of sustainability (Virchow *et al.* 2016).

Another concept being used is that of industrial symbioses, that is, 'the cooperative management and exchange of resource flows—particularly materials, water, and energy—through clusters of companies' (Chertow and Ehrenfeld, 2012). Chertow and Ehrenfeld (2012) discuss the process of the organization of circular economies in industrial symbioses. The concept of industrial symbiosis is also used by others to interpret bioeconomy issues. See, for example, Mouzakitis *et al.* (2017) who use this concept to analyse olive mills wastewater as a trigger for a bioeconomic industrial symbiosis, through valorization into biopolymers and bio-energy production.

The dichotomies, and also the need for the integration of the territorial and industrial views, are also evident in empirical modeling exercises related to the bioeconomy. This is, for example, emphasized by Jonsson *et al.* (2016). The authors propose a modelling approach for the assessment of policy options within the forest-based bioeconomy. They note that the feedback between the forestry dynamics model and the economic model of the global forest-based sector is essential: (i) for assessing the impact of different management regimes on markets of wood-based products on the one hand; and (ii) for understanding the development of forest resources and assessing future harvesting potential on the other.

4.3.4 Research and innovation mechanisms in the bioeconomy

Research and innovation are primary concerns in the bioeconomy. There are two broad and interconnected areas in which the bioeconomy shows specificities: the first is research, in the sense of how new knowledge can be built; and the second is how this new knowledge translates into innovation. These two aspects are interwoven and are hence difficult to disentangle, even if they are sometimes represented as two subsequent steps in a linear process.

In a simplified vision of research and technology uptake, the market and policies play major roles. The market can be seen as a promoter of private research and the main determinant of technology uptake, whereas the public sector, for its part, is a research provider and funder, as well as a promoter of targeted technology uptake (e.g. through subsidies). The transfer of innovation from science to industry, initially thought of as a sort of linear process, has over time been supported by specific extension policies aimed at smoothing and encouraging this one-way flow.

This naïve idea of technology transfer has been widely challenged over time, notably by investigating the number of actors, institutions and mechanisms that provide a bridge between research and industry and considering feedback loops between industry and research (Viaggi, 2015).

Taking the example of the EU, participatory approaches in different steps of the research to innovation processes have become particularly evident in the last couple of decades. The process was boosted by the EU technology platforms at the outset of the Seventh Framework Programme and has become paramount in the context of the new H2020 programme, focusing on the contribution to competitiveness through research and innovation and promoting a multi-actor approach in research projects. A wide set of examples of instruments to support this connection is illustrated in Chapter 7.

A key role in understanding innovation is played by the concept of innovation systems. In the field of activities related to agriculture and food, as well as for other region-based production activities, the role of regional innovation systems is of particular importance. The model of innovation systems is largely attached to the helix concept. The triple helix concept is the one mostly widely used, and involves collaboration between universities, government and industry. In connection to the bioeconomy, several authors have recently proposed an expansion of this concept towards that of a quadruple helix that also includes civil society, or even a quintuple helix system, where the fifth helix represents the environmental setting of a specific region (Grundel and Dahlström, 2016).

The concept of innovation systems has been widely elaborated for agriculture. This is of general interest for the bioeconomy because it is related to innovation challenges when a sector is composed of small firms that are highly heterogeneous as well as geographically spread out. The broad-based literature on agricultural knowledge and innovation systems indicates that this has become increasingly important in agriculture, especially in relation to the bioeconomy (Esposti, 2012). Networks are becoming an increasingly important category in this field. Networks also support a dynamic view of the innovation process, in which different types of actors can get involved at different stages and innovators need to continuously reflect and re-position themselves with respect to their environment. This view of the innovation process also implies that facilitators and innovation monitoring and evaluation methods can play a major role.

A particular difficulty is that of diagnosticating and disentangling success factors in these systems. In a number of cases, project-level organization has become the focus of funding, exploitation and analysis. In this direction, Turner *et al.* (2017) develop a nested multi-level framework to analyse how project actors engage with other actors to configure capabilities and resources to leverage positive path-dependences and weakening hindering ones.

A similar concept used in the field of the bioeconomy as a framework for analysis is that of technological innovation systems (TIS). TISs are understood as: 'a set of networks of actors and institutions that jointly interact in a specific technological field and contribute to the generation, diffusion and utilization of variants of a new technology and/or a new product' (Markard and Truffer, 2008). Innovation systems have three structural elements (Giurca and Späth, 2017):

- actors, which are individuals, firms, government bodies, universities, industry, nongovernmental organizations, bridging organizations and interest groups;
- networks that may include companies, networks, learning networks, policy networks, etc.; and

- institutions, which are legal and regulatory aspects as well as norms and cognitive rules that influence the decisions, activities and learning processes of actors.

An example of TIS with reference to the lignocellulosic biorefinery is shown in Fig. 4.1.

Giurca and Metz (2017) also analysed the wood-based bioeconomy system in Germany using a social network analysis and highlighting that, even with weak linkages, there can be a high level of trust and some commonalities in vision. However, a low frequency of contacts and a lack of common policy vision actually hinders the network's initiatives.

The role of intermediary organizations such as biotechnological clusters is very important due to the complexity of the system and the need to coordinate with diverse regional entities (Kearnes, 2013). As part of the innovation system, the interface between new advanced technologies and primary industry is a specific area in which innovation expertise and capabilities able to make the link are particularly needed.

This will require diverse scientific capability derived from research fields of science and technology currently external to conventional primary industry capabilities (McHenry, 2015).

One of the most relevant phenomena in recent decades, which was also largely supported by the development of the bioeconomy, is the role of entrepreneurship in research and innovation. Entrepreneurship can take different forms, from brokerage of innovation as a specific business activity, to the financing of innovation, up to a sort of 'innovation entrepreneurship'. Specifically, the process of developing 'entrepreneurship' activities and attitudes by researchers is increasingly being promoted. The role of entrepreneurship has attracted attention in the field of biotechnology for at least two decades with the emergence of the term 'bio-entrepreneurship' (Schoemaker and Schoemaker, 1998). The concept includes a wide range of typologies, ranging from researchers developing enterprises to exploit their knowledge, to entrepreneurs investing in life science research and development companies. This concept is now widespread and

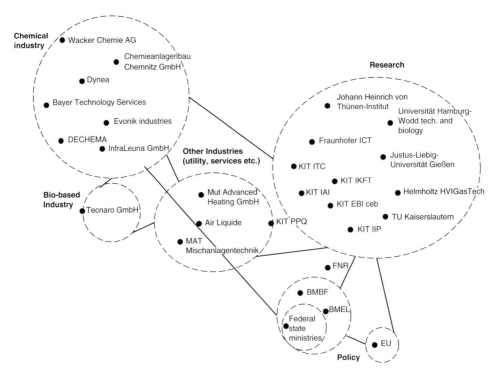

Fig. 4.1. An example of an actor network in the lignocellulosic technological innovation systems in Germany. Source: Giurca and Späth (2017).

increasingly used in specific educational programmes involving university students and researchers (see, for example, Uctu and Jafta, 2015).

This pathway also draws attention to the process of building research objectives and priorities. A specific point concerns the growing role of businesses in building innovation strategies and, through these, guiding the development of oriented applied research. As this approach becomes more important, the building of circular connections between research and business seem to be giving additional weight to business, which is providing a vision for the future and contributing to defining appropriate pathways in this direction.

Another key issue in recent years is the incorporation of collective values (e.g. public goods concerns such as environmental and resource issues) into public research as well as in market strategies by the private sector (and hence also into private research). This can be viewed as the progressive merging of private and public type values. This is witnessed by a process of strategic choice on the part of industry, awareness and related behavioural changes by consumers and the appropriate functioning of markets and marketing, including the communication of values and the transmission of information about products and processes.

Several of these concepts, while not bioeconomy-specific, are of special relevance for the bioeconomy because of its drivers and sustainability objectives. One is the concept of eco-innovation, which also offers examples of the articulated interplay between firm strategies, their economic context and inter-firm relationships. Four eco-innovation drivers can be identified (Rashid et al., 2014): regulatory push, technology push, market pull and firm strategies. In the context of this trend, the connection between vision, research objectives and the impacts of technological innovation is becoming of central importance.

Policy, management and communication instruments are playing a key role here as promoters of change. For example, quality management and certification schemes have been at the core of a wide field of research in recent decades and are increasingly the subject of analysis (see also Chapter 9).

Box 4.2. Complexity of innovation systems

The complexity of innovation systems and the interplay among the different factors is illustrated in a review of biorefinery by Bauer et al. (2017). An important finding is that investing more resources in research and development does not necessarily lead to more success and commercial investments. In spite of the recognized importance of entrepreneurship, there is no agreement on how to facilitate conditions for entrepreneurs and small- and medium-sized enterprises to enter the field of biorefinery. Policies on biorefinery technologies and products have had an important role in creating a market for biofuels and bioenergy; however, for a better development of biorefinery, it will be necessary to incentivize non-energy products. Finally, policy support for biorefinery is strongly connected to legitimacy and social acceptance.

The need for a better use of technology and indeed the governance of technology development leads to the consideration of innovative technology interpretation and management approaches for the bioeconomy. Technology and innovation management and open innovation approaches are considered to be key to the issues of the bioeconomy, though they are not often used as a means of interpretation (Van Lancker et al. 2016). The authors develop a set of guiding principles for the management of innovation processes in the bioeconomy comprised in three key issues: (i) the relevant stakeholder groups and their importance in innovation development; (ii) the innovation network strategy and management; and (iii) organizational features considered prerequisites for collaborative innovation. Five main factors influencing the implementation of an innovation management processes are: (i) many innovations will be radically new and disruptive-type innovations; (ii) bioeconomy innovations are based on a complex knowledge base; (iii) bioeconomy technology will require a large degree of cooperation among actors; (iv) commercialization and adoption of new technologies may be challenging; and (v) the bioeconomy is still dealing with fragmented policy schemes (Fig. 4.2).

The role of product design as a means of controlling for product characteristics from

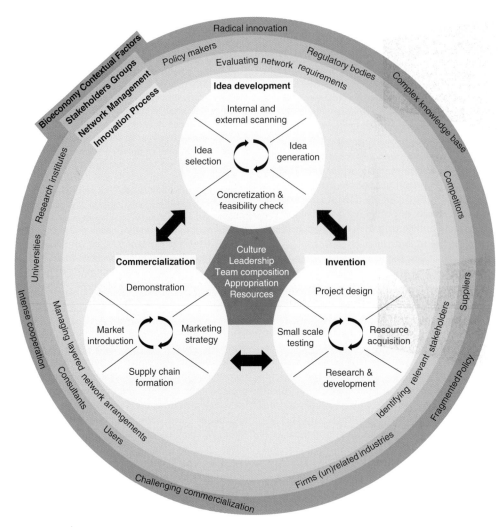

Fig. 4.2. An schematic representation of the innovation process in the Bioeconomy context. Source: Van Lancker *et al.* (2016).

inception is also considered a major and growing topic of current innovation systems. An example of this issue is available in Hildebrandt *et al.* (2017) for bio-based polymers. An area in which the incorporation of societal needs and concerns should be fed into the technology from the very beginning is that of synthetic biology. Indeed, synthetic biology tries to avoid trial and error approaches in the modification of living organisms, in favour of planned design of biological system and components, which tends to combine the technology itself with the decision-making and design process (Fig. 4.3).

4.3.5 Human beings in the bioeconomy

Many bioeconomy issues are addressed by studying the behaviour of human beings. Economics does so often by approaching separately the different roles of human beings in society: consumers, employees, researchers, entrepreneurs and citizens are some of the most frequently addressed. Each of these roles entails specific issues that are treated by the bioeconomy literature. In addition, three very important issues are emerging as specific to the bioeconomy.

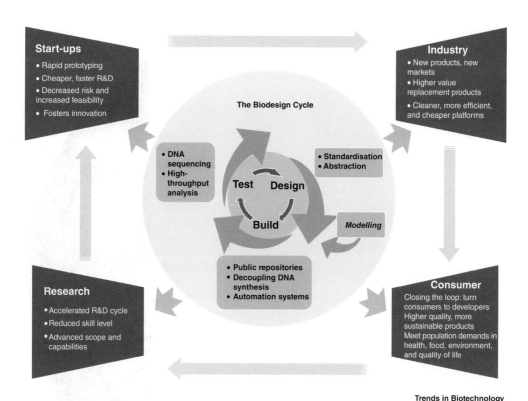

Fig. 4.3. The synthetic biology biodesign cycle and impacts on stakeholder groups in the product supply chain. Source: Flores Bueso and Tangney (2017).

First, all these different roles get to play together in the bioeconomy and need to be treated together in their dynamic interplay among actors in order to understand phenomena. For example, the state of being a citizen, consumer and entrepreneur constitutes a typical mix of conditions on the part of innovative small farmers and bioeconomy entrepreneurs, which are linked to motivation and capacity to develop new ideas. The interplay between consumers and citizens roles in household behaviour has been already highlighted.

Second, the bioeconomy is bringing the physiological side of humans directly into play. In this respect, not only are manpower, ambition and creativity important, but the bioeconomy can also touch upon human biology and physiology directly through implications for health, nutrition and medical solutions. This leads, among others, to entering the field of politics of population control and physiology of motivation and incentives, which is beyond the scope of this book but clearly strictly interwoven.

Third, in spite of the rhetoric correctly emphasizing the need for the bioeconomy to deal with urgent world challenges such as population growth, climate change and the need to reduce dependency on fossil fuels, the system of values and preferences in the bioeconomy is increasingly based on social construction in which information, promissory values, expectations, fears and relationships contribute to define reactions based on rather unstable preferences and positions. This instability and 'plasmability' of problems, preferences and solutions is a key feature of the current stage of the bioeconomy.

4.4 Conceptualizing the Bioeconomy

4.4.1 A socio-ecological value web

This section attempts to formulate a general approach to representing the bioeconomy using the concept of the socio-ecological value web (SEVW) based on the integration of two widely used reference concepts, though at the moment only marginally used for bioeconomy issues. One is the approach of socio-ecological systems and the other is the approach of value webs. Though we use SEVW as a working concept in the book, a much more complete (and maybe telling) definition would be that of a socio-ecological technological value-enhancing web (SETVEW).

Using elements of the above components, a SEVW can be thought of as:

1. A system of biophysical and social components that interact among each other.
2. The linkages among sub-systems, identified by both material and value flows on multiple product chains and socio-ecological systems.
3. A complex system of adaptation solutions from the interaction between its multiple components, including feedback mechanisms.
4. A set of knowledge and technologies.

In contrast to global chain analysis and in common with value webs, SEVW is not product-oriented, though a final product can be one of the agglomeration criteria to distinguish the different components of the system. Agglomeration would rather follow bundles of products or, more generally, bundles of attributes (characteristics), which directly links SEVW with the attribute theory of goods. Compared with value webs, which are mainly supply-related concepts in their origin, SEVW also integrates demand-side components (households, etc.), the institutional context of production and consumption as well as the relevant environmental components.

Compared with SES, SEVW provides a more technical structure of the flows of value within an economy, by explicitly integrating the chain and interchain connections and expanding the concept to the global bioeconomy. As such, it can also be viewed as a series of relationships between different SES and explicitly accounting for the global geographical complexity of production processes.

Similar to SES, the approach emphasizes the interactions and grey areas between social systems and ecological systems. At the same time, it highlights the importance of social and economic interactions in producing value.

Individuals affect their context through direct actions, purchasing actions and contribution to governance decisions. There are individuals and institutions devoted to supply goods. There are institutions and mechanisms to allow for coordination in order to answer to these needs. At least to some extent, this system also incorporates attention to resilience and sustainability.

Both a global bioeconomy SEVW and several partial SEVWs can be identified, for example, linked to one specific final product, or intermediate product or a specific region. However, having in mind the whole system makes it possible to better account for interrelationships that have a global dimension, for example, linked to climate change or to global product value chains and to account for their local interactions.

The framework also acknowledges that value and resilience is built-in, thanks to both resources and human activities.

It explicitly takes into account the role of technology in affecting the functional links among different components of the system, including communication.

Another distinctive factor is that SES are largely oriented towards analysing the sustainability of social-ecological systems, while SEVW tends to integrate concepts that also allow for more prescriptive and predictive diagnoses.

Using this approach also means going beyond the mere dichotomy between markets and state, the reduction of the role of the states (which are increasingly institutions identified by contractual relationship with a geographical reference) and the role of intermediate types of institutions. This also allows the review of the nature and role of the firm. Indeed, one could envisage the SEVW as being navigated by 'amoeba objects' identifiable by institutions such as states, firms, networks, clusters, etc., with each one identified by some stabilizing mission and topological feature (e.g. location), but characterized by a dynamic shape and type of linkages. One qualifying issue is how to define fixes, that is, characters that can allow to identify these entities. These can be identified in bundles of property rights and actions (routines) oriented to ensuring self-sustainability of goals, as well as defined goals. However, fixity is not a value; rather change, asymmetric relationships producing rents and adaptation to change are considered as strong points. More generally, these objects (the nodes of the web) can be identified as mission-oriented bundles of rights and capabilities.

From the spatial and regional/inter-regional point of view, SEVWs allow the understanding of the governance of complex economic systems with flexible boundaries and a number of more or less formalized interconnections. This may accommodate migration of labour/households, multinational corporations and international linkages of small firms, circular economy in a multi-scale and multi-regional sense, sustainability and resilience based on interplay across complementary areas, networks of various kind, flexible specialization based on cheap production locations, and different access to resource and knowledge.

It also may serve to go beyond simplistic dichotomies, such as those of local versus global, production versus consumption and promissory versus real value systems.

The ambition of the SEVW is to explain value creation. Due to its flexible nature, the SEVW can help link value generated by consumer preferences and value generated by increasing biological capacity of production, and is expressed by transactions, assets and property rights. A link with location can also be identified through density of value, capital or density of the web itself.

4.4.2 A transition from fossil to renewable resources

A different view of the bioeconomy technology is that of a technology shift with respect to non-biobased, notably fossil-based, technologies. Following Zilberman *et al.* (2013), the issue of the technology shift from non-renewable to renewable resources can be illustrated as the combination of two topics: (i) the different profile of profitability over time of non-renewable versus renewable-based technologies; and (ii) the coexistence of the use of non-renewable versus renewable resources.

Figure 4.4 illustrates the first point, by depicting the trend of per year profitability over time. Both non-renewable (NR) and renewable biological (RB) technologies have an initial time in which they are not profitable because they require investments in research before they can be actually profitably used.

NR may have a shorter period for this stage, as the technologies used are somehow simpler. When they become profitable, they then initially experience growing profitability, followed by stabilization and then a decrease with the exhaustion of the resource or because harvesting becomes more difficult or extracted product have a lower quality. Eventually profitability drops to zero.

The process of research and development for RB technology is potentially longer due to the greater complexity of knowledge required, including for their implementation. For this reason, it also takes more time to become profitable. Yet, profitability lasts longer and in principle forever due to the renewability of the resources used.

An additional option, however, is to also have non-biological renewable (RNB) (e.g. wind) technologies. These may need even more investment in the initial stages and last forever in the final stage of their life given their renewability. A question is whether these technologies will become more (RNB') or less (RNB") profitable than bio-based technologies.

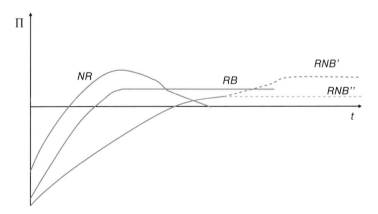

Fig. 4.4. Profitability over time of technologies using non-renewable and renewable resources. NR, non-renewable; RB, renewable biological; RNB, non-biological renewable. Source: modified from Zilberman *et al.* (2013).

Somehow these trends define the space in time and markets of the bioeconomy. Currently there is indeed a coexistence of the three types of technologies, accordingly we are somewhere in the intermediate area of the graph.

To a certain extent, different NR, RB and RNB technologies can coexist in the longer run due to the heterogeneity of their costs and hence also due to the shape of their supply curve (Zilberman *et al.*, 2013) (see Chapter 6). This coexistence would also increase the total welfare for society (higher total production at lower cost) and allow for flexibility in sources of the product. In the light of this, the transitions in Fig. 4.4 can be imagined as a smooth change of market shares between NR, RB and RNB as technology change makes RB, and in turn RNB, less costly and more profitable.

In addition, there are many different technologies within the NR, RB and RNB categories, each with a different time frame, performance and investment requirements. They could replace each other over time or lead to different mixes for the same point in time. Also, each technology may have positive and negative externalities that could modify actual (societal) profitability upward or downward (for positive or negative externalities, respectively). Furthermore, some changes in profitability can occur over time depending on exogenous variables, such as prices of complementary resources or incentives connected to policies.

4.4.3 Exploitation versus investment in anthropized biological capital

This section develops a stylized model of the bioeconomy based on a purely neoclassical approach, focusing on the annual trade-off between exploitation and investment. This is connected to two major distinctive factors of the bioeconomy: (i) the production of economic goods requires the (creation and) maintenance of an anthropized biological capital (ABC); and (ii) consumer utility (demand) is directly affected by both the production of market goods in the current period and the status of the ABC.

These can be seen as two different aspects of capital characteristics. The first point is connected to the idea of ABC already discussed in Chapter 3. The second point derives from the observation that the state of ABC can affect utility. This occurs primarily because the state of ABC determines expectations about the future production of the derived market goods. Another effect is by way of public goods components. It can also include existence values or option values generated by ABC. This may depend on societal risk aversion, acceptability and ethical issues. It is also relevant to note that such effects may, in principle, be both positive and negative in terms of utility. The set of these components can also affect expectations, willingness to pay by consumers and financial market behaviour.

Based on these assumptions, in the short-term the decision maker (society as a whole assumed to take the form of a single decision maker) has the choice between investment in the stock of ABC (in view of future benefits represented by the present value of money invested) or for exploitation effort (Fig. 4.5). The term 'exploitation' is used here as comprising harvesting, cultivation and use in processing, considering the fading distinction between these terms discussed in Chapter 3, section 3.3.

Y_1 is an isoquant of social welfare in the exploitation–investment effort space, i.e. the combination of harvest and investment yielding the same welfare. The shape is due to the expectation

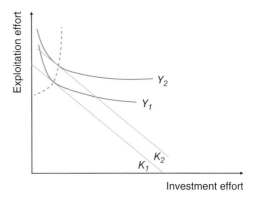

Fig. 4.5. Optimal allocation of investment and harvesting effort with optimal expansion pathway. *Y* is an isoquant of social welfare in the exploitation–investment effort space; *K* is the set of combinations of effort and investment that can be accomplished with the same total cost.

of decreasing marginal utility in both current production and future production possibilities, which implies that getting far from the centre of the figure in one direction or the other, the amount of increasing effort on one of the two activities needed to compensate for the decrease of the other will be more than proportional.

On the contrary, each line K is the set of combinations of effort and investment that can be accomplished with the same total cost, i.e. given some amount of economic resources used in the process. If effort and investment are expressed in physical units, K is determined by the amount of investment and effort each multiplied by its price. If investment and effort are expressed in monetary units, the line is a $45°$ straight line. Given a certain iso-revenue curve, the optimal combination of effort and investment is when the line K is tangent to the iso-yield curve, that is, the point in which that level of revenue can be achieved at the minimum cost. On the opposite, assuming resource constraints (total resources available in a given year), for example, K_1, the highest curve touched would determine the yield achieved and the combination of effort and investment at the point of tangency would be the optimal distribution of effort among those two.

Assuming that this constraint is removed, different optimal combinations may be envisaged for achieving different yields and the optimal one would be the one achieving higher profits (difference between revenue and cost). This expansion pathway shows the optimal combination between effort and investment as a function of the resource available (maximum yield given the cost). This pathway could also be interpreted the other way round (similar to what is usually done in cost functions) as the minimum cost to achieve increasing levels of output from the bioeconomy. In this way, the optimal equilibrium combination of revenue, investment effort and exploitation effort can be represented and ends up yielding the optimal level of revenue production from biological resources in the short and long run.

The framework as described until now, does not consider externalities or public goods explicitly; there are indeed two types of public goods here: those related to the amount of ABC and those related to the exploitation itself. Considering these public goods would modify the optimal level of capital stock and the optimal effort, by modifying the iso-revenue curve. For example, if some amount of public goods is attached to capital stock, a higher level of such capital stock would become optimal. Even more generally, if the iso-revenue is considered as an iso-social utility, representing the flow of utility derived by the combination of exploitation and ABC attributes at a point in time, the framework can be considered as a really comprehensive way of representing the bioeconomy trade-off in a single period.

The framework shown in Fig. 4.5, besides explaining costs and connecting them with anthropized capital stocks, may be the basis for explaining the effects of innovation. Innovation would modify the trade-offs among investments in the next time period. Notably, technology change can allow higher utility with the same resources available, hence moving the Y_1 downwards. This can also be non-neutral with respect to stock-effort of resources, or better biological capital versus other production factors. One possibility is to go in the direction of a flatter right arm, which means that harvest effort may contribute poorly to revenue, while investment will contribute more; this may be the current trend, but innovation can also go in the other directions. As a result, the share of allocation to investment or harvesting changes over time.

This may also depend on external drivers, e.g. climate change, rate of return (which affects future benefits from investment), stock of knowledge (which affects the return on additional investment in innovation).

4.5 Outlook

The excursus carried out in this chapter illustrates some of the different economic viewpoints and interpretations that can be given to the bioeconomy. It also gives an idea of the different views and research agendas that can arise from addressing the bioeconomy, from new management challenges to issues connected to politics and cultural change, to genuinely new economic concepts.

Indeed, this is just a way of sketching these issues, but a number of qualifications of the concepts expressed and of the models illustrated above are possible. This applies to the neoclassical model of exploitation–investment trade-off, mainly because it is purposefully only a primer seeking to extend rather standard economic theory to the bioeconomy. This is also because each dimension of the model is actually a simplification that needs a number of more detailed specifications in order to make it more realistic. This also applies to the SEVW concept, the subcomponents of which and the mechanisms connecting them, need to be better qualified.

This is done in the next three chapters of this book, which are organized in three main areas:

1. The consumer/citizen side behaviour and the structure of demand, its changes, and its connection with the world's challenges will be addressed in Chapter 5.
2. The structure of supply (costs side, including behavioural perspectives or other decision-making mechanisms by firms) will be addressed in Chapter 6.
3. The role of different coordination mechanisms in matching demand and supply, and meeting societal challenges, including markets, public policies and chain coordination mechanisms will be addressed in Chapter 7.

One major cross-cutting issue concerns behavioural assumptions. In the models presented in this chapter we have assumed profit maximizing or utility maximizing behaviour by a 'benevolent dictator' depending on the model. These assumptions also include perfect knowledge of future outcomes, while an intrinsic issue of bioeconomy is uncertainty. This can actually touch upon different aspects of the process and may be more relevant for the future effects linked to investment; this may also be combined with different assumptions about risk aversion. It may also depend on expectations (e.g. about climate change). In some cases, the result of different behavioural assumptions would appear rather straightforward. For example, taking the exploitation–investment trade-off, higher risk aversion could reduce investment in anthropized capital if future outcome are uncertain. However, this topic is usually much less clear in practice. These issues are addressed horizontally through the next chapters.

There remains a most ambitious challenge, however, namely synthesizing the promising ideas developed in terms of new conceptualizations of the bioeconomy and modelling. In this respect, this book is only a first attempt and this challenge will largely be left for further research. This will be illustrated in more detail at the end of each chapter and a synthesis will be provided in the final chapter after exploring further the most recent signals about the future of the bioeconomy.

References

Bauer, F., Coenen, L., Hansen, T., Mccormick, K. and Palgan, Y.V. (2017) Technological innovation systems for biorefineries: A review of the literature. *Biofuels, Bioproducts and Biorefining* 11, 534–548. doi: 10.1002/bbb.1767.

Birch, K. (2017) The problem of bio-concepts: biopolitics, bio-economy and the political economy of nothing. *Cultural Studies of Science Education* 12, 915–927. doi: 10.1007/s11422-017-9842-0.

Birch, K. and Tyfield, D. (2012) Theorizing the bioeconomy: Biovalue, biocapital, bioeconomics or . . . what? *Science Technology & Human Values* 38(3), 299–327. doi: 10.1177/0162243912442398.

Boulding, K.E. (1966) The Economics of the coming spaceship Earth. In Jarrett, H. (ed.) *Environmental Quality in a Growing Economy*, Baltimore, Maryland, USA, Johns Hopkins University Press, pp. 3–14.

Brunori, G. (2013) Biomass, biovalue and sustainability: Some thoughts on the definition of the bioeconomy. *EuroChoices* 12(1), 48–52. doi: 10.1111/1746-692X.12020.

Ceapraz, I.L., Kotbi, G. and Sauvée, L. (2016) The territorial biorefinery as a new business model. *Bio-based and Applied Economics* 5(1), 47–62.

Chertow, M. and Ehrenfeld, J. (2012) Organizing self-organizing systems. *Journal of Industrial Ecology* 16(1), 13–27. doi: 10.1111/j.1530-9290.2011.00450.x.

Coe, N.M., Dicken, P. and Hess, M. (2008) Introduction: global production networks – debates and challenges. *Journal of Economic Geography* 8(3), 267–269. doi: 10.1093/jeg/lbn006.

Cooper, M. (2008) *Life as Surplus*. Seattle, Washington, USA, University of Washington Press.

D'Amato, D., Droste, N., Allen, B., Kettunen, M., Lähtinen, K., Korhonen, J., Leskinen, P., Matthies, B.D. and Toppinen, A. (2017) Green, circular, bio economy: A comparative analysis of sustainability avenues. *Journal of Cleaner Production* 168, 716–734. doi: 10.1016/j.jclepro.2017.09.053.

Dearing, J.A., Wang, R., Zhang, K., Dyke, J.G., Haberl, H., Hossain, M.S., Langdon, P.G., Lenton, T.M., Raworth, K., Brown, S., Carstensen, J., Cole, M.J., Cornell, S.E., Dawson, T.P., Doncaster, C.P., Eigenbrod, F., Flörke, M., Jeffers, E., Mackay, A.W., Nykvist, B. and Poppy, G.M. (2014) Safe and just operating spaces for regional social-ecological systems. *Global Environmental Change* 28, 227–238. doi: 10.1016/j.gloenvcha.2014.06.012.

Esposti, R. (2012) Knowledge, technology and innovations for a bio-based economy: Lessons from the past, challenges for the future. *Bio-based and Applied Economics* 1(3), 235–268.

Festel, G. and Rittershaus, P. (2014) Fostering technology transfer in industrial biotechnology by academic spin-offs. *Journal of Commercial Biotechnology* 20(2), 5–10. doi: 10.5912/jcb631.

Flores Bueso, Y. and Tangney, M. (2017) Synthetic biology in the driving seat of the bioeconomy. *Trends in Biotechnology*, 35(5), 373–378. doi: 10.1016/j.tibtech.2017.02.002.

Georgescu–Roegen, N. (1975) Energy and economic myth. *Southern Economic Journal* XLI, 347–381.

Giurca, A. and Metz, T. (2017) A social network analysis of Germany's wood-based bioeconomy: Social capital and shared beliefs. *Environmental Innovation and Societal Transitions*, 2017 (online). doi: 10.1016/j.eist.2017.09.001.

Giurca, A. and Späth, P. (2017) A forest-based bioeconomy for Germany? Strengths, weaknesses and policy options for lignocellulosic biorefineries. *Journal of Cleaner Production* 153, 51–62. doi: 10.1016/j.jclepro.2017.03.156.

Golembiewski, B., Sick, N. and Bröring, S. (2015) The emerging research landscape on bioeconomy: What has been done so far and what is essential from a technology and innovation management perspective? *Innovative Food Science & Emerging Technologies* 29, 308–317. doi: 10.1016/j.ifset.2015.03.006.

Grundel, I. and Dahlström, M. (2016) A quadruple and quintuple helix approach to regional innovation systems in the transformation to a forestry-based bioeconomy. *Journal of the Knowledge Economy* 7(4), 963–983. doi: 10.1007/s13132-016-0411-7.

Häyrinen, L., Mattila, O., Berghäll, S., Närhi, M. and Toppinen, A. (2016) Exploring the future use of forests: perceptions from non-industrial private forest owners in Finland. *Scandinavian Journal of Forest Research* 32(4), 327–337. doi: 10.1080/02827581.2016.1227472.

Heinonen, T., Pukkala, T., Mehtätalo, L., Asikainen, A., Kangas, J. and Peltola, H. (2017) Scenario analyses for the effects of harvesting intensity on development of forest resources, timber supply, carbon balance and biodiversity of Finnish forestry. *Forest Policy and Economics* 80, 80–98. doi: 10.1016/j.forpol.2017.03.011.

Henderson, J., Dicken, P., Hess, M., Coe, N. and Yeung, H.W.-C. (2002) Global production networks and the analysis of economic development. *Review of International Political Economy* 9(3), 436–464. doi: 10.1080/09692290210150842.

Hildebrandt, J., Bezama, A. and Thrän, D. (2017) Cascade use indicators for selected biopolymers: Are we aiming for the right solutions in the design for recycling of bio-based polymers? *Waste Management and Research* 35(4), 367–378. doi: 10.1177/0734242X16683445.

Jonsson, R., Rinaldi, F., Räty, M. and Sallnäs, O. (2016) Integrating forest-based industry and forest resource modeling. *iForest* 9(5), 743–750. doi: 10.3832/ifor1961-009.

Kearnes, M. (2013) Performing synthetic worlds: Situating the bioeconomy. *Science and Public Policy* 40(4), 453–465. doi: 10.1093/scipol/sct052.

Korhonen, J., Honkasalo, A. and Seppälä, J. (2018) Circular Economy: The concept and its limitations. *Ecological Economics* 143, 37–46.

Mangoyana, R.B., Smith, T.F. and Simpson, R. (2013) A systems approach to evaluating sustainability of biofuel systems. *Renewable and Sustainable Energy Reviews* 25, 371–380. doi: 10.1016/j.rser.2013.05.003.

Markard, J. and Truffer, B. (2008) Technological innovation systems and the multi-level perspective: Towards an integrated framework. *Research Policy* 37(4), 596–615. doi: 10.1016/j.respol.2008.01.004.

McHenry, M.P. (2015) A rural bioeconomic strategy to redefine primary production systems within the Australian innovation system: Productivity, management, and impact of climate change. In *Agriculture Management for Climate Change*. Hauppauge, New York, USA, Nova Science Publishers, Inc.

Mouzakitis, Y., Aminalragia-Giamini, R. and Adamides, E.D. (2017) From the treatment of Olive Mills wastewater to its valorisation: Towards a bio-economic industrial symbiosis. *Sustainable Design and Manufacturing, 2017. Smart Innovation, Systems and Technologies* 68, 267–277. doi: 10.1007/978-3-319-57078-5_26.

Ostrom, E. (2009) A general framework for analyzing sustainability of social-ecological systems. *Science* 325(5939), 419–422. doi: 10.1126/science.1172133.

Pavone, V. and Goven, J. (eds) (2017) *Bioeconomies*. London, UK, Palgrave Macmillan Ltd.

Rajan, K. S. (2006) *Biocapital: The Constitution of Postgenomic Life*. Durham, North Carolina, USA, Duke University Press.

Rashid, N., Jabar, J., Yahya, S. and Shami, S. (2014) Dynamic eco innovation practices: A systematic review of state of the art and future direction for eco innovation study. Asian social science. *Canadian Center of Science and Education* 11(1), 8–21. doi: 10.5539/ass.v11n1p8.

Redman, C.L., Grove, J.M. and Kuby, L.H. (2004) Integrating social science into the long-term ecological research (LTER) network: Social dimensions of ecological change and ecological dimensions of social change. *Ecosystems* 7(2), 161–171.

Rockström, J., Steffen, W., Noone, K., Persson, Å., Chapin, F.S., Lambin, E.F., Lenton, T.M., Scheffer, M., Folke, C., Schellnhuber, H.J., Nykvist, B., De Wit, C.A., Hughes, T., Van Der Leeuw, S., Rodhe, H., Sörlin, S., Snyder, P.K., Costanza, R., Svedin, U., Falkenmark, M., Karlberg, L., Corell, R.W., Fabry, V.J., Hansen, J., Walker, B., Liverman, D., Richardson, K., Crutzen, P. and Foley, J.A. (2009) A safe operating space for humanity. *Nature* 461(7263), 472–475. doi: 10.1038/461472a.

Rose, N. (2001) The politics of life itself. *Theory, Culture and Society* 18(6), 1–30. doi: 10.1177/02632760122052020.

Scheiterle, L., Ulmer, A., Birner, R. and Pyka, A. (2016) From commodity-based value chains to biomass-based value webs: The case of sugarcane in Brazil's bioeconomy. *Journal of Cleaner Production* 172, 3851–3863. doi: 10.1016/j.jclepro.2017.05.150.

Schmidt, O., Padel, S. and Levidow, L. (2012) The bio-economy concept and knowledge base in a public goods and farmer perspective. *Bio-based and Applied Economics* 1(1), 47–63.

Schoemaker, H.J. and Schoemaker, A.F. (1998) The three pillars of bioentrepreneurship. *Nature Biotechnology* 16(Suppl), 13–15.

TEEB (2010) The economics of ecosystems and biodiversity: mainstreaming the economics of nature: a synthesis of the approach, conclusions and recommendations of TEEB. Available at http://www.teebweb.org/publication/mainstreaming-the-economics-of-nature-a-synthesis-of-the-approach-conclusions-and-recommendations-of-teeb/ (accessed 23 May 2018).

Turner, J.A., Klerkx, L., White, T., Nelson, T., Everett-Hincks, J., Mackay, A. and Botha, N. (2017) Unpacking systemic innovation capacity as strategic ambidexterity: How projects dynamically configure capabilities for agricultural innovation. *Land Use Policy* 68, 503–523. doi: 10.1016/j.landusepol.2017.07.054.

Uctu, R. and Jafta, R.C.C. (2013) Bio-entrepreneurship as a bridge between science and business in a regional cluster: South Africa's first attempts. *Science and Public Policy* 41(2), 219–233. doi: 10.1093/scipol/sct049

Van Lancker, J., Wauters, E. and Van Huylenbroeck, G. (2016) Managing innovation in the bioeconomy: An open innovation perspective. *Biomass and Bioenergy* 90, 60–69. doi: 10.1016/j.biombioe.2016.03.017.

Viaggi, D. (2015) Research and innovation in agriculture: beyond productivity? *Bio-based and Applied Economics* 4(3), 279–300.

Virchow, D., Beuchelt, T.D., Kuhn, A. and Denich, M. (2016) Biomass-based value webs: A novel perspective for emerging bioeconomies in Sub-Saharan Africa. In Gatzweiler, F.W. and von Braun J. (eds) *Technological and Institutional Innovations for Marginalized Smallholders in Agricultural Development*. New York, New York, Springer, pp. 225–238. doi: 10.1007/978-3-319-25718-1_14.

Waldby, C. (2002) Stem cells, tissue cultures and the production of biovalue. *Health* 6(3), 305–323. doi: 10.1177/136345930200600304.

van Zanten, B.T., Verburg, P.H., Espinosa, M., Gomez-y-Paloma, S., Galimberti, G., Kantelhardt, J., Kapfer, M., Lefebvre, M., Manrique, R., Piorr, A., Raggi, M., Schaller, L., Targetti, S., Zasada, I. and Viaggi, D. (2013) European agricultural landscapes, common agricultural policy and ecosystem services: a review. Agronomy for Sustainable Development. *EDP Sciences* 34(2), 309–325. doi: 10.1007/s13593-013-0183-4.

Zilberman, D., Kim, E., Kirschner, S., Kaplan, S. and Reeves, J. (2013) Technology and the future bioeconomy. *Agricultural Economics* 44(S1), 95–102. doi: 10.1111/agec.12054.

5

Driving Forces and Demand-side Economics

_____

5.1 Introduction and Overview

This chapter illustrates the demand-side issues in the bioeconomy. The central focus of this topic is individual consumer behaviour. Consumers have been mentioned at several points in the previous chapters, as their individual choices are key to understanding trends and the development of markets and sectors. Yet, consumers are not acting in isolation. On the contrary, they take decisions in a context in which they are called to interact continuously with other consumers and society as a whole. Individual and collective choices may well be promoted by major needs laying in the background but acting as driving forces and enacted through policies, network's actions and information by various opinion groups.

This section seeks to address the entirety of the demand-side issues related to the bioeconomy. It starts with the macro drivers guiding the need for bioeconomy technologies and investigates how they affect the market through individual consumers' and citizens' behaviour. It then focuses on individual behaviour, initially based on rather standard consumer theory, focused mostly on utility as linked to a good's attributes. This is then enriched with behavioural approaches and other considerations. Topics on aggregate behaviour, information and rational ignorance are also touched upon, as well as specific values related to bioeconomy processes and products. Finally, the chapter addresses aggregated behaviour and closes with some outlook, after describing some specific case studies.

5.2 Major Driving Forces and Scenarios

Among the driving forces recognized to provide strong motivations for the current and future development of the bioeconomy and that contribute to shape its internal pathways, the following areas of change play a key role (Pätäri _et al._, 2015; Wesseler _et al._, 2015):

1. Growing population projections, associated with growing per capita incomes, which will be reflected in higher demand for goods of any type, with a special role for food and fuel, and related pressure on land, water.

2. Climate change that will change the productivity of biological systems, in many areas with potential for higher production, in other areas with lower productivity and potential high impact on critical resources such as water.

3. Need for the substitution of fossil fuel, both as a way of dealing with climate change and as a way of addressing the inherent limitations of the use of fossil fuels as a non-renewable resource.

4. Advances in related and complementary technologies, especially in the biological sciences and information and communication technologies.

© D. Viaggi 2018. _The Bioeconomy: Delivering Sustainable Green Growth_ (D. Viaggi)

5. Changes in the organization of industries, including horizontal and vertical integration in agricultural supply chains, increasing inter- and intra-industry exchanges and the increase in the globalization of the economy and product chains.

The outcome of these major drivers clearly points in the direction of higher pressure on biological systems and related natural resources such as land, water and fertilizers. The need to match higher needs with limited resources is emphasizing the research component of the bioeconomy, thanks to its potential for producing more with less input, through technical change.

Projections are often used to depict the future of the drivers listed above. The most recent United Nations population prospects predict that the current world population of 7.6 billion will reach 8.6 billion by 2030, 9.8 billion by 2050 and 11.2 billion by 2100 (United Nations, 2017). The role of different countries and geographical regions will, however, change. India is expected to grow faster and to reach a larger population than China. Africa is expected to grow faster than Asia and to almost reach Asia itself by 2100.

However, future scenarios can take rather, or even totally, different shapes depending on a number of variables: awareness can change individual behaviour, information flows can modify individual awareness, the impact of supply and demand modifies market prices. States may anticipate issues and take different policies, societies may proceed with different political strategies country-wise and worldwide. As a result, there are many different possible futures. The highly dynamic nature of these drivers creates a continuous need for scenario development and technology forecasting. Both the identification of these driving forces and the role they can play for the future bioeconomy are investigated in the literature, as well as the understanding (measurement) of the future needs they entail and the potential and credible strategies for the bioeconomy to address these needs.

Some of the most recent and comprehensive scenarios on world futures are developed and illustrated in Bauer *et al.* (2017), O'Neill *et al.* (2017), Popp *et al.* (2017), Riahi *et al.* (2017) and van Vuuren *et al.* (2017). Riahi *et al.* (2017) illustrate some of these scenarios (indicated as shared socioeconomic pathways [SSPs]) (Table 5.1).

Bauer *et al.* (2017) illustrate simulations on energy use in different scenario applying detailed energy system models, integrated with land-use and macroeconomic models. In a baseline, without climate policies, energy uses would increase until 2100 to between two and almost five times the current use. Climate policies will affect this growth depending on the targets set, but the best option simulated (lowest energy use) yields nonetheless an energy use of between 1.5 and three times the current levels. The authors also note that the scenarios with higher increases in energy use would be characterized by a higher share of fossil fuel use. In all scenarios the role of biomass remains below about one-fifth of total energy production and it is negligible in some scenarios (and even declining in the second half of the 21st century).

Popp *et al.* (2017) illustrate land use in different scenarios. The primary drivers of land use are food demand and bioenergy demand. Future demand for food depends on population and per capita demand, which in turn depends on income, preferences, and food price sensitivity. In the baseline (without climate policies), the change in crop demand will be in a range of roughly between +50% and +100% compared to current levels; the same applies to livestock products. Second-generation dedicated bioenergy crops would change dramatically across scenarios. The baseline scenarios report a range between almost nothing and around 5000 million tonnes worldwide, while the strongest climate policy scenarios would require an increase of up to four times this amount.

The outcome of these different needs results in a broad range of potential land-use futures. Without climate change mitigation, global cropland (including bioenergy crops) decreases by 3 million ha in SSP1 (Sustainability) between 2005 and 2100 and increases by up to 753 million ha in SSP3 (Regional Rivalry). The extremes of pasture land use show an increase of up to 380 million ha in SSP4 (Inequality) and a decrease of 742 million ha in SSP1 (Sustainability). The inclusion of mitigation efforts broadens the range of potential future agricultural area sizes. In the most ambitious mitigation scenarios, total cropland expands by up to 1 413 million ha until 2100 in SSP4 (Inequality) due to the expansion of bioenergy cropland, while pastureland decreases by up to 940 million ha until 2100 in SSP5 (Fossil-fuelled development).

Table 5.1. Summary of different shared socioeconomic pathway (SSP) scenario narratives.

Scenario identifier	Short scenario narrative
SSP1 Sustainability – Taking the Green Road (low challenges to mitigation and adaptation)	The world shifts gradually, but pervasively, towards a more sustainable path. It emphasizes a more inclusive development that respects environmental boundaries, with an increasing commitment to achieving development goals. Inequality is reduced while consumption is oriented toward low material growth and lower resource and energy intensity.
SSP2 Middle of the Road (medium challenges to mitigation and adaptation)	The world follows a path in which social, economic and technological trends do not shift markedly from historical patterns. Development and income growth proceeds unevenly, with some countries making relatively good progress while others fall short of expectations. Global and national institutions work toward but make slow progress in achieving sustainable development goals. Environmental systems experience degradation, although there are some improvements and overall the intensity of resource and energy use declines. Global population growth is moderate and levels off in the second half of the century. Income inequality persists or improves only slowly and challenges to reducing vulnerability to societal and environmental changes remain.
SSP3 Regional Rivalry – A Rocky Road (high challenges to mitigation and adaptation)	A resurgent nationalism, concerns about competitiveness and security, and regional conflicts push countries to increasingly focus on domestic or, at most, regional issues. Policies shift over time to become increasingly oriented toward national and regional security issues. Countries focus on achieving energy and food security goals within their own regions at the expense of broader-based development. Investments in education and technological development decline. Economic development is slow, consumption is material-intensive, and inequalities persist or worsen over time. Population growth is low in industrialized and high in developing countries. A low international priority for addressing environmental concerns leads to strong environmental degradation in some regions.
SSP4 Inequality – A Road Divided (low challenges to mitigation, high challenges to adaptation)	Highly unequal investments in human capital, combined with increasing disparities in economic opportunity and political power, lead to increasing inequalities and stratification both across and within countries. Over time, a gap widens between an internationally-connected society that contributes to knowledge- and capital-intensive sectors of the global economy, and a fragmented collection of lower-income, poorly educated societies that work in a labour intensive, low-tech economy. Social cohesion degrades and conflict and unrest become increasingly common. Technology development is high in the high-tech economy and sectors. The globally connected energy sector diversifies, with investments in both carbon-intensive fuels like coal and unconventional oil, but also low-carbon energy sources. Environmental policies focus on local issues around middle and high income areas.
SSP5 Fossil-fueled Development – Taking the Highway (high challenges to mitigation, low challenges to adaptation)	This world places increasing faith in competitive markets, innovation and participatory societies to produce rapid technological progress and development of human capital as the path to sustainable development. Global markets are increasingly integrated. There are also strong investments in health, education, and institutions to enhance human and social capital. At the same time, the push for economic and social development is coupled with the exploitation of abundant fossil fuel resources and the adoption of resource and energy intensive lifestyles around the world. All these factors lead to rapid growth of the global economy, while global population peaks and declines in the 21st century. Local environmental problems like air pollution are successfully managed. There is faith in the ability to effectively manage social and ecological systems, including by geo-engineering if necessary.

Source: Riahi *et al.* (2017)

The broad range of results from the scenarios highlight a number of technical uncertainties and the difficulty in accounting for all relevant topics in modelling. A pertinent example is that of carbon fertilization: if carbon fertilization is included, the net impact of climate change is likely an increase in yield; if carbon fertilization is not included, the effects are more likely a reduction in yields.

In spite of these uncertainties, the clear message is that there are broad options for the future that will depend to a significant extent on behavioural, policy and political determinants. The development of the bioeconomy and its answers in terms of technologies linked to the use of biological (and related) resources will be shaped by the continuous interplay with these forces.

5.3   Individual Consumers in the Bioeconomy

5.3.1   Utility and product attributes

Individual consumers make choices about the consumption of different goods or bundles of goods and, for this reason, are key actors and drivers of market economies. Consumer behaviour is one of the big focuses of attention in current bioeconomy research, especially concerning food and agriculture. Some of the specific bioeconomy issues related to food (e.g. genetically modified [GM] crops) have dominated food consumption debates in recent years, but the issue is likely to be of even greater importance in the future. Not only does it imply understanding consumer preferences, but it also includes investigating consumer awareness, its changes and the role of consumers in new models of connection between supply and demand (Viaggi, 2016).

In microeconomics, individual consumer behaviour when faced with one or more goods is represented using the concepts of preferences and utility, with different approaches and applications. Notably, the focus of choice mechanism is on the assumption that consumers make decisions by maximizing their utility. An important turning point for the theory of consumer behaviour in relation to food and the bioeconomy is the work by Lancaster (1966) and later related literature. The basic notion introduced by

Lancaster is that a good produces utility for the consumer as a result of the characteristics (attributes) of the good itself.

The notion that utility depends on the attributes of goods is now widely used for the study of consumer behaviour. Attributes can be anything depending on the single consumption problems; however, there are different groups of attributes of particular relevance for the bioeconomy goods:

- price of the goods;
- intrinsic characteristics of the goods (e.g. calories or protein);
- health or nutritional properties;
- the process through which the good is produced;
- effects on (anthropized) natural capital, for example, related to the type of biological resources used (e.g. rare species, land, location);
- attributes related to inputs (renewable/non-renewable); and
- attributes related to ethical issues.

Looking carefully, this concept of attributes can cover any aspect of the various technology features described in Chapter 3 and can be used in a rather flexible way to accommodate changing consumer interests. Some specific attributes have been frequently investigated in bioeconomy studies, such as:

- product features related to packaging and portioning, such as colour, size, functionalities;
- the fact of using genetically modified organisms, and within that, different types of genetic modifications, such as cisgenic versus transgenic (to test to what extent consumers distinguish), including in connection to health, ethical, environmental and public goods aspects;
- environmental benefits associated with production, e.g. the reduction of carbon emissions or the reduction of pesticides, for which consumers have a positive utility;
- the production of food through organic farming techniques, which is appreciated not only through consumers studies, but also increasingly through market trends;
- geographical indications that certify the origin of food or other bioeconomy products;
- quality certifications qualifying the quality of products or processes;

- historical or cultural aspects of the breeds used, of the food or products, often also related to the origin; and
- fairness or ethical features related to production methods.

Several of these attributes contribute to give to the bioeconomy goods the features of credence goods (that cannot be experienced or verified by the consumer) or experience goods, rather than search goods.

An important aspect is that while some of the features above are largely in the form of private goods (defined in economics as being characterized by full rivalry and excludability), others have aspects associated with public goods (non-rivalry and non-excludability). The utility of private goods is totally enjoyed by those having property rights on them, which implies that the preferences related to them are completely expressed on the market. This does not occur for public goods-type features, for which free rider behaviour may occur (i.e. one can derive higher utility by taking benefits of a good without contributing to its production, that is, not paying for it). Indeed, consumers may be more likely to value a good that produces public goods, but have no incentive to express this preference in purchasing actions. There are two consequences of this. First, public goods values are not expressed through the market and hence markets do not lead to the optimal level of production of such goods. This is a motivation for policy intervention in this field. Second, as the value of public goods is not expressed in prices, it may require specific methods of evaluation (such as choice experiments, contingent valuation, etc.). The current consumer literature is largely focused on quantifying utility by understanding consumers' willingness to pay (WTP) for different attributes or for increasing/decreasing levels of these attributes.

Together with public goods-type features, these studies are of special interest for new characteristics, for which market observations are not yet available. Indeed, consumers have shown to be very sensitive to the features of some bioeconomy-related innovations.

Another feature is that these attributes are not independent from each other, but rather interact in the mind of the consumer and this interaction may become more or less evident depending on external conditions.

Box 5.1. Interdependence across attributes

Akaichi *et al.* (2016) consider fair trade, organic and lower carbon footprint bananas in three EU countries and studied their interaction. They concluded that, in the market situation at the time of the study, these three types of food quality were not generally competing against each other. Nonetheless, they were likely to eventually compete for consumers' money in the following cases: (i) if the price of organic foods decreases significantly; (ii) if the price of Fairtrade food products is set higher than consumers' WTP; and (iii) if lower carbon footprint bananas are made available to consumers at a price lower than consumers' WTP.

5.3.2 Behavioural approaches

Together with accounting for multiple attributes and their interaction, utility theory needs to be complemented by a number of considerations to get closer to actual consumer behaviour and answer the different types of consumer concerns in the bioeconomy. In fact, actual decision processes include a much more complex set of aspects. Through a systematic review of genetically modified organism (GMO)-related consumer papers, Frewer *et al.* (2013) identify the following constructs:

- intention and acceptance;
- attitude;
- benefit perception;
- risk perception;
- concerns (including ethical/moral); and
- trust.

While the sum of this can be translated into utility or willingness to pay, it is especially important to know what the determinants are, as well as the components of the process in each step, how they relate to each other and how they are connected to the collective side of consumption. Some of these concepts may require further specifications. For example, a distinction can be made between three conceptual categories of acceptance (Ganzevles *et al.*, 2015): market acceptance, socio-political acceptance and community acceptance.

Behavioural economics focuses on the complex processes leading to decisions. Notably, there is a growing body of literature that emphasizes the importance of behavioural economics in explaining food choice patterns and consumer behaviour in general (Wansink *et al.*, 2009; Zilberman *et al.*, 2015), especially for topics having a higher degree of uncertainty or controversy (e.g. genetic engineering technologies). Zilberman *et al.* (2015) highlight the following aspects of behavioural approaches:

- Loss aversion: economic agents weigh losses more than gains; this can explain the tendency of consumers to overestimate small risks (somehow connected to the lack of evidence assuring them that the technology is completely safe) as well as the establishment of a precautionary principle (that aims to eliminate risks).
- Weighting of probabilities: economic agents tend to assign larger weights to lower probability events, which lead to overestimating their relevance compared to expected utility.
- Framing: economic agents' perception of the outcomes of a choice depends on context; this can be determined by information, in particular in connection with the role of media.

The notion of risk and its behavioural consequences can become very important in a regulatory context in which some of the players have a veto power.

Box 5.2. Framing

Zilberman *et al.* (2015) reported numerous studies suggesting that media have largely introduced the general public to GM technologies, and provided the framing that has affected their opinions. Herring (2008) suggests that that consumers' negative attitudes towards GM technologies in agriculture were due to the context in which they were presented by opponents to GM, as they were framed as hazardous technologies with no benefits for consumers and pushed by profit-seeking corporations and farmers. According to McCluskey and Swinnen (2004), the term 'Frankenfoods' provided a context that induced many consumers to take positions against GM technologies. According to Herring and Paarlberg (2016), risks attached to GM technologies are due to political use of framing, linking biotech to 'corporate control, hubristic science, and irreversible genetic pollution'.

5.3.3 Heterogeneity

A major topic in interpreting consumer behaviour is heterogeneity, in spite of the fact that very often consumers are implicitly regarded as homogeneous. Heterogeneity may depend on personal characteristics, individual socioeconomic characteristics (e.g. income) or other preference components.

This heterogeneity can translate into different behavioural characteristics and can ultimately be measured with different WTP for one or more specific attributes. Heterogeneity can be also represented through the way consumers respond to different groups of attributes and to what extent consumers are really interested in making trade-offs among attributes. For example, some consumers may choose primarily based on price, while others care mainly for health factors.

There are several other examples in which consumers show opposite behaviour on the same feature. With regard to pesticides, for example, Hamilton *et al.* (2003) show that a majority of consumers are not willing to pay for pesticide-free food; yet 15% of consumers are willing to pay an additional price for pesticide-free food. In the same study, the authors found that 40% of Californian citizens are willing to vote for banning pesticides; however, some of the people willing to vote to ban pesticides are not willing to pay more for pesticide-free food. This suggests that this decision is driven by environmental concerns rather than food quality or health concerns, and that WTP is highly affected by the perceived allocation of rights.

Heterogeneity is very relevant in terms of marketing as it is the basis for market segmentation, but also for understanding political and regulatory processes. In this context, heterogeneity not only concerns areas clearly distinguished by law or geography (e.g. different countries), but also groups defined by their roles (e.g. non-farmers versus farmers). Intra-country and intra-group heterogeneity is also very relevant as is the case for overall group positioning as shown, for example, in the GE literature (Zilberman *et al.*, 2015). This can also lead to differences between political positioning and individual consumer behaviour.

5.3.4 Information and goodwill

Information is a key determinant of behaviour. In general, preferences are constructed over time, taking into account information on outside factors or consumers' personal experience. Information may also be connected to personal education as well as information provided with the product, for example, through labelling or certifications, or from marketing actions or other means.

Information can, in principle, act in two different ways:

- In the first instance, better information about goods increases the level of consistency between utility expectation and actual utility. In other words, it increases the utility of the consumption of a given product through better predictability of the benefits derived from the choice of the goods.
- Also, information can modify actual utility by producing a different connection between the good and the constructs producing utility (e.g. the notion that something is not causing ethical issues, pain for animals, global warming, etc.). This would translate in different WTP for a good.

A widely investigated topic is the role of the media in affecting GE technologies and food in general. Curtis *et al.* (2008) discuss this topic, emphasizing that the media may tend to highlight negative news. They also noted that different biases in the media with respect to GE, as well as different levels of access to media between countries, result in different consumer attitudes.

There is clear evidence that information can affect opinions and behaviour. Thorne *et al.* (2017) provide an example of GM potatoes, for which information showed to be able to modify WTP and willingness to buy. However, this also depends on the interaction with other features, for example, education or the fact of having children.

New information may be particularly relevant when consumers do not have a strong opinion about a topic or product. One key issue with respect to consumers and information is that in most cases there is no information available to judge a product or the information available is not comprehensible for the average consumer.

This is related to behaviour when confronted with uncertainty.

On the other hand, there are limitations on the role of information due to the fact that information burden may be too high and consumers may decide on purpose to remain ignorant or partially uninformed (rational ignorance). For example, very detailed labels may imply a higher disutility due to reading compared to the benefits (utility) from higher information, as the information they carry does not in fact tell much to the reader and it would be too costly to obtain all of the background information necessary for interpretation. This raises the issue of how consumers deal with choices using proxy information.

Information and rational ignorance is also connected to trust, as higher trust in the counterpart may actually be a substitute for better information. One way to analyse the evolution of these attitudes, and to link information and decision making, is to apply the concept of goodwill towards a good or technology (Zilberman *et al.*, 2015). Goodwill can be viewed as a stock variable. As such it evolves over time and hence each consumer can have their own goodwill at any moment that reflects their assessment of the benefits and risks of a good or technology. The goodwill one feels towards a technology is updated over time on the basis of new information concerning the merits or characteristics associated with the good or technology. It is expected that the decisions that individuals will support about the good or technology (e.g. in voting for or against it) will depend on their level of goodwill.

5.3.5 Drop-in versus new products

The term 'drop-in product' is used in the bio-based literature to identify products that are absolutely the same as existing products, except that the process leading to their production is different. Bio-polymers that have the same characteristics as those derived from fossil fuels are one example. Another example is that of biofuel molecules that can be used as drop-in fuels for existing engines (Petrovič, 2015).

The main feature of these products is that they already have a (well-known) market and are well-known or non-distinguishable to the eye of the consumer. This also implies that consumers have no difficulty in expressing their preferences about the product itself. Differences with existing products then focus on:

- the cost of the new products (which may indeed change); and
- the process (which can, for example, have improved features in terms of impact on public goods provision).

This can put the consumers in a totally different position compared with the new products. New products may bring a higher information burden for consumers and preferences may be more uncertain for consumers themselves, so that reaction to their introduction is also highly uncertain. Drop-in products raise much fewer problems; however, they may require improved communication if they have features that could make the product more appealing to consumers (e.g. green features due to a share of biofuel in fuel) but that are not evident to the consumers.

5.3.6   Stability/instability

As a result of the above, consumer preferences and WTP change over time. Several examples have been studied in the literature on GMOs as well as novel foods. While some cases show substantial instability, in other cases good stability is identified. Connor and Siegrist (2016) use survey data from 2008 and 2010 to show that risk and benefit perceptions of gene technology applications are moderately stable. The topic is made more complex by the fact that different components of preference may vary differently. For example, risk perception may vary differently from perception of benefit due to additional information gained by the consumer on this specific topic.

Stability is not an issue per se, but should rather be considered together with information events or in the context of the interplay with policy, contingent drivers and mega changes. Some alleged changes in demand are driven by higher information or awareness, and also by regulation drivers. An interesting case is that of wood construction. Toppinen *et al.* (2016) found that the emphasis on sustainability is mainly driven by changing regulations reflecting societal needs rather than the push for sustainability arising directly from changing individual consumer preferences.

5.4   Aggregate/Collective Consumer Behaviour and the Bioeconomy

5.4.1   Market demand

Similar to individual demand, aggregate demand can also be studied in different ways and with progressively higher behavioural complexity. The first view is to consider the standard economic representation of demand, whereby aggregate demand is the sum of individual consumer demand, expressed in terms of quantity as a function of some exogenous variable, primarily price. In practice, the aggregated demand can be depicted as a function in which the quantity demanded by a population depends on a number of variables, including the price of the goods, the price of alternative goods, the size of the population, income and income distribution, habits and tastes, information, education, age and household composition.

The price of the goods is the primary variable that affects demand, against which the classical demand curve is usually represented. This is relevant as it connects consumer preferences with cost functions through market functioning and the related price/quantity equilibriums, as we will see in Chapter 7. When considering reactions to market prices, the parameter used is price elasticity. Price elasticity can be defined as the ratio between the relative change of quantity and the relative change of prices (causing the quantity to change). Food or other goods matching primary needs usually have low elasticity, meaning that consumption decreases little when prices increase and grow little when prices decrease.

Prices of other goods are also important. Cross-elasticity can be defined as the ratio between the relative change in quantity of a product and the relative change in the price of another product. This measure serves to identify, based on the sign, products that are substitutes or complements. From the point of view of the bioeconomy, this has a special interest in order to understand the interplay between fossil-based and bio-based products. For example, if the products are substitutes, the increase in oil prices will cause an increase in the demand for bio-based substitute for the production of energy.

As much of the bioeconomy discourse is driven by the comparison between a growing population and finite resources, human population numbers are a major driver of demand. However, this may need to be qualified based on a number of other features. Income and income distribution are also very relevant, as a higher income population is expected to pay more, consume more and, importantly, to consume differently. In particular, higher income populations may be expected to be more sensitive to soft or credence attributes, such as those related to risks (e.g. in GM crops) or to public goods and negative externalities attached to the production process of the goods, including attention to environmental issues and future generations, as well as for rights related to living organisms.

Habits and tastes are also classical issues in food demand and translate into heterogeneity with respect to world population. Heterogeneity is visible among countries, geographical regions, cultures and religions.

Information is also an important determinant, linked to both knowledge capital by individuals and specific information messages related to a specific product.

Age, education and household composition may be very important determinants of individual and, taking the average values, of aggregated consumption; for example, having a high share of families with children (especially buying food for them) affects the expectation of highly healthy food, and avoidance of risks even at high cost (see examples from the GMO literature).

5.4.2 Information and heterogeneity in collective actions

Information is at the core of the bioeconomy, not only as it affects individual behaviour, but also as a way of coordination and alignment in collective actions. Besides information attached to products, information is often advocated as a means of informing consumers of the potential benefits of innovative consumption models or of the dangers of existing models. As mentioned,

this is also connected to instruments such as the use of labels, certification and information in marketing campaigns.

Information in collective action may be linked to different issues, which are in turn related to the existence of different players with contrasting (or simply different) interests. The most common case remains the interaction between different interest groups, or between the public, firms or consumers (see also policy instruments), or between companies and consumers, or among different actors in the value chain.

Different interests can also be identified in the same actor group. The main point is that very motivated subgroups can be active, even by way of strong actions, in providing information for the others or trying to build a consensus around their positions. This has been witnessed in anti-GMO campaigns, for example, in environmental communication or in the action of local food networks.

Indeed, for some parts of the bioeconomy, promotion of changes in lifestyles is very important to achieve transitions; for example, with regard to energy, several authors support the idea that, besides substituting plant biomass energy in place of fossil fuel energy, we should restructure our lifestyles and relocalize the production and consumption of food and biomass (Smolker, 2008).

5.4.3 Indirect demand

Demand for intermediate bio-based products or raw materials for the bio-based industry are derived by the primary (consumer) demand function and by the production costs of the processing stage. In other words, the demand for raw materials will be determined by the profit in the downstream sector. When more than one step of the product chain is involved, this applies to every single stage.

This is indeed a well-known topic for production factors in production economics. It is very relevant for the bioeconomy due to the high and growing number of potentially separable stages in production in which building blocks are produced, processed and traded. To some extent this can also be expanded to the demand for investments (capital) and research and innovation.

There are three major issues related to these points:

- Demand depends on downstream demand elasticity, as well as on input elasticity in the downstream market (i.e. the production function).
- Taking into account market reactions in the downstream market, the changes in demand of the upstream market are always less elastic than it appears from the downstream change because of price adaptation in the market.
- Technology in the downstream sector changes the demand curve of the derived product, including potential rebound effects for technologies increasing efficiency.
- Substitution between bio-based and non-bio-based goods in both the product markets and the input markets makes things even more complicated by introducing the issue of cross-elasticity in indirect demand markets.

The relationships across the bioeconomy are further complicated by the fact that they are not perfectly competitive markets, due to both oligopolistic/monopolistic structures and information issues.

5.5 Consumer–Citizens–Communities

A phenomenon of noteworthy importance in current trends, and of particular importance specifically for the bioeconomy, is the hybridization of the role of consumer and of that of citizen. Hence, rather than 'just' consumers, we have a bundle of people caring, to varying extents, for their consumption and for decisions taken about private and public goods together. The consumer–citizenship concept highlights this merging and the contradiction between the two terms (Johnston, 2008). This is to some extent an individual behaviour issue. However, it is in fact largely connected to the perception by individuals of their role in societies and to collective behaviour.

One major contradiction among the two roles concerns the role of equity, which is a cornerstone of democratic citizenship, and not a priority in consumer behaviour *per se* (Jubas,

2007). As a result, a consumer expressing their point of view as a citizen when taking consumption decisions can value equity more than it would be from their point of view as an individual when they are playing their role as a citizen. This may actually apply to a number of value-related issues or public goods.

When political or citizen-related views are explicitly expressed through the market, by purchase decisions, this phenomenon is called 'dollar voting' or 'voting with your dollar' or 'voting with your fork', in the case of food.

An early work highlighting this trend concerns food (Johnston, 2008) and notes that indeed the development of the consumer–citizenship concept is strongly related to food. Notably, this approach is especially important when food-environmental issues are concerned, which is a key topic in the bioeconomy (e.g. in the food versus energy debate). Examples concern, for example, the banning of chemicals (Guthman and Brown, 2016).

A relevant case in which this concept is witnessed in practice is that of community supported agriculture (CSA), also referred to as community-shared agriculture. CSA is an alternative, locally based economic model of agriculture and food distribution characterized by the fact that growers and consumers share the risks and benefits of food production. A basic model is that CSA members pay at the beginning of the growing season for a share of the anticipated harvest; once harvesting begins, they periodically receive shares of production. Examples of concepts and value behind it can be found in O'Kane (2016).

Another area of interest is that of community-based consumption initiatives, where not only individual actions count, but collective action is determined largely through the use of social media and shared opinions and information. Food networks are growing in importance as drivers of opinion, also thanks to tools that allow communication, exchange and the identification of products.

Framing and trust views of single bioeconomy products also related to the opinions that citizens and actors in general have of the bioeconomy (Fresco, 2015; Lynch *et al.*, 2016; Sleenhoff and Osseweijer, 2016). Specifically, trust in science is a major issue for a highly innovation and research-intensive sector such as the bioeconomy.

Box 5.3. Trust in science and views of the bioeconomy

According to Fresco (2015), there is a lack of belief in human learning and a general distrust in science, at least in Europe. Hence, to meet future needs, the biggest challenge is to bridge the gap between the sciences and society and to engage society in the development of science.

A major issue is that of citizens' perceptions of the bioeconomy as a whole and its components. Lynch et al. (2016) explore Dutch citizens' arguments for and against three selected bio-based innovations: bioplastics, bio-jet fuels and small-scale biorefineries. Citizens are generally in favour of bio-based technologies as they recognize the contributions to economic growth and sustainability. They are also aware of the negative effects, such as high costs, food shortages or deforestation. Their acceptance and support for bio-based technologies increases when they feel more engaged with these technologies and when they see direct personal benefits.

Sleenhoff and Osseweijer (2016) found that individual members of the public foresee distinct ways and levels in how they can engage with the transition to a bio-based society and that these often do not concur with views expressed by stakeholders' representatives.

The picture of communities/stakeholder involvement is evolving in different ways. Two major drivers are the attention to sustainability and the potentiality for interaction due to the availability of digital technologies (Lock and Seele, 2017).

5.6 Selected Cases

5.6.1 Organic food

Organic production has been at the forefront of market differentiation in recent years and one of the most relevant cases of market segmentation. Organic quality is now well established and connected to certification systems. Organic production has seen steady market growth in several countries. The relevance of this market segmentation in terms of demand is illustrated by the difference in price of organic versus non-organic agricultural products. In 2015 the price of organic corn was about three times the one of conventional corn, with the difference increasing over time, at least since 2010. The implications for market segmentation in the US are discussed with respect to GM and non-GM, and non-organic crops in Greene et al. (2016).

From a consumer's perspective, a number of studies have estimated WTPs for organic food. Interestingly, organic seems to be an easy message to transmit despite the complexity of production prescriptions behind it and the scientific doubts about several statements concerning its health benefits and, in some cases, the benefits for the environment. Organic, especially in the early times of its development, was also attached to strong ideological positions and to strong networks involving producers and consumers.

5.6.2 Genetically modified organisms

GMOs are, to date, one of the most debated bioeconomy-related issues. The discussions around GMOs have led to a major differentiation of positions across different countries and consumer groups. The diverging opinions have resulted in strong market segmentation and coexistence issues. Consumer behaviour related to GMOs is largely driven by unknown (or non-existent) risks and information received by consumers. Despite opposition in some areas, GMOs have been rather well-accepted in many parts of the world (Bhullar and Bhullar, 2009).

Bennett et al. (2013) provide a broad overview of the evolution of GMO technology and related issues. The authors note that GM crops have been on the market for more than 15 years and, at the time of the study, were adopted on more than 170 million ha, almost equally shared between developed countries (48%) and developing countries (52%). First-generation GMOs have reduced the input and management costs and hence mainly benefited the producers (though this could be discussed based on actual profit distributions given different market power). Instead, second-generation GM crops are expected to provide benefits to consumers, for example, with enhanced food quality parameters. There are also third-generation GM crops that are being developed specifically for industrial purposes. This evolution will also change consumer perceptions.

According to Bennett *et al.* (2013), the economic and environmental impacts of GM crops can now be summarized with some certainty, and the analysis indicates that, on balance, many benefits have accrued from the adoption of such crops. There continues to be many ethical issues that are being debated, and many are being resolved through institutional interventions. At the same time, there is now a greater awareness that potential risks need to be addressed appropriately (Bhullar and Bhullar, 2009).

A review of acceptance by consumers and its interplay with the regulatory context is provided by Lucht (2015), highlighting that important determinants of consumer attitudes are the perception of risks and benefits, knowledge, trust and personal values. The paper also highlights different trends in different regions of the world, notably an increasingly negative environment for agricultural biotechnology in Europe and a growing discussion in the US, including a push towards the labelling of GMO food. China sees a careful development taking societal discussions into account. Frewer *et al.* (2013) note that both risk and benefit perceptions increased over time. They found that risk perception was higher in Europe than North America and Asia, while the opposite held for benefits and moral concerns.

Lucht (2015) highlights that new breeding techniques address some consumer concerns with transgenic crops or may show positive attributes from the consumer's perspective. A relevant question in this direction is whether consumers actually differentiate between types of GM technologies. An example is cisgenic (containing only DNA sequences from sexually compatible organisms) versus transgenic (including DNA sequences from sexually incompatible organisms), with the former being likely more acceptable to consumers. Indeed, Delwaide *et al.* (2015), based on a survey in different countries of the EU (Belgium, France, the Netherlands, Spain and the United Kingdom in 2013), showed that consumers would be willing to pay much more to avoid consumption of transgenic rice (WTP between €10 and €29) rather than to avoid cisgenic (WTP between €3 and €15). They would also be willing to pay a positive amount for environmental benefits brought by GM crops in most countries (between €0.9 and €6.9), with the exception of France. In a similar study, Shew *et al.* (2016) found that Indian consumers do not differentiate between cisgenic and transgenic rice in terms of their willingness to pay. However, consumers were willing to pay a premium price for GM crops that avoid fungicide use. About three-quarters of the respondents were willing to consume GM rice. Connor and Siegrist (2016) showed that people distinguish between medical, plant and food applications and applications involving animals when evaluating the risk of gene technology. When evaluating the benefits, participants also take consumer-related benefits into account, such as enhancement of functional properties. Frewer *et al.* (2013) highlight that plant-related or 'general' applications are better accepted compared to animal-related GM. The heterogeneity of consumers suggests that there is actually a large majority that does not have a strong preference towards GE (Zilberman *et al.*, 2015).

Thorne *et al.* (2017) investigate consumer WTP for GM potatoes in Ireland, also by considering the effect of information. First, the majority of participants preferred conventional potatoes to GM potatoes, but favourable information (about potential economic and health benefits) about GM increased acceptance and preference for GM potatoes. After being informed, up to two-thirds of participants indicated that they would choose GM potatoes at a 20% price discount. Only 14% would still reject GM potatoes whatever the price. Notably, a lower valuation of GM products was associated with higher levels of education, greater familiarity with GM, and the presence of children in the household.

5.6.3   Bioenergy

Studies on WTP for some types of bioenergy usually relate to two categories. The first is the case in which bioenergy is not established yet and the study focus mainly on the acceptability of its establishment. The second one tries more directly to estimate the monetary value attributed by consumers to the role of biofuels in the fight against climate change, so it is an estimation of its public good component. In fact, the two aspects are inter-related.

Some broad reviews are now available on this topic (Ma *et al.*, 2015; Soon and Ahmad, 2015; Sundt and Rehdanz, 2015; Oerlemans *et al.*, 2016). They emphasize that differences in valuation results may depend more on valuation methods than on actual determinants and that non-bio-based forms of energy are often preferred to bio-based. Several recent examples go, however, in the opposite direction, that is, of showing a preference for bio-based solutions.

Lim *et al.* (2017) assess the public's WTP a premium for the introduction of a bioethanol mandate (see Chapter 7). They estimated a premium price for ethanol of about US$0.26 per litre, which is about 15.6% of the retail price of gasoline. This can be interpreted as WTP for the external benefits due to bioethanol. Other examples in the literature show a similar attitude. Kim *et al.* (2016) estimate a WTP of US$2.5 per household per year for the next 10 years for biogas mandate in Korea. Susaeta *et al.* (2010) investigated selected states in the US and estimated the WTP for payment for CO_2 emission reduction and biodiversity in ethanol blends finding a WTP of US$0.48–1.17 per gallon. Susaeta *et al.* (2011) study woody biomass-based electricity, which can lead to reductions in CO_2 emissions and improved forest habitats in the Southern United States, finding a WTP of US$0.049 per kWh or US$40.5 per capita per year.

Acceptance issues also apply to biogas and even to the evolution towards biorefineries. Ganzevles *et al.* (2015) provide an example related to the BioBased Economy Park at Cuijk, in the Netherlands, including the integration of biopower into a broader smart-use scheme. The results showed that the extension from bioenergy towards smart biomass use does not necessarily increase the acceptance of the project, though it modifies the configuration of social acceptance issues. In particular, the social acceptance of smart biomass use is 'fuzzier, more open to recursive patterns and more dependent upon inter-firm trust'. Features of biomass use that can affect acceptability are: the type of bioenergy, the sector that takes the initiative, the greenfield character of the project and the complexity of the energy scheme.

5.6.4 Recyclable materials

Klaiman *et al.* (2016) use discrete choice experiments to assess consumer WTP for packaging materials and recyclability of a beverage product packaging. Consumer WTP was different across different packaging materials; notably it was highest for plastic packaging, followed by glass, cartons and aluminium. The authors also analyse the effectiveness of indirect questioning in addressing issues of social desirability bias (the tendency of an individual to provide answers in a way that is biased towards their perception of a socially acceptable answer) as well as the effects of information on consumer behaviour. The empirical results reveal that indirect questioning results in WTP values for packaging recyclability that are 60% lower than those obtained from direct questioning. They also find that information from a video treatment had a significant and positive effect on consumer preferences and demand for packaging recyclability.

5.7 Outlook

The understanding of consumer behaviour is a central topic for the bioeconomy. However, it remains a rather open and challenging issue. On the one hand, the current literature highlights the importance of consumer choices as market drivers, also in relation to the wide role of market instruments (see Chapter 7). On the other hand, it highlights the instability, contingency and sometimes unreliability of consumer attitudes toward specific attributes of bioeconomy products. Key to this interplay is the role of information in affecting consumer preferences and the interaction between individual and collective behaviour.

Even more, changes in lifestyle represent one of the key strategies to make the bioeconomy real, which adds to marketing actions as accepted and promoted ways of affecting consumer behaviour. This is a major challenge for demand-side management of the bioeconomy (and economic studies) as it implies explicitly considering consumers preference as a variable that can be willingly modified (to some extent).

Difficulties are more relevant when looking at perspective studies on new products. The ability of studies to detect properly the demand side, and especially WTP, is often questionable when based on self-reported preferences by consumers. Self-reporting suffers from the basic human tendency to present oneself in the best possible way and from several methodological biases depending on the specific technique used (Klaiman *et al.*, 2016). However, revealed preferences also suffer from uncertainty and instability, and this is even more the case for collectively expressed preferences.

In spite of this, consumer needs remain central in the bioeconomy. This poses at least two complementary challenges: one for societies, to envisage improved institutional solutions to promote aware and constructive demand management as well as coordination between private demand and societal challenges; and one for researchers, to better understand the expression of values and preferences in complex SEVWs, including collective and informational embeddedness. This issue, interestingly, takes rather different pathways in different countries, especially following opposite trajectories in income, but also diverging lifestyles.

References

Akaichi, F., de Grauw, S., Darmon, P. and Revoredo-Giha, C. (2016) Does fair trade compete with carbon footprint and organic attributes in the eyes of consumers? Results from a pilot study in Scotland, The Netherlands and France. *Journal of Agricultural and Environmental Ethics* 29(6), 969–984. doi: 10.1007/s10806-016-9642-7.

Bauer, N., Calvin, K., Emmerling, J., Fricko, O., Fujimori, S., Hilaire, J., Eom, J., Krey, V., Kriegler, E., Mouratiadou, I., Sytze de Boer, H., van den Berg, M., Carrara, S., Daioglou, V., Drouet, L., Edmonds, J.E., Gernaat, D., Havlik, P., Johnson, N., Klein, D., Kyle, P., Marangoni, G., Masui, T., Pietzcker, R.C., Strubegger, M., Wise, M., Riahi, K. and van Vuuren, D.P. (2017) Shared socio-economic pathways of the energy sector: Quantifying the narratives. *Global Environmental Change* 42, 316–330. doi: 10.1016/j.gloenvcha.2016.07.006.

Bennett, A.B., Chi-Ham, C., Barrows, G., Sexton, S. and Zilberman, D. (2013) Agricultural biotechnology: Economics, environment, ethics, and the future. *Annual Review of Environment and Resources* 38, 249–279. doi: 10.1146/annurev-environ-050912-124612.

Bhullar, G.S. and Bhullar, N.K. (2009) Acceptance of GMOs worldwide: A consumer and producer perspective. In Bhowmik, P.K. and Basu, S.K. (eds) *Advances in Biotechnology*. Emirate of Sharjah, United Arab Emirates, Bentham Science Publishers, pp. 279–297. doi: 10.2174/978160805090110901010256.

Connor, M. and Siegrist, M. (2016) The stability of risk and benefit perceptions: A longitudinal study assessing the perception of biotechnology. *Journal of Risk Research* 19(4), 461–475. doi: 10.1080/13669877.2014.988169.

Curtis, K.R., McCluskey, J.J. and Swinnen, J.F.M. (2008) Differences in global risk perceptions of biotechnology and the political economy of the media. *International Journal of Global Environmental Issues* 8(1/2), 77. doi: 10.1504/IJGENVI.2008.017261.

Delwaide, A.-C., Nalley, L.L., Dixon, B.L., Danforth, D.M., Nayga, R.M., Van Loo, E.J. and Verbeke, W. (2015) Revisiting GMOs: Are there differences in European consumers' acceptance and valuation for cisgenically vs transgenically bred rice? *PLoS ONE* 10(5), e0126060. doi: 10.1371/journal.pone.0126060.

Fresco, L.O. (2015) The new green revolution: Bridging the gap between science and society. *Current Science* 109(3), 430–438.

Frewer, L.J., van der Lans, I.A., Fischer, A.R.H., Reinders, M.J., Menozzi, D., Zhang, X., van den Berg, I. and Zimmermann, K.L. (2013) Public perceptions of agri-food applications of genetic modification – A systematic review and meta-analysis. *Trends in Food Science & Technology* 30(2), 142–152. doi: 10.1016/j.tifs.2013.01.003.

Ganzevles, J., Asveld, L. and Osseweijer, P. (2015) Extending bioenergy towards smart biomass use Issues of social acceptance at Park Cuijk. The Netherlands. *Energy, Sustainability and Society* 5(1), 22. doi: 10.1186/s13705-015-0053-9.

Greene, C., Wechsler, S.J., Adalja, A. and Hanson, J. (2016) Economic issues in the coexistence of organic, genetically engineered (GE), and non-GE crops. *Economic Information Bulletin No. (EIB-149)*. United States Department of Agriculture, Economic Research Service, 41 pp.

Guthman, J. and Brown, S. (2016) I will never eat another strawberry again: the biopolitics of consumer-citizenship in the fight against methyl iodide in California. *Agriculture and Human Values* 33(3), 575–585. doi: 10.1007/s10460-015-9626-7.

Hamilton, S.F., Sunding, D.L. and Zilberman, D. (2003) Public goods and the value of product quality regulations: the case of food safety. *Journal of Public Economics* 87(3–4), 799–817. doi: 10.1016/S0047-2727(01)00103-7.

Herring, R.J. (2008) Opposition to transgenic technologies: ideology, interests and collective action frames. *Nature Reviews Genetics* 9(6), 458–463. doi: 10.1038/nrg2338.

Herring, R. and Paarlberg, R. (2016) The political economy of biotechnology. *Annual Review of Resource Economics* 8(1), 397–416. doi: 10.1146/annurev-resource-100815-095506.

Johnston, J. (2008) The citizen-consumer hybrid: Ideological tensions and the case of whole foods market. *Theory and Society* 37(3), 229–270. doi: 10.1007/s11186-007-9058-5.

Jubas, K. (2007) Conceptual con/fusion in democratic societies: Understandings and limitations of consumer-citizenship. *Journal of Consumer Culture* 7(2), 231–254. doi: 10.1177/1469540507077683.

Kim, H.-Y., Park, S.-Y. and Yoo, S.-H. (2016) Public acceptability of introducing a biogas mandate in Korea: A contingent valuation study. *Sustainability* 8(11), 1087. doi: 10.3390/su8111087.

Klaiman, K., Ortega, D.L. and Garnache, C. (2016) Consumer preferences and demand for packaging material and recyclability. *Resources, Conservation and Recycling* 115, 1–8. doi: 10.1016/j.resconrec.2016.08.021.

Lancaster, K.J. (1966) A new approach to consumer theory. *Journal of Political Economy* 74(2), 132–157.

Lim, S.-Y., Kim, H.-J. and Yoo, S.-H. (2017) Public's willingness to pay a premium for bioethanol in Korea: A contingent valuation study. *Energy Policy* 101, 20–27. doi: 10.1016/j.enpol.2016.11.010.

Lock, I. and Seele, P. (2017) Theorizing stakeholders of sustainability in the digital age. *Sustainability Science* 12(2), 235–245. doi: 10.1007/s11625-016-0404-2.

Lucht, J.M. (2015) Public acceptance of plant biotechnology and GM crops. *Viruses* 7(8), 4254–4281. doi: 10.3390/v7082819.

Lynch, D.H.J., Klaassen, P. and Broerse, J.E.W. (2016) Unraveling Dutch citizens' perceptions on the bio-based economy: The case of bioplastics, bio-jetfuels and small-scale bio-refineries. *Industrial Crops and Products* 106, 130–137. doi: 10.1016/j.indcrop.2016.10.035.

Ma, C., Rogers, A.A., Kragt, M.E., Zhang, F., Polyakov, M., Gibson, F., Chalak, M., Pandit, R. and Tapsuwan, S. (2015) Consumers' willingness to pay for renewable energy: A meta-regression analysis. *Resource and Energy Economics* 42, 93–109. doi: 10.1016/j.reseneeco.2015.07.003.

McCluskey, J.J. and Swinnen, J.F.M. (2004) Political economy of the media and consumer perceptions of biotechnology. *American Journal of Agricultural Economics* 86(5), 1230–1237. doi: 10.1111/j.0002-9092.2004.00670.x.

O'Kane, G. (2016) A moveable feast: Exploring barriers and enablers to food citizenship. *Appetite* 105, 674–687. doi: 10.1016/j.appet.2016.07.002.

O'Neill, B.C., Kriegler, E., Ebi, K.L., Kemp-Benedict, E., Riahi, K., Rothman, D.S., van Ruijven, B.J., van Vuuren, D.P., Birkmann, J., Kok, K., Levy, M. and Solecki, W. (2017) The roads ahead: Narratives for shared socioeconomic pathways describing world futures in the 21st century. *Global Environmental Change* 42, 169–180. doi: 10.1016/j.gloenvcha.2015.01.004.

Oerlemans, L.A.G., Chan, K.-Y. and Volschenk, J. (2016) Willingness to pay for green electricity: A review of the contingent valuation literature and its sources of error. *Renewable and Sustainable Energy Reviews* 66, 875–885. doi: 10.1016/j.rser.2016.08.054.

Pätäri, S., Tuppura, A., Toppinen, A. and Korhonen, J. (2015) Global sustainability megaforces in shaping the future of the European pulp and paper industry towards a bioeconomy. *Forest Policy and Economics* 66(C), 38–46. Elsevier. doi: 10.1016/j.forpol.2015.10.009.

Petrovič, U. (2015) Next-generation biofuels: A new challenge for yeast. *Yeast* 32(9), 583–593. doi: 10.1002/yea.3082.

Popp, A., Calvin, K., Fujimori, S., Havlik, P., Humpenöder, F., Stehfest, E., Bodirsky, B.L., Dietrich, J.P., Doelmann, J.C., Gusti, M., Hasegawa, T., Kyle, P., Obersteiner, M., Tabeau, A., Takahashi, K., Valin, H., Waldhoff, S., Weindl, I., Wise, M., Kriegler, E., Lotze-Campen, H., Fricko, O., Riahi, K. and Vuuren, D.P. van (2017) Land-use futures in the shared socio-economic pathways. *Global Environmental Change* 42, 331–345. doi: 10.1016/j.gloenvcha.2016.10.002.

Riahi, K., van Vuuren, D.P., Kriegler, E., Edmonds, J., O'Neill, B.C., Fujimori, S., Bauer, N., Calvin, K., Dellink, R., Fricko, O., Lutz, W., Popp, A., Cuaresma, J.C., KC, S., Leimbach, M., Jiang, L., Kram, T., Rao, S., Emmerling, J., Ebi, K., Hasegawa, T., Havlik, P., Humpenöder, F., Da Silva, L.A., Smith, S., Stehfest, E., Bosetti, V., Eom, J., Gernaat, D., Masui, T., Rogelj, J., Strefler, J., Drouet, L., Krey, V., Luderer, G., Harmsen, M., Takahashi, K., Baumstark, L., Doelman, J.C., Kainuma, M., Klimont, Z., Marangoni, G., Lotze-Campen, H., Obersteiner, M., Tabeau, A. and Tavoni, M. (2017) The shared socioeconomic pathways and their energy, land use, and greenhouse gas emissions implications: An overview. *Global Environmental Change* 42, 153–168. doi: 10.1016/j.gloenvcha.2016.05.009.

Shew, A.M., Nalley, L.L., Danforth, D.M., Dixon, B.L., Nayga, R.M., Delwaide, A.-C. and Valent, B. (2016) Are all GMOs the same? Consumer acceptance of cisgenic rice in India. *Plant Biotechnology Journal* 14(1), 4–7. doi: 10.1111/pbi.12442.

Sleenhoff, S. and Osseweijer, P. (2016) How people feel their engagement can have efficacy for a bio-based society. *Public Understanding of Science* 25(6), 719–736. doi: 10.1177/0963662514566749.

Smolker, R. (2008) The new bioeconomy and the future of agriculture. *Development* 51(4), 519–526. doi: 10.1057/dev.2008.67.

Soon, J.-J. and Ahmad, S.-A. (2015) Willingly or grudgingly? A meta-analysis on the willingness-to-pay for renewable energy use. *Renewable and Sustainable Energy Reviews* 44, 877–887. doi: 10.1016/j.rser.2015.01.041.

Sundt, S. and Rehdanz, K. (2015) Consumers' willingness to pay for green electricity: A meta-analysis of the literature. *Energy Economics* 51, 1–8. doi: 10.1016/j.eneco.2015.06.005.

Susaeta, A., Alavalapati, J., Lal, P., Matta, J.R. and Mercer, E. (2010) Assessing public preferences for forest biomass-based energy in the southern United States. *Environmental Management* 45(4), 697–710. doi: 10.1007/s00267-010-9445-y.

Susaeta, A., Lal, P., Alavalapati, J. and Mercer, E. (2011) Random preferences towards bioenergy environmental externalities: A case study of woody biomass based electricity in the Southern United States. *Energy Economics* 33(6), 132–139. doi: 10.1016/j.eneco.2011.05.015.

Thorne, F., Fox, J.A.S., Mullins, E. and Wallace, M. (2017) Consumer willingness-to-pay for genetically modified potatoes in Ireland: An experimental auction approach. *Agribusiness* 33(1), 43–55. doi: 10.1002/agr.21477.

Toppinen, A., Röhr, A., Pätäri, S., Lähtinen, K. and Toivonen, R. (2016) The future of wooden multistory construction in the forest bioeconomy - A Delphi study from Finland and Sweden. *Journal of Forest Economics* 31, 3–10. doi: 10.1016/j.jfe.2017.05.001.

United Nations (2017) *World Population Prospects. The 2017 Revision. Key Findings and Advance Tables.* New York, USA, United Nations.

Viaggi, D. (2016) Towards an economics of the bioeconomy: four years later. *Bio-based and Applied Economics* 5(2), 101–112.

van Vuuren, D.P., Riahi, K., Calvin, K., Dellink, R., Emmerling, J., Fujimori, S., KC, S., Kriegler, E. and O'Neill, B. (2017) The shared socio-economic pathways: Trajectories for human development and global environmental change. *Global Environmental Change* 42, 148–152. doi: 10.1016/j.gloenvcha.2016.10.009.

Wansink, B., Just, D.R. and Payne, C.R. (2009) Mindless eating and healthy heuristics for the irrational. *American Economic Review* 99(2), 165–169. doi: 10.1257/aer.99.2.165.

Wesseler, J., Banse, M. and Zilberman, D. (2015) Introduction special issue 'The political economy of the bioeconomy'. *German Journal of Agricultural Economics* 64(4), 209–211.

Zilberman, D., Graff, G., Hochman, G. and Kaplan, S. (2015) The political economy of biotechnology. *German Journal of Agricultural Economics* 64(4), 212–223.

6

Supply-side Economics

6.1 Introduction and Overview

Supply-side economics covers the issue of how production systems supply goods by reacting to incentives coming from markets or policy. It includes understanding the relationships between the input and output of a production process, how production reacts to market signals (prices), the way this is organized into production units (companies) and the manner in which companies take decisions about production.

This chapter is primarily based on technology and its representation as developed in Chapter 3, and explains how this translates into the production of goods and services. To a large extent, the supply-side economics of the bioeconomy can be addressed using standard concepts, such as the decision making by firms based on production function. As discussed in Chapter 3, traditional production functions in agriculture, for example, take a form that depends to a significant extent on the behaviour of living organisms. Production factors are usually identified in three separate groups:

- nature, especially land but also other raw materials;
- labour, which is human effort exerted in the production process; and
- capital, which is constituted by goods and services produced and used in the production process.

Knowledge can be added to this as a special case of capital, or taken as a separate production factor.

Firms are usually classified as belonging to the primary, secondary and tertiary sector based on their connection with nature. However, the supply of bio-based products relates to a number of interconnected sectors, including the harvesting of natural resources, the production of feedstocks (as products or by-products) and their transformation, as well as the recovery of wastes usable in the processes. Thus, the bioeconomy, given its characteristics, actually covers the whole chain from nature to consumers, including a number of feedback loops, as previously discussed. For this reason, the interconnections among all factors and stages are especially important.

Key sectors for the supply of biomass are agriculture and forestry as they determine the overall production possibilities. In addition, a number of specificities are arising downstream for the bioeconomy, such as the hot topic of the re-use of wastes and by-products.

Furthermore, the supply-side economics of the bioeconomy is interwoven with specificities of technology uptake, entrepreneurship, chain management and financing. One of the key issues besides supply is that it is achieved through the involvement of a number of companies on the same product chain and that different chains are interwoven among each other.

© D. Viaggi 2018. *The Bioeconomy: Delivering Sustainable Green Growth* (D. Viaggi)

This concept has already been highlighted in Chapters 3 and 4, but is so important here that a large part of this chapter is devoted to processes and industrial organization within the supply system. The chapter artificially stops at the interface with consumers, though it also highlights issues related to the connection with consumers and society that will be more comprehensively addressed in the next chapter. This section also does not consider public goods and externality dimensions that will be rather addressed in Chapter 7.

6.2 Bioeconomy Production, Production Functions and Supply

6.2.1 Choices in the attribute space

In this section we pick up technology representation from Chapter 3, to represent the basic production choices in the bioeconomy, beginning with the choice of product characteristics. This subsection investigates how the combination of attribute levels, i.e. the product design, is affected by changes in prices or utility. Each attribute may have different effects on utility and combinations of attributes may have a utility that is not the mere sum of the utility of individual attributes.

Let us assume (contrary to Chapter 3) that continuous combinations of attributes are allowed by technology as in Fig. 6.1.

The figure depicts various options of utility achievable from the two attributes. It is worth noting that utility here can also be intended as the willingness to pay (WTP) by consumers for the goods or by the WTP by firms using the product in the downstream steps of the chain, or as market prices. Each line B may be thought of as the combination of attributes that yield the same level of utility/WTP. Different lines, for example, Ba_1 and Ba_2, represent different levels of utility.

Different shapes depict different configurations of utility of the possible combinations of the two attributes. Line B_0 represents a case of linear utility in the two attributes. Lines Ba_1 and Ba_2 have higher utility for more focused goods (closer to purer attributes); this can be seen by the fact that the given level of utility can be obtained by a relatively low level of each attribute taken in higher isolation. Following the line, it may be seen that the same utility can be obtained by growing combinations of both attributes, which means that one compensates for the other. Beyond the point when the tangent of the function is vertical or horizontal, they start to substitute for each other, that is, the same utility level can be kept by increasing one attribute when the other is decreasing. In cases like this, a process (or research development) can yield a benefit by detaching attributes from each other (assuming they are together for example in a complex feedstock).

The other line (Bb_1) has the opposite behaviour, with higher utility for combinations of

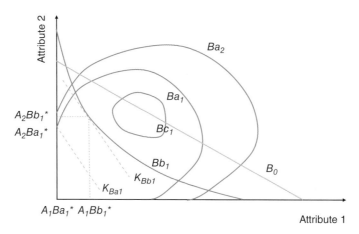

Fig. 6.1. Different goods and utility (willingness to pay) in the attribute space. Different lines, for example, Ba1 and Ba2, represent different levels of utility. See the text for further details.

attributes. Moreover, high presence of an attribute can be as negative as low presence, so circular iso-utility lines are also possible (*Bc*). This means that utility does not necessarily grow with the levels of attributes (e.g. excessive concentration can be seen as being negative as negative as a too low concentration).

The question is now to determine which conditions yield the best economic results in terms of profit or net utility. This can be done by adding an iso-cost curve to the graph, which makes it possible to identify the minimum cost for each combination of attributes. We can assume a linear iso-cost curve K, i.e. the cost is proportional to the amount of two compounds. The optimal combination of attributes is given by the point in which each utility level touches the lower cost (the iso-cost line closer to the origin). The best solution would be in an intermediate point for iso-utility curve Bb_1 $(A_1Bb_1{}^*, A_2Bb_1{}^*,$ with cost level K_{Bb1}) while it would be in one of the extremes (intercepts) for Ba_1 and Ba_2. In the figure, for curve Ba_1, it is the point $(A_1Ba_1{}^*, A_2B_1a^*,$ with cost level K_{Ba1}). The slope of the iso-cost line depends on the relative cost of obtaining a unit crease in the level of each attribute. So, if the cost of obtaining a higher level of attribute A_1 would increase, it would become optimal to obtain products with a lower level of that attribute.

More realistically, higher levels of pureness may have higher costs. In this case, the cost curve is concave towards the origin.

Costs, as well as benefits, can also be completely discontinuous, and not necessarily related to the levels of attributes; in this case the problem of a profit-maximizing decision-maker would be to empirically compare costs and benefits of each combination of attributes to identify the highest positive difference among these parameters.

6.2.2 Input–output

As mentioned, the relationships between input and output in a single stage of the production process are traditionally represented in economics in three ways:

1. Input–output relationships (in a simplified way through one factor-one product function), which allow for the identification of the optimal level of use of a variable input and the related optimal level of the corresponding output.

2. Input–input relationships (in a simplified way through the relationship among two production factors), which allow for the identification of the optimal combination of input, i.e. the optimal technology.

3. Output–output relationships (in a simplified way the product-product function), which make it possible to understand the production frontier and identify the optimal combination of output.

Taking into account these relationships, pofit maximization entails the following conditions (Debertin, 2012):

- the ratios of the values of the marginal products to the respective input prices must be the same for each input in the production of each output;
- the marginal rate of substitution between each pair of inputs must be equal to the corresponding inverse price ratio; and
- the rate of product transformation among each pair of products must equal the slope of the iso-revenue line or inverse output price ratio.

In order to sketch economic choices, we further elaborate on some of the relationships above by illustrating them graphically.

Figure 6.2 illustrates the choice of the optimal level of input. The top of the figure reports two examples of revenue functions obtained by production functions discussed in Chapter 3, multiplied by product price. The bottom part represents the corresponding marginal revenue functions. The optimal level of input (and consequent optimal production level) are given by the intersection between the marginal revenue from production and the prices of the input (respectively $X_A{}^*$ and $X_B{}^*$). Moving from technology A to technology B would reduce the use of input because of the change in shape; however, a simple move of the production function A upward would actually increase the use of input by increasing the marginal revenue. On the other hand, moving the price of the input upwards would decrease the optimal level of use of the

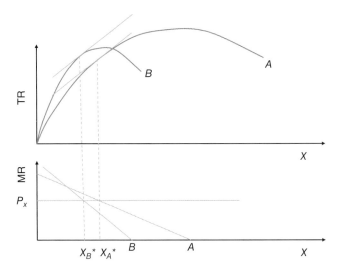

Fig. 6.2. Optimal level of input and production. See the text for details.

input and the optimal level of production. This explains for example the concern for the price of feedstocks.

Figure 6.3 illustrates the optimal combination of input and output (6.3(a) and 6.3(b) respectively). In Fig. 6.3(a), X_{1A}^* and X_{2A}^* are the optimal amount of input to produce the amount given by the isoquant A, given the slope of the iso-cost curve, determined by the prices of inputs X_1 and X_2.

A change in technology would move the isoquant to B, with a new optimal combination of input X_{1B}^* and X_{2B}^*. A change in prices would instead change the slope of the iso-cost curve, moving the optimal combination of input on a different point of the same curve, notably reducing the use of the input the price of which increases in relative terms. This explains the move from fossil to bio-based input when the price of fossil resources increases. Fig. 6.3(b) describes the optimal combinations of output for the three production frontiers A, B and C. The straight lines are iso-revenue lines, that is, all combinations of the two products yielding the same revenue. Iso-revenue lines closer to the origin imply lower total revenue. In the figure, iso-revenue lines have the same slope, that is, prices of goods produced are the same. If one of the prices changes, the iso-revenue line rotates yielding a different optimum. An interesting case is C compared to A. Assume Y_1 is bioenergy and Y_2 is a crop. By increasing the productivity of bioenergy production, the effect is not only the increase of the bioenergy production, but, if the change is

strong enough, there is also a reduction of the crop production.

Once having identified the minimum cost combination for each level of production of a good, this can be mapped on different levels of production through cost functions. Cost functions express the total and marginal costs of each level of production achieved by a firm and are used to explain the level of output of the firm as a function of market price. The essential functioning of these mechanisms is illustrated in Fig. 6.4.

Curves A, B and C illustrate the marginal costs of different firms, using different technologies or having different resource endowments. This first illustrates heterogeneity of costs among firms and for different amounts of production that will be discussed more in detail later on. It also illustrates different production behaviour as a function of costs. For example, given price P_1, the optimal level of production of firm C is Y_{P1C}^*, higher than B, while firm A would stay out of the market as its lowest cost is above the market price. This hence expresses the entry and exit decision from the market. The change in production as a reaction to price generates the supply curve of individual firms. For example, if prices move from P_1 to P_2, production by firm C moves from Y_{P1C}^* to the higher level Y_{P2C}^*.

The aggregated supply function is the horizontal sum of individual firms supply function and is expressed by curve $A+B+C$. This curve expresses the amount of production

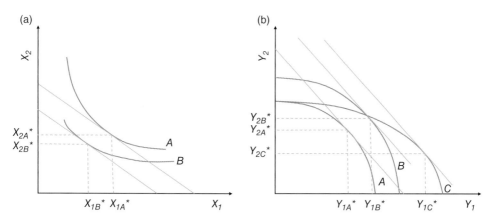

Fig. 6.3. Optimal combination of input and of output. See the text for details.

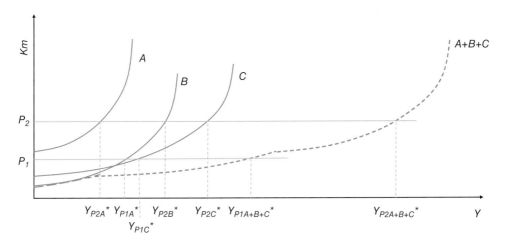

Fig. 6.4. Cost functions and the supply curve. Curves *A*, *B* and *C* illustrate the marginal costs of different firms, using different technologies or having different resource endowments. See the text for details.

of good *Y* as a function of price and represents the market supply function of a good that can be used to express the interaction with demand on a market (Chapter 7). Another point is that different costs may be associated with firms using different technologies with respect to environmental impacts or concerning the use of bio-based versus non-bio-based inputs. This will also be further discussed in Chapter 7.

6.2.3   Harvesting and anthropized biological capital

While this is a general representation of the production process, for harvesting biological

resources an additional key topic is the impact of harvesting on capital stock and its effect on the ability of the capital stock to produce a yield (affecting the potential harvest). This, in turn, depends not only on the stock, but also on other variables, such as, principally, effort by firms (e.g. fishing effort represented by number and size of boats or similar). Figure 6.5 presents revenues as a function of harvesting effort.

R represents the revenue derived from the multiplication of harvest by prices; *K* represents the cost; both are represented as a function of the exploitation effort. The highest profit, hence optimal effort, is represented by the maximum distance between *R* and *K*, in case of monopoly or sufficient coordination

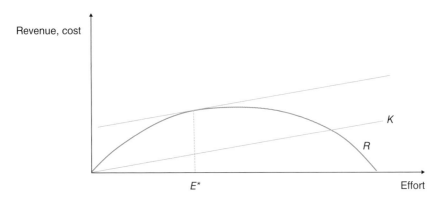

Fig. 6.5. Revenue and costs as a function of harvesting effort. *R* represents the revenue derived from the multiplication of harvest by prices; *K* represents the cost; both are represented as a function of the exploitation effort. See the text for details.

among users (point E^*). This level is also associated with some level of stock; high effort would be associated with low stocks and the other way around; so, at the equilibrium, there should be an optimal level of stock (normally different from the maximum and the minimum) and an associated optimal level of harvesting.

6.3 More on Behaviour

6.3.1 Risk and behavioural economics

The models illustrated in section 6.2 are based on an assumption of profit maximization on the part of firms. Indeed, this is a very simplified assumption that provides a clear-cut reasoning, but which also has limitations with respect to interpreting the bioeconomy.

At least, together with profit maximization, some consideration should be paid to risk aversion. This is connected to the fact that bioeconomy technologies are often characterized either by relevant uncertainty of effects or by potentially long and complex pathways of consequences within the production process itself and beyond it (i.e. in the environment or at the consumption stage).

There are several risk dimensions that are relevant for the supply side of the bioeconomy, in addition to those related to consumers and citizens discussed in the previous chapter. Abbati de

Assis *et al.* (2017) identify the following sources of risk, the first three of which are common to several new ventures:

- volatile markets (downstream and upstream);
- political or policy instability;
- potential upcoming disruptive technologies;
- additional specific risk for bioeconomy activities coming from competing with mature, highly efficient markets dominated by fossil-based companies;
- uncertain feedstock availability;
- limited technological data and information on new processes; and
- uncertain market conditions for newly developed products.

However, the literature goes beyond adjustments of profit-maximizing behaviour to account for risk and, in analogy to consumer behaviour, argues for the need for approaches more directly linked to behavioural economics in order to properly interpret bioeconomy decisions by firms.

Behavioural considerations are usually very important for small-scale actors, such as farmers and small companies. Loss aversion is clearly a paramount issue for small farmers, especially low-income producers, due to the fact that loss has significant relevance and can easily lead to dramatic consequences for the farm and the household. Distorted weighting of events may also occur due to the fact that small actors have more difficulty in accessing information. Framing is also important, especially in areas in which farmers are at the centre of multiple actors' systems with different views and bringing to bear sometimes strongly biased starting points of view, such as farmers' organizations, actors in the product chain and local authorities in the areas where farmers work. This also leads to the fact that the individual characteristics of actors in the supply side are a partial explanation of different attitudes.

Yet this is also an issue for larger companies. Examples related to GM crops are provided by Zilberman *et al.* (2013). The authors argue that some of the objections against new GE crops came from input manufacturers (e.g. pesticides producers) that were being replaced by GE traits. Loss aversion behaviour, in particular, may suggest that manufacturers of these inputs may engage in vigorous (at times excessive) battles to protect their 'territory' due to concerns regarding losses. Indeed, the intensity of the actions seen by the various agents involved in the GE sector concerning regulation may reflect the extreme implications of certain regulatory decisions on their losses.

Risk and behavioural considerations are not only individual issues, but rather relate to a significant extent to collective behaviour through communication and different forms of coordination.

6.3.2 Technology uptake and entrepreneurship

Technology change is a key issue in the bioeconomy. This requires an understanding of technology uptake and innovation behaviour linked to entrepreneurship. This is also linked to both the heterogeneity of operators and an understanding of the productivity (and finally supply) of the sector. This has both an individual dimension and a dimension linked to collective behaviour and policy, also discussed through networking instruments and similar in Chapter 7.

Heterogeneity and timing of technology adoption is an issue at all stages of the chain, though technology uptake by farmers may be the most commonly discussed example. Indeed, primary production in the bioeconomy is largely related to innovation adoption by farmers and, being the sector with the smallest firms, more often farmers are facing the problem of adoption of exogenously developed technologies, in contrast to larger companies that have a more active role in technology development. Specific issues include:

- profitability and economic impact (linked to benefit distribution due to patenting and companies' input strategies);
- risk issues (production/market) and distribution among stages of the chain; and
- social vision and implications (including employment, etc.).

Zilberman *et al.* (2015a) emphasize the heterogeneity of farmers with respect to the adoption of GE crops:

- Among conventional farmers, it may be expected that there are groups that benefit from a trait (and hence may be expected to support it), while groups that may not benefit from the same trait or could even lose out due to the price effect of its adoption by those who benefit; hence, they have reasons to not adopt and even to oppose the trait at the policy making level.
- Some organic growers in the US oppose GE on ideological grounds, while others do so on practical grounds. For example, they may require strong segregation policies because the income of organic farmers may suffer due to contamination from neighbouring GE crops (hence affecting the selling price).
- Their attitudes may also vary by crop or trait depending on the above, or other, reasons.

The heterogeneous and differing interests of farmers are more evident when major technology debates arise, but are actually a feature of any innovation. To some extent similar debates are also witnessed concerning bioenergy plants in some countries, though this is more related to controversies over large facilities in relation to citizens rather than consumers or chain organization.

Box 6.1. Motivation for technology adoption

Decisions to convert to biofuel crops were investigated by Kelsey and Franke (2009). The main motivations for conversion were overall profitability (the predominant reason for converting cropland to dedicated biofuel crops such as switchgrass) and patriotic reasons. The main barriers were: (i) lack of markets (biorefineries); and (ii) (lack of) information about biofuel crop production. The authors recommended an educational campaign regarding biofuel crop production best practices.

Case *et al.* (2017) investigate the decision-making processes underlying the use of processed and unprocessed organic waste-based fertilizers by farmers, which is a key strategy to support a circular bioeconomy. Among a sample of Danish farmers surveyed, they found that future interest in using processed manures and urban waste-derived fertilizer was greater than the current use. Farm and farmer characteristics affecting future interest include: farming activity, farmer age, farm size and conventional/organic farming. However, a large percentage (40%) of farmers did not have access to processed forms of organic fertilizer, particularly those interested in using processed manure (35% of respondents). The most important barriers to the use of organic fertilizer were: unpleasant odour for neighbours, uncertainty of nutrient content, and difficulty in planning and use. The most important advantage was improved soil structure.

This also highlights the fact that (similar to the consumers) the issue of information is key to adoption decisions. In fact, the difficulties in taking clear-cut decisions at the farm level match the state of knowledge from an academic point of view, which is far from being conclusive. In a review of 99 papers in peer-reviewed journal articles published between 2004 and 2015 on the topic of the uptake of GM crops, Fischer *et al.* (2015) found that two-thirds of publications are based on previously published empirical evidence, indicating a need for new empirical investigations, especially with regard to the social impacts of GM crops in agriculture. Economic impact studies are the most frequent and mainly show that GM crops provide economic benefits for farmers. In contrast, social impacts have been less studied and present less straightforward results. In fact, they show that benefits of GM crops vary significantly depending on the political and regulatory setting. They highlight, in particular, that intellectual property rights and private industry dominance limit farmer access to available GM technology and the benefits of adoption. The authors also highlight that wellbeing is (frequently) discussed in the literature, but rarely investigated empirically, with, in addition, largely contradictory and inconclusive evidence.

Retailers can also have different strategies towards technology uptake. Some retailers position themselves in the market by identifying with organic or 'environmentally friendly' practices; therefore, they are likely to support the regulation of GE and, for example, the introduction of labelling.

The biotech industry itself has shown a diversity of attitudes towards GE. Zilberman *et al.* (2015a) note that while almost all biotechnology companies strongly support policies in favour of the introduction of GE technologies, larger companies are more likely to support stricter regulatory procedures than smaller companies. One economic reason is that, facing more demanding regulatory procedures, large companies will have an advantage because they will have more resources to invest in regulatory efforts and in the introduction of new technologies.

Academic communities may also have different positions towards specific innovations and GE is again a pertinent example of this divide. It may be expected that biotechnology researchers will have incentives to strongly support the technology, while others, notably environmental science groups, may tend to oppose it. Indeed, university research provides the basis for both eco-oriented technologies rejecting GE and biotech promoting it, with a number of intermediate positions. While in the past these were largely driven by ideological motivations, research is becoming more diversified and accommodating different (and sometimes contradicting) technology options.

As in most areas characterized by innovation, entrepreneurship plays a major role. This is evident from policy actions in the field of the bioeconomy that are clearly pushing for entrepreneurship. The main problem is to what extent economics itself is able to deal with entrepreneurship, given the personal and unpredictable features of such phenomena. Entrepreneurship is linked to individual attitudes and potential, ability to take risk and to envisage unexpected solutions. Entrepreneurship is also linked to business models and mechanisms.

Box 6.2. Factors affecting success of small to medium-sized enterprises in the bioeconomy

---

Jernström *et al.* (2017) studied small to medium-sized enterprises (SMEs) in south-east Finland with the objective of analysing the main factors challenging new SMEs willing to develop business opportunities in the bio-based economy. Key factors influencing the successfulness of SME companies turn out to be customer value-added, collaboration in research and development, and supply chains, knowledge of markets, products and processes.

Entrepreneurship is, however, not just a matter of the features of a single individual, but is also related to the economic and social environment. In fact, the state of the art of the bioeconomy is to a large extent a matter of system behaviour, through the complex interaction of multiple agents creating the environment for individual choices. The grounded innovation framework is illustrated in Cavicchi *et al.* (2017) as a promising way of understanding adoption choices taking into account these kinds of features.

6.4   Space, Heterogeneity and Circularity in Supply

6.4.1   Transportation costs and spatial location

Spatial distribution of land and seas, and hence of biomass production, provides a major role to transportation costs in the economics of the distribution of bioeconomy activities in space. Similar to market structures aimed at coping with the temporal and composition variability of biomass, logistics is key to coping with spatial variability and distribution (Lamers *et al.*, 2016).

This is connected to spatial location of facilities and with the profitability of local versus imported provision. To give an idea about the relevance of transportation costs, Trishkin *et al.* (2017) show that wood chips from Finland and Russia had a transportation cost ranging from 19% to 24% of the total supply costs, where the highest level applies for cross-border transportation.

With reference to location (distribution in space) in connection with production capacity, Bojesen *et al.* (2014) analyse the Danish biogas sector. In this case study, the authors show that large-scale biogas plants have 16% lower transportation costs than small biogas plants, and hence minimize transportation costs, which is considered to be the single most important production cost factor.

The issue is addressed in several other papers. In the case of biorefinery, it is well known that the size of the facilities depends on the distribution in the territory. This is to a large extent an empirical issue, depending on different area conditions. As an example, Blair *et al.* (2017) study the development of forest-based biorefinery in two provinces of Canada (British Columbia and Ontario), comparing a centralized and a distributed approach. They found that the distributed model could be attractive in both cases. Yet, while biorefinery development is likely to follow a distributed pathway in Ontario, in British Columbia, a centralized approach may be successfully implemented given the identification of two potentially relevant locations in that province.

The literature highlights strategic supply system design from other viewpoints, among which risk, beyond mere logistic costs (Lamers *et al.*, 2015a, 2015b).

Box 6.3. Alternative feedstock supply systems

According to Lamers *et al.* (2016), the vision of the future feedstock supply system is a network of distributed biomass processing centres (depots) and centralized terminals. Depots are located close to the sources of biomass, while shipping and blending terminals are located in strategic logistical hubs with easy access to bulk transportation systems. Lamers *et al.* (2015a) assess three distinct depot

Continued

Box 6.3. Continued.

---

configurations ranging from conventional pelleting to sophisticated pre-treatment technologies. Depot processing costs are likely to range from US$30 to US$63 per dry metric tonne, depending upon the specific technology implemented and the energy consumption. The authors conclude that the benefits of integrating depots into the overall biomass feedstock supply chain will outweigh depot processing costs.

According to Lamers *et al.* (2016), a fundamental part of initiating (pilot-)depot operations is to establish the value proposition to the biomass grower, in order to mobilize biomass production potential and create a market push that will de-risk and accelerate the deployment of bioenergy technologies. In this way, depots produce value-added intermediates that are fully fungible in both a companion and the biorefining market.

Lamers *et al.* (2015b) provide a holistic evaluation of operational and production costs for a biorefinery supply chain in which traditional cellulosic biorefineries are compared with an advanced feedstock supply system based on a network of depots. Processing operations at the depot increase feedstock supply costs initially, but then enable wider system benefits including supply risk reduction, industry scale-up, conversion yield improvements, and reduced handling equipment and storage costs at the biorefinery. The cost reductions per litre of gasoline equivalent (LGE), is between US$-0.46 to US$-0.21 per LGE for biochemical and US$-0.32 to US$-0.12 per LGE for thermochemical conversion pathways.

The importance of transportation costs may have implications in different directions. Gkritza *et al.* (2012) address the issue of biofuels and transportation infrastructure and noted that the massive growth in ethanol production and distribution will create additional transport demand and competition, with similarly significant growth in other major freight categories.

Correll *et al.* (2014) find that novel supply chain designs for bioenergy and bio-based products can result in logistical cost savings of 2–38%. They also highlight that this is supported by a differentiated plant species supply basis, challenging the prevalent assumption that monocultures are preferable from a cost perspective.

Due to transportation costs, complex biorefineries can only be financially viable when a high degree of feedstock concentration is available, because the plant material is extremely voluminous prior to processing. Accordingly, for logistical reasons, the farming intensity of special plants increases in the vicinity of biorefineries. However, the literature also reports externalities from this concentration such as, for example, significantly increased pollen levels in neighbouring urban areas and subsequently an increased risk of allergies, and related increases in costs to national health systems (Vochozka *et al.*, 2017).

The international trade of biomass feedstock and intermediates also requires adequate logistical infrastructure and this need can be partly satisfied through existing supply chain infrastructure, originally constructed for other goods (Searcy *et al.*, 2016). While this is generally true, differential issues may exist for the three main possible states of bio-based products. Existing solid handling infrastructure is well suited to integrate biomass intermediates such as conventional or torrefied pellets. Infrastructure designed for the petroleum industry can serve liquids. However, some problems can arise from corrosive substances due to high oxygen levels. Natural gas grids are already commonly used for biomethane in most of Europe, but limitations exist such as high production costs, pipeline access and the lack of quality standards.

6.4.2 Heterogeneity and dynamics in production

Production and performance (i.e. efficiency, costs, impacts) in the bioeconomy are heterogeneous and varied in time and space.

There are a number of reasons for heterogeneity in production across space and across different firms. First, bioeconomy activities are based on original biological resources of a different nature (e.g. plants, fish, algae, etc.), from different ecosystems and located worldwide, potentially in any area where there is life. Most of this heterogeneity concerns the primary production of biomass. Sources of biomass from waste may be even more differentiated in terms of costs, availability and properties of the biomass, due to the wide differentiation of sources. In addition, biomass tends to have an inherent variability with respect to its properties and its ability to change over time due to the fact of being made of living matter, which keeps transforming even after whole organisms are no longer living.

Firms managing these resources are different in size and organization in order to use different sources of biomass in different conditions. For example, harvesting fish or growing corn or managing a forest requires completely different inputs and organization. In addition, structural characteristics may differ due to historical or legal reasons, or simply due to the social context and path dependency. A pertinent example is that of the variety of organization solutions for farms, ranging from small family farms, sometimes linked to self-consumption, to highly professional large farms.

Heterogeneity may be high also within the same group, as discussed above. For example, small individual companies, or family farms, have heterogeneous behaviour widely driven by the personal characteristics of the people involved. Typical determinants of behaviour are age, gender, education and family composition. These variables, in addition, have a differentiated effect depending on the decision to be taken and may have a substantial role when major choices have to be made such as radical change in technology.

Companies in the downstream sectors are differentiated as they are often specialized in one or a bundle of products or compounds. A clear macroscopic difference found is among food, energy and biomaterials. Within each of these industries there are a number of very specific pathways.

Companies using bio-based material and interfacing with consumers may be less specialized (e.g. big beverage companies using bio-based plastics) and have different types of organization in that they are more of a kind of services industry organization.

The research and innovation sector is, in turn, a very varied environment composed of both private and public organizations, branches of multinationals and small start-ups. The last type, in particular, may be very different case-by-case depending on the specific business model identified.

There are three main consequence of this heterogeneity:

1. First, the technology in different cases (firms, areas) may be radically different, which is clearly represented by the difference in crop yield or production performance worldwide.
2. Second, when facing the same stimulus in terms of market or technology options behaviour may be very different, and, as a consequence, there can be scope for a number of different legitimate strategies for apparently the same management issue, which, by the way, to some extent contributes to diversification and resilience; technology uptake is a major case (see section above).
3. Third, there is a lot of scope for positive chain/network interaction, complementarity, and so on, while, at the same time, there is the risk of unbalanced relationship along the chain, e.g. market power.

The same performance parameters may also change over time. There are three main reasons for caring about dynamics in the bioeconomy.

The first is that production and biomass availability is linked to short-term growth and yields, as well as to biomass dynamics beyond the life of the producing organisms. Examples of this type are related to crops grown during the season and biomass preservation issues.

Second, production of the basic feedstocks, but to some extent also of intermediate products, is linked to biological dynamics, such as population dynamics. Examples of this kind are forest growth or fish growth. This also involves times for research; for example, creation and testing of new varieties may require time due to the need to grow new organisms for selecting and checking for their characteristics.

The third relates to the fact that the bioeconomy is characterized by a continuous innovation process, so changes are expected all the time; this also implies that the life of a product or of a patent as viable business may be rather short or contingent on substitutes. Also, as innovations are connected to each other, the timing some solution would become relevant to the market may depend on the finding of some complementary solution. In a changing environment, timing is essential to guarantee benefits from research and policy.

Box 6.4. The cost of waiting

Zilberman *et al.* (2015b), using a real option approach (i.e. considering the benefits and costs of postponing a decision) estimate that restrictions to the adoption of GM in corn, rice and wheat are causing a (net present value) damage of above US$300 billion to US$1.22 trillion, depending on assumptions about impacts and interest rates. Considering that waiting also allows for better knowledge of the technology, they estimate that the value of the information gained in one year must be at least between US$27 billion and US$82 billion to justify the one-year delay in the introduction of GMOs for these crops. The conclusion of the authors is that 'precaution is very costly'.

6.4.3   Bioeconomy supply and circular economy

As mentioned, a major focus of the bioeconomy is circularity. Recycling can be interpreted as a special way of renewable resource use. This explains the strong emphasis on recycling and re-use. From an economic point of view, the main concern is the optimal level of recycling/re-use (or more generally of circularity). da Cruz *et al.* (2014) identify different types of impacts of recycling, including institutional, financial/economic, environmental and social, so that all of these dimensions should be accounted for in discussing the optimal level of recycling.

There are two main ways of addressing optimal recycling rate. One is from the point of view of waste management. Kinnaman (2014) studies recycling in municipal waste management using the concept of 'optimal recycling rate', intended as the rate of recycling that minimizes the overall social costs involved. These social costs are the net value that results from the sum of all operational costs and revenues associated with waste and recycling programmes, all costs associated with preparing and storing recyclable materials for collection (household costs), all costs associated with waste disposed at landfills or incinerators, and all external benefits associated with the provision of recycled materials. The paper provides a model representing this issue and tests it using data from Japanese municipalities, using external costs and benefits from Europe and the US. It ends up suggesting an optimal recycling rate of 36% for Japan.

The second one is from the point of view of the optimal source of a specific raw material, in which provision of a raw material can be obtained both from extraction or from recycling. An illustration of the economics of recycling in this second meaning is given in Fig. 6.6.

The figure is somehow a variant of the heterogeneous cost model illustrated above. It illustrates the use of a resource that can be obtained either through extraction/mining (supply S) or from recycling, through either technology RS_1, with high costs, or technology RS_2, much cheaper. Given the cost of extraction/mining, the technology based on RS_1 is not profitable, i.e. extraction is providing the resource at lower costs given the demand and supply curves. When RS_2 is available, the opposite situation occurs, in which recycling is now of interest. Note that a similar situation could also happen if demand of the good has grown or because the natural resource has become scarcer.

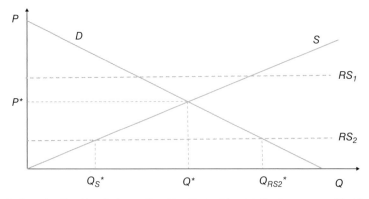

Fig. 6.6. Illustration of optimal level of recycling. See the text for details. Source: modified from Vollaro *et al.* (2016).

The figure illustrates the point that the recycling technology is not necessarily fully replacing the technology based on mining of the resource. On the contrary, in general, there will be an optimal level of distribution of the resource provision between mined (Q_S^*) and recycled ($Q_{RS2}^* - Q_S^*$). Optimal recycling levels are found when the marginal social cost of recycling reaches the marginal social cost of supply by mining or harvesting. Note also that in this case, the availability of the recycling technology has two effects. First, there is an increase in the total amount of resource used; second, there is a welfare gain from the demand side. If the demand side reflects an industry, this would also imply higher profits with potential rebound effect (Vollaro *et al.*, 2016).

In addition to this, recycling interacts with dynamics and status quo; in particular, there is an additional constraint in the recycled part, which is the amount used in the previous period and the maximum efficiency of recycling technology. As the stock of extractable resources declines over time and the stock of used products increases over time, the importance of recycling increases (Zilberman *et al.*, 2013).

A problem is that of failures in the markets of recyclable materials. These may be due to environmental externalities, but also to imperfect and asymmetric information and technological and consumption externalities. Nicolli *et al.* (2012) review the nature of such failures and how they may affect markets for certain recyclable materials. They also discuss how these failures can be overcome by technological innovation and the role for policy measures in this innovation in the area of plastic packaging.

Another key issue for waste and other recycled material (similar but with different problems compared to primary production) is logistic. Costs may include fixed and variable costs per vehicle (transport), personnel cost, container or bag costs as well as emission costs (estimated as about 15% in Groot *et al.* (2014) for urban wastes). Costs may depend on the way waste is collected and managed and hence depend on facility choices and network design.

6.4.4 Bioeconomy and primary production

Because the bioeconomy is dependent on the availability of primary biomass production, this topic attracts very high attention in the literature and policy, with two main aspects: on the one hand, limitations to biomass production entail limitations in the growth of the bioeconomy; on the other hand, this need stimulates actions, in terms of land mobilization or research and innovation in order to ensure higher productivity of land. Several works have addressed this issue about food. As an example for energy, Lewandowski *et al.* (2016) discussed options and needs for improvement of miscanthus production. Similarly, Sikkema *et al.* (2017) discussed the need for wood production and the strategy to ensure increased wood production in Europe. Papers on this topic largely emphasize issues related to total availability (including uncertainty in estimates) and cost (Hennig *et al.*, 2016). Actually, primary production responds to economic stimuli in the form of supply–demand interaction, so that higher feedstock production can be obtained at sufficiently high prices.

While primary production has implications for the bioeconomy, the development of the bioeconomy and the evolution of needs expressed towards primary production is affecting agriculture, forestry and fisheries. Therond *et al.* (2017) provide a broad view of the different farming systems and the conditions driving their features. They propose a new classification of farming systems based on the two dimensions of external input versus ecosystem services and relationships based on market prices versus territorial embeddedness. The authors also discuss the connection with bioeconomy technologies, finding that all models identified require adapted cultivars, farm machinery and information and communication technology. They also note that all models identified can help develop the bioeconomy, however each one is better connected with a different type of bioeconomy. Input-based agriculture models are more oriented towards developing a global bioeconomy, with a stronger focus on processing and upgrading biological raw materials, and large-scale supply chains; on the contrary, agriculture models more integrated within territorial contexts are more compatible with a bioeconomy mainly focused on a local or regional level, characterized by sustainable management of the nexus between food, non-food and natural resources as well as by a high level of circularity. In this perspective, the bioecology vision of sustainability prevails (Bugge *et al.*, 2016). The authors also discuss the co-existence among these models at different levels and how they can be complementary in reaching societal goals.

6.5 Supply System Organization

6.5.1 Biorefinery, production facilities and scheduling

The features of bioeconomy technologies linked with the functioning of demand and interrelated markets, bring to the issue of production organization at the facility and at the chain level. This is particularly important as the connection between energy and biomaterials production typical of a biorefinery is also at the core of the interwoven set of processes typical of the bioeconomy. In this section we focus on the facility issue and in the next to the chain level.

Sammons Jr. *et al.* (2007) argue that though most of the fundamental processing steps involved in biorefining are well known, there is a need for a methodology capable of evaluating the combination of processes (or the integrated processes) in order to identify the optimal set of products and the best route for producing them. The authors propose a mathematical optimization-based framework enabling the inclusion of profitability measures and other techno-economic metrics along with process insights obtained from experimental as well as modelling and simulation studies. This topic, in a way, is a special case of the more general issue of optimal product variety, pricing and scheduling decisions in a production facility. An example of general treatment of this topic is given by Chen *et al.* (2017).

However, optimization assuming static and deterministic conditions is only part of the story. Most often, facilities have to match with variable feedstock availability and market conditions. The concept of flexible production facility is largely developing in the direction of matching these issues. Flexible production facility stands for a flexible system able to adapt to changes, in order to stay competitive facing process uncertainty, raw material variability, market uncertainty and fluctuating demands (Swartz *et al.*, 2015).

In order to assess the benefits of process integration a number of different effects may need to be considered and measured through different

performance parameters, such as carbon balance, profit and flexibility.

> **Box 6.5.** Advantages of flexibility
>
> Brambila-Paz *et al.* (2013) address the case of multipurpose project plants producing bioethanol and sugar together, in contrast to plants producing either bioethanol or sugar. The analysis shows that a multipurpose project is better than a single-purpose project under unstable price scenarios because the critical value is lower. The project value increases by US$6 million when considering the real option of producing sugar to bioethanol or vice versa. The paper concludes by arguing in favour of flexibility in biorefinery using first-generation raw materials such as sugarcane. An example of process combination and flexibility is also discussed in Francavilla *et al.* (2016), who propose an innovative solution for microalgae cultivation ('third-generation' feedstock) in combination with anaerobic digestion and composting technologies

The advantage of integrated processes is also connected to feedstock availability and in turn has an impact on the WTP for biomass and on the profitability of the downstream sectors. In this sense, this issue is already strongly connected to chain organization and to the connection with the territory and the spatial aspects of biomass provision.

> **Box 6.6.** Process integration and integrated techno-economic assessments
>
> A case of techno-economic assessment bringing together several issues and methods is provided by Höltinger *et al.* (2014) for green biorefinery (GBR), aimed at providing alternative utilization pathways for surplus grassland areas and using them to produce bioenergy, biomaterials, livestock feed and organic acids. In order to deal with the techno-economic assessment of this solution, the authors developed a spatially explicit, mixed integer programming model. The model maximizes total producer surpluses of GBR supply chains subject to resource endowments by selecting optimal plant locations and sizes. The impacts of uncertain model input parameters on model outputs are analysed by Monte-Carlo simulations and regression analysis.

6.5.2 Bioeconomy and system organization: from value chain to value web

Several issues related to the development of the bioeconomy have implications for chain organization, that is, for the way different firms operating along the same product chain interact. First, chain organization is a widely studied topic in some bioeconomy sectors, such as food. Chain organization issues are at the core of food industry because of the complexity of the chain and of the different structural characteristics and consequent market power of the actors. The most evident feature, in this direction, is the structural difference between the myriad of small farms (and the high number of small companies in the food processing industry) and the big players of the food processing and big retailers. The other relevant feature is the increasing number of points in which technology allows separability in the process, which encourages specialization along the chain, but at the same time, increases the effort for coordination, bringing the trade-off between transaction cost and operational costs. At the same time, the diversification of chains and the distribution worldwide increase the number of different options for each step in the technological process, which further change their role depending on market and regulatory evolution, so that suitable (flexible enough) chain organization is a key for competitiveness and sustainability. Finally, new technologies and better economic connections are making links across chains more and more important, pushing attention beyond chains and towards the concept of value web.

6.5.2.1 Chain organization and flexibility

Chain organization issues are now widely discussed in the literature on biorefinery. In particular, a wide review carried out by Espinoza Pérez *et al.* (2017) identifies the main models of chain organization, related to the evolution of biorefinery chains (Fig. 6.7).

The first phase is characterized by an individual input processed in a plant with a fixed transformation technology and leading to a determined end product. Phase II is characterized by a flexible processing technology, which allows different end products using the same raw material. Finally, phase III uses a variety of raw materials through a flexible transformation technology to obtain several end products. The move from the first to the third phase can potentially allow lower costs and higher flexibility facing uncertain market conditions, also considering competing claims on biomass and agricultural land (e.g. food versus energy debate). However, it also implies higher chain coordination issues, due to the higher complexity of the interplay among different actors and markets.

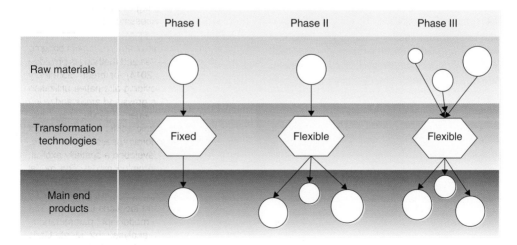

Fig. 6.7. Three phases of biorefinery chain organization. Source: Espinoza Pérez *et al.*, (2017).

Flexibility at the level of chain complements the notion at the level of processing
plant and is often related to flexibility of crops
and commodities. According to Borras *et al.*
(2016) 'flex crops and commodities' are those
having 'multiple uses (food, feed, fuel, fibre,
industrial material, etc.) that can be flexibly
interchanged while some consequent supply
gaps can be filled by other flex crops'. Examples
of flex crops include soya (feed, food, biodiesel),
sugarcane (food, ethanol), oil palm (food, biodiesel, commercial/industrial uses) and corn
(food, feed, ethanol). Examples of emerging
flex crops are cassava, coconut, sugarbeet,
rapeseed and sunflower. Other commodities
with similar features are trees for timber, pulp,
biomass, ethanol and carbon sequestration
purposes.

One dimension of the awareness of these
issues is an increasingly proactive approach
through the field of chain design. Espinoza Pérez
et al. (2017) provide a typology of decision making at three levels of analysis: strategic, tactical
and operational (Fig. 6.8), together with a specific set of tools used to model and optimize the
biorefinery supply chain.

An example of flexibility and optimization
in integrated forest biorefinery is provided by
Swartz *et al.* (2015). The flexibility of feedstock
or intermediaries has several effects:

- In a static view, it affects prices, welfare and
  welfare distribution among sectors producing original products.
- From a dynamic perspective, it can allow
  flexibility to adapt to the availability of different sources or to shocks in case of, for example, the unexpected failure of a source of
  biomass.
- It can also encourage economies of scale in
  the platform level of the chain.
- It may more easily allow for the use of small-
  scale feedstocks that would not be able to
  build their own chain.

6.5.2.2   Interplay of chains

One emerging issue in view of the bioeconomy
development is the need and ability of actors to
work across different product chains and of
different chains to interact through technical
relationships, for example, through flows of
products. This is largely determined by the way
different stages of the chain (namely agriculture
and processing) interact among each other. As
an example, Shortall *et al.* (2015) highlight the
move of agriculture towards a responsible multipurpose sector. Based on stakeholder interviews,
they identify three different pathways:

- a biorefinery concept framed within an industrial agricultural paradigm that envisages

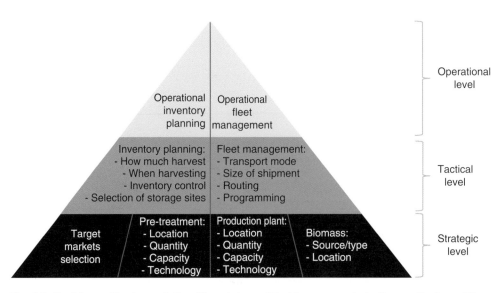

Fig. 6.8. Decision-making issues in the different steps of the bioeconomy chain. Source: Espinoza Pérez
et al. (2017).

large-scale production stimulated by biotechnology innovation;

- an 'alternative agriculture' paradigm, which envisions sustainable multipurpose biomass production in terms of on-farm nutrient and energy cycling and local, smaller scale production; and
- a potential overlapping solution based on the concept of quality industrial biomass production.

Another topic concerns chain organization related to the growing emphasis on by-products and waste re-use. This not only has implications for chain organization and management, but also has implications for cost calculation in a chain perspective. One key aspect of chain interaction occurs at the farm level, especially for farming companies accessing energy markets. This phenomenon is identified by some authors as an example of an industry convergence process. Industry convergence is related to convergence of scientific fields and technologies,

and is a case in which a firm's capability to yield flexible and dynamic responses to the context is key for its future success (Golembiewski *et al.*, 2013).

6.5.2.3   Value web

The bioeconomy, driven by the increased interconnection between different value chains, tends to be growingly depicted as a system or a web, under the concept of a 'bio-based value web' (Virchow *et al.*, 2016).

The concept of a "biomass-based value web", which was developed as an extension of the value chain concept with the aim to capture the links within and between value chains that arise from the cascading and joined use of biomass' is combined with 'the "national innovation system" (NIS), which serves to identify the different types of actors involved in the biomass value web and the linkages between them' by Scheiterle *et al.* (2016) to analyse the sugar cane sector in Brazil.

Box 6.7. Value webs in Africa and Brazil

Virchow *et al.* (2016) use the emerging bioeconomy value web for Africa to describe the transformation of African agriculture from a food-supplying to a biomass-supplying sector, including the production of feed, energy and industrial raw materials. It is expected that the emerging bioeconomy will intensify the connections between biomass production, processing and trading. To depict these increasingly complex systems, adapted analytic approaches are needed, among which those based on the concept of biomass-based value webs. Scheiterle *et al.* (2016) seek to analyse the position of Brazil as the world's leader in sugarcane production, using the lens of value webs. The paper shows that the development of Brazil's international competitiveness in sugar and ethanol was based on political incentives. These incentives resulted, among others, in a strong network of institutions focusing on these two products. The authors argue that for Brazil to become a front-runner in the future bioeconomy, the existing innovation network needs to be expanded. Some avenues are found in the integration of national and international private sector organizations, but policies and funding for risky investments are also needed to encourage collaboration among companies.

Box 6.8. Integrated modelling

Panteli *et al.* (2016) provide a modelling and optimization exercise on biorefining chain systems. The authors use an integrated approach and model all of the entities involved across the technology chain. The approach is motivated by the fact that the success of biorefinery is linked to the ability to achieve a full exploitation of biomass and its macro components (cellulose, hemicellulose and lignin) for the production of bio-based products and platform chemicals. This includes the identification of the most promising pre-treatment process that fractionates biomass into cellulose, hemicellulose and lignin (usually accounting for the highest cost share of the entire biorefining system), addresses the overwhelmingly high capital costs of second-generation technologies and accounts for uncertainty on the market (price, demand).

6.5.3 Business models

The development of the bioeconomy and its organization are also linked to the emergence of new business models in various fields and sub-sectors. A business model describes the way an organization creates, delivers and captures value. The emergence of the business model concepts is on the one hand linked to higher level of detail and options in describing business, but also to the wider concept of shared value produced by companies for a large number of stakeholders (Lüdeke-Freund *et al.*, 2017).

One key area of application is the topic of innovation development and exploitation, in which the mechanism remains based on patenting and exploitation rights on patents, but also involves the development of ideas able to bring the innovation to deliver to final users and consumers. A frequent example concerns the use of new patents through spin offs. As this may be a dynamic and often uncertain business, this goes largely through small start-up and spin-off companies that sell targeted services. Also, the acquisition of patents through merging and similar processes applies.

Know-how embedded in personal knowledge is increasingly an issue, together with the need for companies to employ specific expertise. This results in mobility of personnel, as well as start-up creation and services delivery.

Other examples are seen in biorefinery or bioenergy production, where new operators take the lead in matching new technical solutions, regulation interpretation and sometimes difficult relationships with local administration, and offer, for example, ready-to-go projects to investors. This is also a way of ensuring access to land, for which also new contract solutions are emerging.

Growing business models are those connecting innovation to chain and market management. An example is that of producer organizations in the farming sector. Producer organizations decide which innovation to bring in their area, identify the market for the products of that innovation (e.g. a new crop breed), support farmers through financial incentives for its adoption and provide incentive pricing for the product.

The term bio-business is also being used. See, for example, the work with different case studies from Brazil in Sousa *et al.* (2016).

6.5.4 Investment and financing

Investment of venture capital in the biotechnology and bioeconomy industries is relevant both as an indicator of economic attractiveness and as an indicator of innovation effort. Similar to other sectors, investment in bioeconomy relates to the attractiveness of capital compared to competing sectors. This, in turn, is determined by profitability and risk.

Investment involves both production-related bioeconomy sectors and those related to research investment and innovation financing, which are in a way the most typical of the bioeconomy as a sector based on innovation (Festel and Rammer, 2015). Financing innovation often means a long time period between initial investment and return on investment, as well as highly uncertain outcomes. Also, investment may be needed in different distributed components of the network of actors. In this respect, investment is highly related to risk taking especially in the highly varied world of new technologies that may work or not depending on the complex dynamics of consumers and regulatory actions (as well as technology robustness itself).

A study by the European Investment Bank (2017) investigates the access to finance in bio-based industries and the blue economy, highlighting that bio-based industries projects still face difficulties in accessing private capital. The main funding gaps concern bio-based industry projects scaling up from pilot to demonstration projects and in bio-based industries moving from demonstration to flagship/first-of-a-kind and industrial-scale projects.

Actually, the promising growth potential of the bioeconomy attracts actors of the financial market, but the funding is hindered by perceived high risks and information asymmetries. Notably, not only markets (demand), but also regulatory framework conditions are perceived as the most important drivers and incentives but also present the biggest risks and challenges for bio-based project promoters and financial market investors. In addition, access to funding may be particularly difficult for small companies and investment conditions may be different across countries and regions (Dammer and Carus, 2016).

In this context public policies have clearly a role, but this role should be strengthened and made more consistent, with a focus on risk reduction.

6.6 Outlook

The core problem of the supply side of the bioeconomy is to produce cheap bio-based goods or attributes through innovation, optimization of processes, web organization and cost-minimizing spatial distribution. Besides this apparently straightforward view, the supply side of the bioeconomy is a complex and highly dynamic system. The current direction is towards increasing its degree of heterogeneity and organizational complexities. The core concepts of bioeconomy supply are the breaking down of the process and its consequent multiplicity of connections, the interplay between chain and spatial organization and the overall move towards value webs as the key interpretation concept. This reflects also in changing and complex impacts on supply in terms of prices and distribution.

Most likely this trend will also make the sector less and less understandable and predictable by economists, hence with a huge variety of potential futures depending on technology, internal organizational changes and the constraints and incentives coming from demand and regulation. The most puzzling aspect is the interconnection between economic behaviour and technology. Reimer *et al.* (2017) analyse this issue through a fishery model exploring the interplay between incentives and technology and highlighting that basing ex-ante analysis on observing technologies tends to represent the current state of incentives rather than real technological possibilities in a different institutional setting.

This topic, as many in economics of the bioeconomy, requires more evidence. However, evidence will always have limitations on three grounds:

- the difficulty of having conclusive economic information in a wide variety of production conditions and in which industrial competition is based on an accurate management of confidential information;
- because of the importance of distributional factors linked to evolution of property rights and regulation; and
- the relevance of soft factors linked to decision-making, such as social aspects and risk considerations, as well as the role of entrepreneurship and creativity in some steps of the chain.

References

Abbati de Assis, C., Gonzalez, R., Kelley, S., Jameel, H., Bilek, T., Daystar, J., Handfield, R., Golden, J., Prestemon, J. and Singh, D. (2017) Risk management consideration in the bioeconomy. *Biofuels, Bioproducts and Biorefining* 11(3): 549–566. doi: 10.1002/bbb.1765.

Blair, M.J., Cabral, L. and Mabee, W.E. (2017) Biorefinery strategies: exploring approaches to developing forest-based biorefinery activities in British Columbia and Ontario, Canada. *Technology Analysis and Strategic Management* 29(5), 1–14. doi: 10.1080/09537325.2016.1211266.

Bojesen, M., Birkin, M. and Clarke, G. (2014) Spatial competition for biogas production using insights from retail location models. *Energy* 68. doi: 10.1016/j.energy.2013.12.039.

Borras, S.M., Franco, J.C., Isakson, S.R., Levidow, L. and Vervest, P. (2016) The rise of flex crops and commodities: implications for research. *Journal of Peasant Studies* 43(1), 617–628. doi: 10.1080/03066150.2015.1036417.

Brambila-Paz, J.J., Martínez-Damián, M.A., Rojas-Rojas, M.M. and Pérez-Cerecedo, V. (2013) La bioeconomia, las biorefinerías y las opciones reales: El caso del bioetanol y el azúcar. *Agrociencia* 47(3), 281–292.

Bugge, M.M., Hansen, T. and Klitkou, A. (2016) What is the bioeconomy? A review of the literature. *Sustainability* 8(7), 691. doi: 10.3390/su8070691.

Case, S.D.C., Oelofse, M., Hou, Y., Oenema, O. and Jensen, L.S. (2017) Farmer perceptions and use of organic waste products as fertilisers: A survey study of potential benefits and barriers. *Agricultural Systems* 151, 84–95. doi: 10.1016/j.agsy.2016.11.012.

Cavicchi, B., Palmieri, S. and Odaldi, M. (2017) The influence of local governance: Effects on the sustainability of bioenergy innovation. *Sustainability* 9(3), 406. doi: 10.3390/su9030406.

Chen, P., Xu, H., Li, Y. and Zeng, L. (2017) Joint product variety, pricing and scheduling decisions in a flexible facility. *International Journal of Production Research* 55(2), 606–620. doi: 10.1080/00207543.2016.1229065.

Correll, D., Suzuki, Y. and Martens, B.J. (2014) Logistical supply chain design for bioeconomy applications. *Biomass and Bioenergy* 66, 339–345. doi: 10.1016/j.biombioe.2014.03.036.

da Cruz, N.F., Simões, P. and Marques, R.C. (2014) Costs and benefits of packaging waste recycling systems. *Resources, Conservation and Recycling* 85, 1–4. doi: 10.1016/j.resconrec.2014.01.006.

Dammer, L. and Carus, M. (2016) Study on investment climate in bio-based industries in the Netherlands. In Snyder, S.W. (ed.) *Commercializing Biobased Products: Opportunities, Challenges, Benefits, and Risks*. London, UK, Royal Society of Chemistry Publishing, pp. 315–335.

Debertin, D.L. (2012) *Agricultural Production Economics*. Available at www.uky.edu/~deberti/ap/ap.pdf (accessed 25 May 2018).

Espinoza Pérez, A.T., Camargo, M., Narváez Rincón, P.C. and Alfaro Marchant, M. (2017) Key challenges and requirements for sustainable and industrialized biorefinery supply chain design and management: A bibliographic analysis. *Renewable and Sustainable Energy Reviews* 69, 350–359. doi: 10.1016/j.rser.2016.11.084.

European Investment Bank (2017) *Study on Access-to-Finance. Conditions for Investments in Bio-Based Industries and the Blue Economy*. Kirchberg, Luxembourg, European Investment Bank.

Festel, G. and Rammer, C. (2015) Importance of venture capital investors for the industrial biotechnology industry. *Journal of Commercial Biotechnology* 21(2), 31–42. doi: 10.5912/jcb685.

Fischer, K., Ekener-Petersen, E., Rydhmer, L. and Edvardsson Björnberg, K. (2015) Social impacts of GM crops in agriculture: A systematic literature review. *Sustainability* 7(7), 8598–8620. doi: 10.3390/su7078598.

Francavilla, M., Intini, S. and Monteleone, M. (2016) Designing an integrated technological platform centered on microalgae to recover organic waste and obtain multiple bioproducts. *European Biomass Conference and Exhibition Proceedings*, pp. 294–299, doi: 10.5071/24thEUBCE2016-1CV.4.16.

Gkritza, K., Nlenanya, I. and Jiang, W. (2012) Bioeconomy and transportation infrastructure impacts: A case study of Iowa's renewable energy, green energy and technology. In Gopalakrishnan, K., van Leeuwen, J. and Brown, R.C. (eds) *Sustainable Bioenergy and Bioproducts: Value Added Engineering Applications*. London, UK, Springer Ltd, pp. 173–188. doi: 10.1007/978-1-4471-2324-8_9.

Golembiewski, B., Sick, N. and Leker, J. (2013) Agriculture and energy industry in the setting of an emerging bioeconomy: Are there any signs of convergence on the horizon? Contributed to the PICMET '13 Conference 'Technology Management in the IT-Driven Services', San Jose, California, USA.

Groot, J., Bing, X., Bos-Brouwers, H. and Bloemhof-Ruwaard, J. (2014) A comprehensive waste collection cost model applied to post-consumer plastic packaging waste. *Resources, Conservation and Recycling* 85, 79–87. doi: 10.1016/j.resconrec.2013.10.019.

Hennig, C., Brosowski, A. and Majer, S. (2016) Sustainable feedstock potential - A limitation for the bio-based economy? *Journal of Cleaner Production* 123, 200–202. doi: 10.1016/j.jclepro.2015.06.130.

Höltinger, S., Schmidt, J., Schönhart, M. and Schmid, E. (2014) A spatially explicit techno-economic assessment of green biorefinery concepts. *Biofuels, Bioproducts and Biorefining* 8(3), 325–341. doi: 10.1002/bbb.1461.

Jernström, E., Karvonen, V., Kässi, T., Kraslawski, A. and Hallikas, J. (2017) The main factors affecting the entry of SMEs into bio-based industry. *Journal of Cleaner Production* 141, 1–10. doi: 10.1016/j.jclepro.2016.08.165.

Kelsey, K.D. and Franke, T.C. (2009) The producers' stake in the bioeconomy: A survey of Oklahoma producers' knowledge and willingness to grow dedicated biofuel crops. *Journal of Extension* 47(1), 1RIB5.

Kinnaman, T.C. (2014) Determining the socially optimal recycling rate. *Resources, Conservation and Recycling* 85, 5–10. doi: 10.1016/j.resconrec.2013.11.002.

Lamers, P., Roni, M.S., Tumuluru, J.S., Jacobson, J.J., Cafferty, K.G., Hansen, J.K., Kenney, K., Teymouri, F. and Bals, B. (2015a) Techno-economic analysis of decentralized biomass processing depots. *Bioresource Technology* 194, 205–213. doi: 10.1016/j.biortech.2015.07.009.

Lamers, P., Tan, E.C.D., Searcy, E.M., Scarlata, C.J., Cafferty, K.G. and Jacobson, J.J. (2015b) Strategic supply system design: A holistic evaluation of operational and production cost for a biorefinery supply chain. *Biofuels, Bioproducts and Biorefining* 9(6), 648–660. doi: 10.1002/bbb.1575.

Lamers, P., Searcy, E. and Hess, J.R. (2016) Transition strategies: Resource mobilization through merchandisable feedstock intermediates. In Lamers, P., Searcy, E., Hess, J.R. and Stichnothe, H. (eds) *Developing the Global Bioeconomy: Technical, Market, and Environmental Lessons from Bioenergy*. London, UK, Academic Press, pp. 165–186. doi: 10.1016/B978-0-12-805165-8.00008-2.

Lewandowski, I., Clifton-Brown, J., Trindade, L.M., Van Der Linden, G.C., Schwarz, K.-U., Müller-Sämann, K., Anisimov, A., Chen, C.-L., Dolstra, O., Donnison, I.S., Farrar, K., Fonteyne, S., Harding, G., Hastings, A., Huxley, L.M., Iqbal, Y., Khokhlov, N., Kiesel, A., Lootens, P., Meyer, H., Mos, M., Muylle, H., Nunn, C.,

Özgüven, M., Roldán-Ruiz, I., Schüle, H., Tarakanov, I., Der Weijde, T., Wagner, M., Xi, Q. and Kalinina, O. (2016) Progress on optimizing miscanthus biomass production for the European bioeconomy: Results of the EU FP7 project OPTIMISC. *Frontiers in Plant Science* 7(NOVEMBER20). doi: 10.3389/fpls.2016.01620.

Lüdeke-Freund, F., Massa, L., Bocken, N., Brent, A. and Musango, J. (2017) *Main Report: Business Models for Shared Value*. Available from https://nbs.net/p/main-report-business-models-for-shared-value-4122f859-2499-4439-824e-7535631a14ed (accessed 23 May 2018).

Nicolli, F., Johnstone, N. and Söderholm, P. (2012) Resolving failures in recycling markets: The role of technological innovation. *Environmental Economics and Policy Studies* 14(3), 261–288. doi: 10.1007/s10018-012-0031-9.

Panteli, A., Giarola, S. and Shah, N. (2016) A generic MILP modelling framework for the systematic design of lignocellulosic biorefining supply chains. In Computing and Systems Technology Division 2016 - Core Programming Area at the 2016 AIChE Annual Meeting.

Reimer, M.N., Abbott, J.K. and Haynie, A.C. (2017) Empirical models of fisheries production: Conflating technology with incentives? *Marine Resource Economics* 32(2), 169–190. doi: 10.1086/690677.

Sammons Jr., N., Eden, M., Yuan, W., Cullinan, H. and Aksoy, B. (2007) A flexible framework for optimal biorefinery product allocation. *Environmental Progress* 26(4), 349–354. doi: 10.1002/ep.10227.

Scheiterle, L., Ulmer, A., Birner, R. and Pyka, A. (2016) From commodity-based value chains to biomass-based value webs: The case of sugarcane in Brazil's bioeconomy. *Journal of Cleaner Production* 172, 3851–3863. doi: 10.1016/j.jclepro.2017.05.150.

Searcy, E., Lamers, P., Deutmeyer, M., Ranta, T., Hektor, B., Heinimö, J., Trømborg, E. and Wild, M. (2016) Commodity-scale biomass trade and integration with other supply chains. In Lamers, P., Searcy, E., Hess, J.R. and Stichnothe, H. (eds) *Developing the Global Bioeconomy: Technical, Market, and Environmental Lessons from Bioenergy*. London, UK, Academic Press, pp. 115–138. doi: 10.1016/B978-0-12-805165-8.00006-9.

Shortall, O.K., Raman, S. and Millar, K. (2015) Are plants the new oil? Responsible innovation, biorefining and multipurpose agriculture. *Energy Policy* 86, pp. 360–368. doi: 10.1016/j.enpol.2015.07.011.

Sikkema, R., Dallemand, J.F., Matos, C.T., van der Velde, M. and San-Miguel-Ayanz, J. (2017) How can the ambitious goals for the EU's future bioeconomy be supported by sustainable and efficient wood sourcing practices? *Scandinavian Journal of Forest Research* 32, 551–558. doi: 10.1080/02827581.2016.1240228.

Sousa, K.A., Santoyo, A.H., Rocha, W.F., De Matos, M.R. and Silva, A.D.C. (2016) Bioeconomia na Amazônia: Uma análise dos segmentos de fitoterápicos & fitocosméticos, sob a perspectiva da inovação. *Fronteiras* 5(3), 151–171. doi: 10.21664/2238-8869.2016v5i3.p151-171.

Swartz, C.L.E., Wang, H. and Mastragostino, R. (2015) Operability analysis of process supply chains-toward the development of a sustainable bioeconomy. *Computer Aided Chemical Engineering* 36, 355–384. doi: 10.1016/B978-0-444-63472-6.00014-8.

Therond, O., Duru, M., Roger-Estrade, J. and Richard, G. (2017) A new analytical framework of farming system and agriculture model diversities. A review. *Agronomy for Sustainable Development* 37(3), 21. doi: 10.1007/s13593-017-0429-7.

Trishkin, M., Goltsev, V., Tolonen, T., Lopatin, E., Zyadin, A. and Karjalainen, T. (2017) Economic efficiency of the energy wood chip supply chain from Russian Karelia to Finland. *Biofuels* 8(4), 411–420. doi: 10.1080/17597269.2016.1225648.

Virchow, D., Beuchelt, T.D., Kuhn, A. and Denich, M. (2016) Biomass-based value webs: A novel perspective for emerging bioeconomies in Sub-Saharan Africa In Gatzweiler, F.W. and von Braun J. (eds) *Technological and Institutional Innovations for Marginalized Smallholders in Agricultural Development*. New York, Springer, pp. 225–238. doi: 10.1007/978-3-319-25718-1_14.

Vochozka, M., Stehel, V. and Maroušková, A. (2017) Uncovering a new moral dilemma of economic optimization in biotechnological processing. *Science and Engineering Ethics* [Epub ahead of print]. doi: 10.1007/s11948-017-9925-z.

Vollaro, M., Galioto, F. and Viaggi, D. (2016) The circular economy and agriculture: New opportunities for re-using phosphorus as fertilizer. *Bio-based and Applied Economics* 5(3), 267–285.

Zilberman, D., Kim, E., Kirschner, S., Kaplan, S. and Reeves, J. (2013) Technology and the future bioeconomy. *Agricultural Economics* 44(S1), 95–102. doi: 10.1111/agec.12054.

Zilberman, D., Graff, G., Hochman, G. and Kaplan, S. (2015a) The political economy of biotechnology. *German Journal of Agricultural Economics* 64(4), 212–223.

Zilberman, D., Kaplan, S. and Wesseler, J. (2015b) The loss from underutilizing GM technologies. *AgBioForum* 18(3), 312–319.

7

Matching Demand and Supply: Markets, Policies and Beyond

7.1 Introduction and Overview

This chapter investigates how demand and supply can meet each other or, in more general terms, what mechanisms can be put in place to ensure that supply satisfies societal needs. This matching, in the current economic system, is largely driven by markets. However, markets do not always work properly and policies or other governance instruments need to be devised to meet societal needs. There are several reasons for this to happen in the bioeconomy. First, the bioeconomy is rooted on new technology developments, based on the role of research, in which public funding and research institutions have a major role. Second, the bioeconomy includes a number of cases in which public goods and externalities may emerge and this is a typical case in which private interests do not necessarily meet societal needs. Several markets for bio-based products are characterized by high production costs compared with alternative (e.g. fossil based) products; however, their potential superiority is advocated based on environmental considerations. An example is that of biodegradable polymers (Dietrich *et al.*, 2017). Moreover, bioeconomy products may result in potential risk perceptions, for example, with regard to health, which has traditionally yielded a strong work of regulation by public bodies. Third, the bioeconomy sector is in its inception and this is arguably a stage in which public intervention may be needed to stimulate the development of new markets and to provide

guidance to the direction technology can take. Some bioeconomy priorities are linked to envisaged long-term goals that do not match with short-term profitability, which may also require strong policy support to get industries started (e.g. in the case of bioenergy). Finally, uncertainties may amplify investors' perception of risk and reduce investments or hinder funding availability.

Besides markets and policy, an increasing role is played by intermediate types of economic relationships (e.g. contracts, networking and other forms of integration). In addition, the bioeconomy entails a specific focus on a number of global issues. All these topics are embedded and interwoven in the larger institutional setting, hosting processes related to governance, participatory decision making and information.

Indeed, one feature of modern economies, and especially of the bioeconomy, is that the distinction between markets and policy (i.e. the private and the public sphere of action) is increasingly fuzzy and a number of intermediate institutional forms are developing. In addition, markets and policy support each other in providing incentives focused on societal needs. For example, a functioning market needs appropriate regulation, product definition and enforcement of rules. For these reasons, markets, policies and intermediate instruments are treated here in the same chapter, acknowledging the continuum between them and the fact that markets and policy are part of the wider category of mechanisms

 © D. Viaggi 2018. *The Bioeconomy: Delivering Sustainable Green Growth* (D. Viaggi)

and institutional arrangements the purpose of which is to ensure the functioning of the economy.

A final point is that these different instruments are accompanied by continuous institutional innovation, especially in connecting demand and supply (or even more progressively dismantling the clear-cut distinction between these two components). This will be discussed in section 7.7, Outlook.

7.2 Markets for Bioeconomy Products

7.2.1 Market functioning and externalities

The simpler representation of a market for a bioeconomy product is that of a perfectly competitive market situation depicted by the interaction of demand (D) and supply (S) in Fig. 7.1.

The equilibrium price P^* and quantity Q^* are determined by the intersection between the demand and the supply curves. This form of market is characterized by many buyers and many sellers, homogeneous products and perfect information. This works for private goods, in which the effects of production or consumption decisions are fully borne by the respective decision makers.

However, bio-based products, in light of the properties discussed above, often provide public goods or involve externalities linked to production or consumption (e.g. reducing CO_2 emissions, benefiting future generations, etc.). The figure considers a bio-based product involving the production of a public good (or positive externality) or, alternatively, a bio-based product that could rather produce 'public bad' (or a negative externality).

If the private good also produces a joint public good, and assuming the value of the public good is proportional to that of the private good provision, the total social benefit from the production of the private good and of the associated public goods will be represented by the supply curve S_{pe}. This is lower than S by the amount represented by the public good provision, as we assume this could be subtracted from the production costs to yield the actual social cost of the good provision. It is worth noting that, in this case, the optimal amount of the private good (Q_{pe}^*) is higher and the equilibrium price for society is lower (P_{pe}^*). The opposite occurs if the bio-based good produces a negative externality (Q_{ne}^* and P_{ne}^*).

The problem is that, given the nature of externalities, private actors will not have the incentive to push the market towards the optimal level (either Q_{pe}^* or Q_{ne}^*); rather, the market will stick to Q^* unless instruments are envisaged to provide incentives to actors in the appropriate direction. One is the establishment of clear property rights, between those causing and those affected by the externality. Another way is linked to information and certification (or similar instruments) that encourage consumers to recognize the features of the product and to pay the 'right' amount for it (e.g. through a premium price in case of positive externality). Finally, the state could intervene with dedicated instruments,

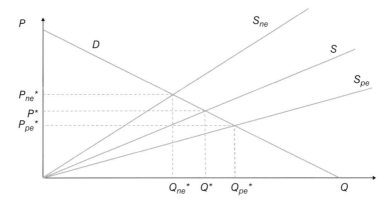

Fig. 7.1. Market of a bio-based products and the role of a public good/bad provision. *D*, demand; *P*, price; *Q*, quantity; *S*, supply.

such as taxes or subsidies (or other instruments), internalizing the value of the public good. This is discussed in detail in section 7.3.

Before doing that, however, it needs to be highlighted that there are several variants of this market representation, in addition to public goods. Two relevant points for the bioeconomy are market structure and information. The existence of one seller only characterizes the monopoly structure, which yields higher prices and a lower quantity of products exchanged, at the expense of consumers and society as a whole. Intermediate situations are possible, such as oligopoly or monopolistic competition, which could be envisaged across small processing units in a given territory. Another reason for differences in market functioning is linked to imperfect information by one part, e.g. the consumer not knowing relevant characteristics of the product (asymmetric information). The effect of this occurrence is a mismatch between willingness to pay (WTP) by consumers and their actual utility from the purchase of partly unknown goods, leading to sub-optimal solutions. This may occur in a number of cases and can require specific solutions, such as labelling or suitable contracts. Indeed, it is very common in the food chain.

Box 7.1. Relationships between private and public goods in the bioeconomy

According to Dietrich *et al.* (2017), biodegradable polymers such as polyhydroxyalkanoates (PHAs) can reduce pollution. Indeed, life-cycle analysis and the environmental footprint show that PHAs can contribute to greenhouse gas emission reduction targets, waste reduction, green jobs and innovation. However, they have a higher price due to high production costs. Specific support policies may hence be needed to boost the dissemination of these products. In addition, due to high variability in the industrial production of PHAs in terms of feedstock, energy sources, polymer properties, and so on, the choice of optimization criteria influences the design of new production processes and it is hence necessary to direct the technical design of sustainable PHA production.

Opposite considerations are available in Lewandowski *et al.* (2016) who report costs per unit of CO_2 from the production of miscanthus, with negative results (i.e. profitability of carbon fixation). Lee (2016) provides an economic evaluation of different biohydrogen technologies.

7.2.2 Connections across markets

The interconnection across markets is not an exclusive feature of the bioeconomy, but is emphasized by the fact that bio-based products tend to substitute existing products, propose new products and/or create links between different value chains. It is also amplified by technology trends towards platform products and flexible products, facilities and chains. In this section we consider three different cases. The first is a case with multiple supply options for the same good. The second concerns the opposite case of different industries using the same production factors. The third concerns vertical interconnection among markets.

In the bioeconomy these cases are not exclusive, but rather occur at the same time. Notably, both the first and second issue apply and are emphasized for platform products that, by definition, can be developed from multiple sources. This means that the supply curve will be derived from the composition of the different feedstocks in their market, reflecting the marginal costs of each upstream feedstock. In analogy, overall demand is affected by several downstream production or consumption processes.

7.2.2.1 Multiple supply options for the same goods

The case of multiple sources of a bio-based product is of general interest, but for some bioeconomy sectors, in particular energy, it is of special interest as it questions the existence and the market share for the bioeconomy itself. This case is of special importance also for drop-in bio-based products, that is, bio-based products that are exactly the same as traditional fossil-based products, but are developed from a process based on biological sources. The link between the demand for a product and the option of producing it with different technologies is illustrated in Fig. 7.2 taking the example of energy. Assume energy can be produced from non-renewable sources (fossil) (S_{NR}), bio-based sources (S_R) and renewable non-bio-based sources (S_{NBR}). Needless to say, each supply curve is represented by a variety of different sources. Taking the example of energy, both renewable and non-renewable energy can come from a variety of sources (e.g. oil, coal versus wood, biogas, bioethanol, etc.).

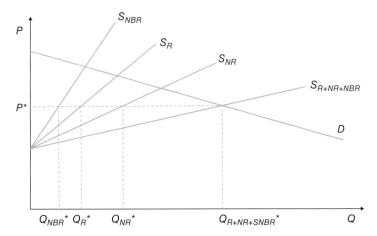

Fig. 7.2. Market with fossil, bio-based and renewable non-bio-based supply. *D*, demand; *P*, price; *Q*, quantity; S_{NR}, non-renewable supply; S_R, bio-based supply; S_{NBR}, renewable non-bio-based supply. Source: modified from Zilberman *et al.* (2013).

There are sources with different costs within each group and subgroup, due, for example, to location and quality of the raw material, which explain the shape of the supply curve and justify the existence of different sources at the same time.

The total supply (given by the horizontal sum of the three curves above) is represented by the $S_{R+NR+NBR}$ curve and the market equilibrium determines the market price and the total market size. Assume that the quantity demanded/supplied of a good is represented by Q and the good faces a demand D. Assuming that at a certain price it is profitable to produce some level of each of the three types, all of them will have some positive value and the equilibrium will be given by the intersection of D and $S_{R+NR+NBR}$, yielding the equilibrium price P^*. At this price, Q_R, Q_{NR} and Q_{NBR} represent the optimal allocation of production of renewable, non-renewable resources and non-biological renewable, respectively. The space for each type of production is determined by the relative product costs that shape the supply curve, and in particular by the slope of the curve. The figure depicts a situation in which S_{NR} has the lower costs and the highest share of the market, followed by S_R and S_{NBR} with lower shares of the market. An expansion of S_{NBR} (e.g. due to technological change resulting in the downward movement of the supply function) would increase the overall supply possibilities and the equilibrium amount of goods, as well as

causing a decrease in market price. As a result, the share (and the absolute amount) of both bio-based and fossil products would drop. This example illustrates the case of, for example, bio-based energy, the market space of that is squeezed between fossil fuel markets and non-bio-based renewable sources, such as eolic energy.

Variants of this framework can be devised with the intercept of either S_R, S_{NR} or S_{NBR} on the vertical axis higher than the demand. This would mean that the use of renewable, non-renewable or non-biological renewable resources, respectively, is possible but not economically relevant at that point in time. It is indeed still a widespread condition that production of bio-based goods is possible, but not economically profitable for society.

A modified version of this model can account for externalities, as represented in Fig. 7.3.

In this figure the existence of a positive externality reducing the social cost of the bio-based product modify the optimal social outcome. In this case we have assumed that this is strong enough to make the bio-based product less costly for society than the non-bio-based product. If measures are taken to achieve the optimal level of production of the bio-based goods (e.g. through labelling or a subsidy), the actual market behaviour with respect to bio-based products can follow curve S_{Rpe}; this also has the potential of creating a further displacing effect on the

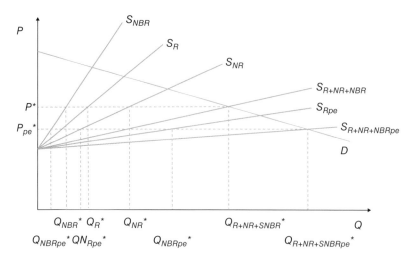

Fig. 7.3. Positive externalities (internalised) reduce the cost of bio-based products. *P*, price; *Q*, quantity; *S*, supply.

market when leading to the aggregate supply curve $S_{R+NR+NBRpe}$:

- the social optimal would be an even higher level of exchange of the good;
- the overall price would be lower;
- as a result of the above, the share of the *NR* product would decrease (due to the lower price of the achieved market equilibrium); and
- the share of the bio-based product would increase.

Some variants of this model are possible. Notably, there may be cases in which the presence of an externality compensates high costs at low levels of production and results in the product being introduced to the market when this positive externality is considered through public policies or by consumers. Second, this could be a dynamic process, as the internalization and adaptation of production are usually not simultaneous. Finally, by increasing profits for the bio-based industry, there is more capacity to invest in and expand the industry, hence leading to positive feedback towards substitution of non-bio-based products.

It is worth noting that there are two other ways of telling the same story. One is to envisage a comparative negative externality for the non-bio-based product. This would most likely require 'opposite' policy instruments, such as taxes, or negative labels, with the same effects on the mechanics of the short term, but likely less able to encourage a specific direction of innovation in

the bio-based industry (due to the lack of the potential positive loops envisaged for the other instruments).

Another way of telling the same story is to attach a positive value to the demand side of the bio-based product only, with a higher WTP for it.

To be rigorous, this choice would depend on the connection of the public good with the product itself or with the production process, and would be likely reflected in the instrument used to achieve internalization, for example, the bio-based product with labelling that ensures higher utility would be represented though a higher WTP, while a subsidy to production due to positive externalities would, for its part, translate into a reduction of costs.

Of course, the whole story is reversed if there is a negative externality related to the bio-based product, which is indeed as it was perceived in some cases (such as genetically modified organisms [GMOs] or biorefinery encouraging the simplification of agricultural systems, etc.)

7.2.2.2 Multiple demands for the same product

The opposite case is that of multiple demands for the same good, as illustrated in Fig. 7.4.

Different demands (e.g. D_1 and D_2) sum horizontally to determine the total demand (D_{1+2}) that, in turn, determines the market equilibrium. The different demands interact among each other. For example, consider the food demand of different

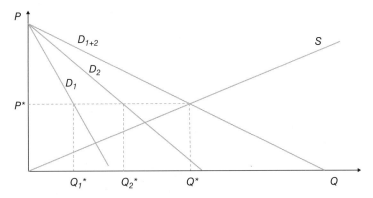

Fig. 7.4. Multiple demands for the same good. *P*, price; *D*, demand; *S*, supply; *Q*, quantity.

countries in a given international market. An increase in food demand in one area would push up the market price for all others. The same applies for biomass in general; for example, the bioenergy industry and food compete for the same products or the same basic resources, such as land. Policy instruments affecting demand would also affect these relationships. For example, a subsidized bioenergy industry would result in an increase in the market price of biomass, hence also affecting the food market (meaning higher prices for food). This has been in the spotlight as part of the discussion about the energy–food nexus (Lochhead *et al.*, 2016). The problem can also be seen from the opposite perspective. For example, from the non-food bioeconomy point of view, more (food) biomass demand increases prices and inhibits investment in biofuels or new technologies (Millinger and Thrän, 2016). Even more widely, Lewandowski (2015) highlights the importance of managing competing claims on biomass for food, feed, fibre and fuel production and of securing a sustainable biomass supply in a growing bioeconomy and provides recommendations in this direction.

This mechanism is particularly relevant for intermediate steps in the chain (e.g. platform products that can be used in several markets). It is similarly the case for the connection between markets of different products sharing the same resource. This applies, for example, to land, which is the basic resource for biomass production. Assuming two goods competing for the same land (e.g. wheat and a biofuel crop), the relationships across products at a given point in time may be presented, following Millinger and Thrän (2016) as:

$$p_a = \frac{p_b Y_b - c_b + c_a}{Y_a}$$

Where: p_a and p_b represent product price, Y_a and Y_b are yields and c_a and c_b are production costs, respectively of products *a* and *b*. The relationship shows that prices in equilibrium (i.e. equal to minimum average costs) would be determined by the production costs of the goods, plus the opportunity cost given by the profit of the alternative crop.

Several considerations may emerge from this relationship. First, depending on the alternative crops with an economic value in an area, the benchmark may be different (usually it would be a mix of crops). Second, increasing prices for food crops due to higher demand would counteract the emphasis on the need for bioenergy production, by increasing the opportunity cost of land. Finally, providing incentives in one market (e.g. for bioenergy) would increase the opportunity cost of the land for alternative uses and hence increase the market price for the other crop.

Box 7.2. Context variables and connections across markets

Mertens *et al.* (2016) provide a case study in which context variables and interconnections across markets are analysed for biogas production in Flanders, Belgium. The authors investigate the effect of market context on the purchase of local biomass for anaerobic digestion. The timing (i.e. when a firm enters the market) is the main factor explaining difficulties in obtaining a stable supply of maize silage for biogas plants. A price increase for maize silage occurs for farmers in competition with a biogas plant (especially in the case of a maize silage deficit in the market). The different institutional arrangements used do not make any relevant difference.

7.2.2.3 Vertical connection and derived demand

Vertical connections among markets concern the case in which different markets along the same product chain, or along interconnected chains affect each other. This is the case for derived demand. For example, taking the previous sub-section, if the market relates to an intermediate product or a production factor, the demand will be from an industry and will be determined by the profits that the industry derives from the product under analysis. This will, in turn, depend on the price of the final product and on the production costs, and hence the technology used.

7.2.2.4 Extensions and qualifications

As long as the destructuring of biomass increases, including interchain connections and flexibility based on platform products in a global market, the types of interactions illustrated above expand in an exponential way, leading to the need to account explicitly for increasing linkages.

Concepts such as that of 'platform products' are also amenable to producing monopolies when largely diffused and protected by intellectual property right (IPR). This highlights, in general, the fact that the issues above would often be cast in a non-competitive market context, with different results in terms of overall price and social outcomes.

Moreover, these connections may increase uncertainty in local markets due to the growing number of potentially unexpected interactions. Finally, transmission does not only include price levels, but also price stability and instability.

The sum of increased flexibility and efficiency at the global level and potential increases in instability at the local level have resulted in a push for investments in solutions for managing these relationships, such as vertical integration. This is rarely a win–win solution, as it implies transaction costs that detract from the overall increase in efficiency. This will be further discussed in section 7.4.

7.2.3 Segmentation and coexistence issues

There are a number of cases in which the bioeconomy implies the coexistence in the same broad market of types of products that are different and sometimes incompatible from the point of view of consumers, regulation and supply. This applies to different technologies (e.g. organic, conventional, biodynamic, etc.) and is managed through market segmentation. The basis for market segmentation is the heterogeneity of consumer behaviour (see Chapter 5). However, it becomes a coexistence issue and a topic related to chain organization when the different types of products are not compatible in terms of production technologies and/or may interact during the production/processing stages, including negatively affecting each other's properties, which ultimately requires having distinct chains or production locations.

The most evident case of this topic is the coexistence between GMO production, conventional non-GMO production and organic production. Greene *et al.* (2016) investigate the coexistence issues of these three technologies in the US. The development of GM crops has now reached about half of the cultivated area (above 90% for soybeans, corn, cotton, sugar beet and canola) and is about to expand to a number of widely cultivated and consumed crops, such as potatoes. The expansion of GM crops has stimulated the growth in the market for non-GMO crops. However, coexistence requires traceability or certification, and the prevention of interaction during processing, and contamination during the production stage (e.g. due to pollination). At the same time, the market for organic products has grown, and now accounts for around 1% of cultivated area in the US. Organic, in turn, requires compliance with a series of prescriptions, which include among others the prohibition of using GMOs. This latter point is presently under discussion, and its revision would change the situation significantly.

Coexistence produces benefits to society through a better segmentation of the market and an expression of higher WTP for products with well-defined characteristics for consumers with different preferences. On the other hand, it also results in a number of cost items, including:

- damage costs due to contamination of non-GM conventional or organic by GM crops;
- additional costs for actions to prevent contamination in the field (e.g. buffer strips);

- costs for additional transport, conservation, processing lines to avoid contamination; and
- costs for traceability, certification, and separate commercialization.

Greene *et al.* (2016) found that in 2014, 1% of all US certified organic farmers reported that they experienced economic losses (amounting to US$6.1 million) due to GM commingling during the period 2011–2014. In the case of soybeans, the top practices used by farmers to reduce the risk of commingling include the use of buffer strips (which also reduce the risk of pesticide drift) and delaying planting by organic corn producers to reduce the risk that their crops pollinate at the same time as GM crops.

The allocation of coexistence costs connected to regulation is non-neutral in terms of economic efficiency. Moschini (2015) shows that an equal distribution of costs among non-GM and GM producers is more efficient than allocating costs on GM producers only.

7.2.4 New markets, expanding markets

As mentioned in Chapter 5, market issues are very different between products that replace existing ones (totally or partially) as compared to products that are new to the market. For products that do not have a market yet, assuming markets make sense from a social point of view, the issue is how to smoothly create new markets. This also depends on the degree of novelty of the product and on the existence of (partially different) substitutes. This is a clear policy concern in most of the policy documents available and in research and innovation funding. It is also a clear consequence of the political will to boost the bioeconomy.

Creating new markets usually involves a high degree of uncertainty and high marketing costs for companies, as well as uncertainty and lack of knowledge on the part of consumers. In order to boost market development, these issues need to be overcome, which may be achieved through policy intervention. A number of instruments are envisaged in bioeconomy strategies, such as public procurement, production/consumption incentives, labelling and infrastructure development, most often at the same time (see also section 7.3).

Box 7.3. Factors affecting new markets for plastics

The complexity of the relationships across products is exemplified by the case of bioplastics in Lettner *et al.* (2017). The paper considers four biopolymers: polylactic acid (PLA), polyhydroxyalkanoate (PHA), lignin and cashew nut shell liquid. The sales volumes of all four biopolymers depend on the price and on marketing activities. The price of PLA and PHA is affected by the process costs. In contrast, prices of cashew nut shell liquid and lignin-based novel bio-based plastic materials are more influenced by further technological innovations.

7.3 Policies

7.3.1 Overview

As mentioned, the bioeconomy is the object of a number of policies and policy intervention is widely advocated to lift barriers to bioeconomy development. In the work of the European Investment Bank (2017), the first key recommendation for the development of the bioeconomy is to 'Establish an effective, stable and supportive regulatory framework for BBI at the European Union (EU) level'. According to the German Bioeconomy Council (2015b), about half of the countries that have adopted a bioeconomy strategy also plan to create an enabling policy framework for bioeconomy development. However, only eight countries have dedicated bioeconomy policy strategies (EU, Finland, Germany, Japan, Malaysia, US, South Africa and the West Nordic Countries). These countries adopt a holistic perspective to leverage the full potential of renewable biomass.

Another group includes countries with holistic regional bioeconomy strategies: Australia (South Australia), Belgium (Flanders), Canada (British Columbia, Alberta, Ontario), Germany (Baden-Wurtenberg, North Rhine-Westphalia) and UK (Scotland).

Four main policy strategies are identified in the other cases: research and innovation strategies with a focus on the bioeconomy; green and blue economy; bioenergy strategies; and high-tech and biotechnology strategies.

An overview of the main policy tools is provided by the German Bioeconomy Council (2015a, 2015b), which also provides some classification of policy instruments. In this classi-fication, the instruments cover five main areas: promoting innovation; infrastructure; commercialization; demand-side instruments; and policy framework conditions. Interestingly, in most of the countries, all or most of these areas are covered, highlighting the need for an integrated view of the bioeconomy. Most of the countries have a clearer focus on research and innovation, though with different instruments. The literature also corroborates this observation highlighting that policy efforts have been largely concentrated on research, while less has been done in terms of support to bioenergy and biomaterials (Philp, 2015). However, there are also opposite country examples, especially with respect to bioenergy.

Infrastructure is the less frequent area of policy, though this is notably very relevant in the countries with the most high-tech strategies. Existing infrastructure is largely devoted to linking research with education. Commercialization is focused, for the most part, on making the link between innovation development and the process of bringing such innovation to the market. Demand-side instruments focus on increasing demand for new products through public procurement and using labels to identify and qualify products. Policy framework conditions are also very important. The most frequent case is that of green taxes (aimed at discouraging competing products rather than developing demand) together with some regulatory cases. However, most bioeconomy-specific initiatives relate rather to participation, policy coherence and legislation for the approval of new technologies.

These instruments may have been used differently depending on contexts, products and sub-sectors within the bioeconomy. Notably, several examples in the literature highlight differences among different sub-sectors of the bioeconomy, but some also highlight that overall support of the bioeconomy will help the development of individual sub-sectors (Turley, 2015).

The OECD secretariat (OECD, 2017) lists a number of areas of future policy intervention for the bioeconomy, distinguishing different subgroups, closely connected to the simplifying distinction between supply and demand:

- demand side: mandates and targets, public procurement, standards and certification for bio-based products, fossil carbon taxes and emissions incentives, reforms of fossil-fuel subsidies;

- supply-side policy measures: supporting R&D and commercialization of bioproduction, tax incentives for industrial R&D, technology clusters, small and medium enterprises and start-up support, supporting local and international access to feedstocks, support to production facilities such as financing demonstration and full-scale biorefineries, support to integrated biorefineries; and,

- crosscutting (mix of supply- and demand-side policy measures): developing metrics and agreeing definitions and terminology, skills and education initiatives with industry for workforce training and public–private partnerships (PPPs) for creating and maintaining specialist training facilities.

7.3.2 Role of regulation

Regulation is the basis and often the principal instrument of public policy. It can take either the narrow meaning of command and control instruments often used in environmental economics, or the broader meaning of public rule setting as a support for private actors in defining what is feasible and what the permitted technologies are.

These topics have a significant role for the bioeconomy due to the preponderant role of regulation on topics that are sensitive for consumers and citizens due to public opinion and that sometimes touch ethical, equity or quality of life issues, as well as living organisms. In this respect, a clear regulatory status of the bioeconomy and the connection with agriculture is advocated as a need (Twardowski et al., 2017).

An important component concerns regulating research and innovation activities. Examples of actions in this area (coming mostly from the GM literature) include:

- deliberate limitations to carrying out research in a given field, e.g. prohibition of research on GMOs;

- procedures, exclusions and timing for patenting new solutions; and

- procedures for permission at the stage of production or commercialization of new products.

Countries take different approaches in this respect and the stringency of these norms change over time.

Box 7.4. Procedures and timing for GM authorization

Smart *et al.* (2017) describe the process for GM authorization. In the EU, the approval process starts with an application in a given member state, followed by a scientific risk assessment. It ends with a political decision-making step (risk management). From 1996 to 2015 the overall time trend for approvals in the EU had an overall mean completion time of 1763 days, first decreasing and then flattening over time. The US approval system begins with a scientific (field trial) step and ends with a 'bureaucratic' decision-making step. From 1988 until 1997 the overall completion time decreased, with a mean approval time of 1321 days; from 1998 to 2015, the mean approval time was 2467 days, with an almost stable trend.

Regulation as a hampering factor and the need to establish dialogue with consumers and society as a whole, is particularly relevant (e.g. for the plant breeding industry) (Małyska and Jacobi, 2017). Potential hurdles in permission and regulation are also often understated in projections, leading to over-optimistic expectations with regard to the impact of new technologies, at least with respect to the timing of their adoption (Chapotin and Wolt, 2007). On the other hand, regulations, or the potential need for regulation, serve to flag potential problems to researchers and innovation developers from the point of view of societal perceptions. Awareness of potential regulatory problems can then be key to anticipate potential issues. For example, Chapotin and Wolt (2007) highlight that, in the crop sector, this makes it possible for developers to look for solutions to avoid applications that are unlikely to meet regulatory approval or gain market and public acceptance, as well as design crops that have a stronger appeal for stakeholders and greater potential for widespread adoption.

Another economic issue, stemming in part from the evolution in governance/legislation and the existence of more sophisticated technologies, is the growing complexity of property rights in research results. This results in a wider distribution of the use of research results and the appropriation of their benefits. The participation in the exploitation, and the potential for alternative uses of knowledge, also become greater while, at the same time, requiring more flexible ways of collaboration.

Another broad area of regulation concerns property rights over resources and the sharing of benefits. Access to classical production factors, such as land, is a common topic in the study of property rights. One well-known action is the Nagoya Protocol, which is a supplementary agreement of the 1992 Convention on Biological Diversity. The objective of the Nagoya Protocol is the fair and equitable sharing of benefits arising from the utilization of genetic resources, thereby contributing to the conservation and sustainable use of biodiversity. Needless to say, these actions have a cost in terms of administrative burden, which could potentially damage the monitoring and collection of biodiversity, or even research.

A less specific but nonetheless important area of regulation concerns the prohibition of the use of competing (e.g. fossil-based) products or technologies to support the development of new bio-based markets. For example, the prohibition of purely fossil-based products in some areas (e.g. plastic bags in Italy) is a strong signal in favour of bio-based substitutes as it forces actors to use these substitutes.

7.3.3 Innovation promotion, funding and orientation

A major action area for public bodies in the bioeconomy is that of support for research and innovation. This takes several forms and depends largely on different country approaches (see, for example, Schütte (2017) for reflections on bioeconomy and innovation policy in Germany). This topic has largely to do with public funding of research. However, it is not just a matter of money. On the contrary, a major reconfiguration of organizational, administrative and operational aspects related to the bioeconomy in education and research institutions is also needed to achieve the greatest benefits from breakthroughs in research. Moreover, research funding rules and technology transfer mechanisms may need important changes to meet both local needs and global challenges (Beachy, 2014).

7.3.3.1 *Promotion/funding of basic research and applied research*

A basic role is to promote research through direct research funding in public institutions or support to private research initiatives. Focusing on relatively high technology readiness level (see Chapter 3) implies a greater role for private actions through private research, collaboration and partnership between public and private research (see section 7.3.3.3). Indeed, private sector investment in research has grown over the past decades. However, the topic of private versus public research in the bioeconomy goes beyond this, as there are interdependencies between public and private research and much of the foundational research supporting private sector activities is still generated by the public sector research and funding (Beachy, 2014)

Due to their role, the relevance of public institutions as funders of key enabling technologies (KET) or crosscutting technologies is paramount compared with applications, in which the role of private companies is prevailing. It is also important to mention that this may go far beyond the bioeconomy as defined in this book and link it to, for example, medicine. This implies economies of scale in research and also increasing crosscutting collaboration and interest.

Another factor in determining the role of public versus private research is the public good nature or the public interest as a driver of research. Beachy (2014) lists several good reasons for a public effort in funding research in the agriculture and bioeconomy sectors, such as the high rate of return on investments in research and development and the (already discussed) urgent societal challenge of providing sufficient food for a growing population and combatting climate change. However, public support of research in agriculture has increased at a much slower pace than other sectors such as health and energy; accordingly, it is advocated that greater support to public research is required for this sector (Beachy, 2014).

Public policies are also oriented towards promoting some specific features of research. One is interdisciplinarity, or even transdisciplinary, as the bioeconomy is typically problem-oriented and needs to deal with a number of issues at the same time, ranging from genetics to social sciences. The integration of economics and behavioural sciences with more hard sciences is actually one of the challenges of current research pathways and promoted vigorously by current funding schemes, as well as non-academic actor involvement. Another feature is international collaboration. Since similar problems tend to be distributed worldwide (and touch upon genuinely global issues, such as climate change), it is important to achieve a critical mass or exploit synergies between the various strengths of research in different countries. This topic touches the highly geographically interconnected supply chains of the bioeconomy.

7.3.3.2   Pilot and demonstration plants

Having achieved a sufficiently high level of technology readiness level, the promotion of pilot and demonstration plants is a widely used strategy to spread new technologies. These plants have two main functions. First, they allow testing the real-life functioning of processes, thus demonstrating that the solutions work. Second, they support information sharing through site visits, which is more informative compared with generic communication about new technologies, in particular when these new technologies have potentially uncertain performance, are not robust or simply where trust needs to be built among operators.

The importance of trials, pilots and demonstration plants is illustrated in Fevolden *et al.* (2017), and relates to the concepts of learning by doing, which includes both learning through R&D and learning through manufacturing, and learning by using. Knowledge can be embodied, when it enters the design of goods, or disembodied, when it concerns manufacturing and utilization.

7.3.3.3   Clusters and public–private partnerships

One feature of the innovation processes described above is to have different actors interacting and contributing with different roles and in different stages. Two widely used policy instruments to ensure consistency in this process and to improve collaboration are clusters and PPPs. Clusters and PPPs are different, yet overlapping, instruments due to the existence of weak and multiple definitions.

Clusters are groups of actors, including companies and public institutions, ideally comprising industries, research and investors, working on the same issues or chains. This approach is promoted in bio-based industries to ensure coordination, identification of common strategies and cooperation on innovation projects. When geographically clustered, they may also include concepts such as industrial districts. Clusters are often the object of specific support by public policies seeking to create clusters of companies and other stakeholders, which specialize in a particular technology to build an 'industrial ecosystem' (Philip and Winickoff, 2017).

Box 7.5. Bioeconomy clusters

According to Stadler and Chauvet (2017), the cluster policy introduced in France in 2005 to boost innovation in the bioeconomy and to develop innovation platforms allowed for industrial demonstration and feasibility at the TRL from 5 to 9. The paper illustrates the industries and agro-resources (IAR) cluster based in the Hauts-de-France and Grand Est region, and completely dedicated to the bioeconomy. IAR has sought to build a complete innovation ecosystem in the bioeconomy by developing relationships between actors along the full value chain and facilitated investments in new platforms and programmes through PPPs. Kircher *et al.* (2016) use the model of the German cluster CLIB2021 to discuss the role of cluster organizations and professional cluster management in gaining a competitive advantage; the paper further cites examples of how cluster management supports the formation of an industrial consortium. Auer *et al.* (2016) provides the example of a bioeconomy cluster that combines applied research in interdisciplinary teams and innovative industrial companies in the wood industry, fostering the conversion process of forest resources into building materials, base chemicals, plastics and new materials.

Among the lessons learned, it should be mentioned that this approach requires a long-term strategy and support, in which the role of regional and national public partners (as well as a wide range of stakeholders) is essential.

PPPs are joint initiatives including both companies and public institutions that provide funding to be used in common for research projects, as well as devising their research topics and agendas. A pertinent example of such a PPP is the Bio-based Industries Joint Undertaking (BBI JU) promoted by the EU. BBI JU is a PPP between the European Commission and the Bio-based Industries Consortium, an industry-led private not-for-profit organization representing the private sector across the bio-based industries. The PPP model is now considered a successful way of promoting research and innovation, reducing risk for investment in research and work as a catalyst for private initiatives in the EU bioeconomy (Mengal *et al.*, 2017).

While collaboration may be organized around any issue, sustainability is a special field for inter-organizational collaboration due to the wide range of societal issues involved. Nuhoff-Isakhanyan *et al.* (2016) investigate available knowledge on sustainability of inter-organizational collaboration in bio-based businesses, while considering the three pillars of sustainability: environmental, economic and social.

Furthermore, different forms of inter-organization networking and collaboration exist to boost innovation. Lopes (2015) argues for the development of eco-industrial parks together with reliable policies for carbon pricing and revenue reinvestments in disruptive technologies to boost the transition from a fossil to a bio-based economy.

7.3.3.4 Patenting and openness in science

A major issue in the markets for new technologies is the distribution of benefits from innovation, through patenting and royalty regulation and promotion. A patent is a type of IPR through which a sovereign state grants to an inventor a set of exclusive rights. This occurs for a limited period of time in exchange for detailed public disclosure of an invention. In spite of the fact that patents in some biotechnological fields are a controversial issue, patenting is widely used as an instrument to protect and exploit intellectual capital and the number of patents is increasingly sharply over time in the field of biotechnology (García *et al.*, 2013). The role of patents is to provide incentives to research and innovation by providing remuneration for the investment needed in order to arrive at an invention. In spite of the number of patents, patenting in the biotech industry is often considered to be a low profitability venture that is sometimes hindered by long waiting times for patenting and the uncertainties of downstream markets (Cota *et al.*, 2016).

García *et al.* (2013) discuss external and internal factors affecting the exploitation of biotechnological patents. External factors are: (i) the market needed for biotechnological products and services; (ii) freedom to carry out analyses prior to entering the market; and (iii) efficiency in prosecution by patent offices. Internal factors are the characteristics that the patent's owner must take into consideration in order to have a strong, broad subject matter in the granted patent. To this end, the experimentation needed to obtain an adequate scope of the subject matter in claims is regarded as a critical issue in the exploitation of a patent.

IPR issues are also emerging in fields in which they were previously less relevant. For example, Myking *et al.* (2017) discuss IPR issues related to somatic embryogenesis (SE) (a way in which an almost unlimited number of genetically identical plants (clones) can be produced from a single mother plant/seed). The authors highlight how different IPR systems may possibly differ in how well sufficient genetic variation can be encompassed by protection claims. Accordingly, they argue that process patents are most applicable, and that genetic variation should be explicitly considered in future patenting claims.

A relevant debate is the contrast between patenting and open science. Indeed, open science is largely promoted to allow for the widest range of benefits from new knowledge (Levin and Leonelli, 2017). This does not merely contrast with the use of patents; on the contrary, a careful use of open science requirements in public funding together with patenting can contribute to maximizing societal benefits from research and innovation.

7.3.3.5 *University spin-offs and other strategies from research to application*

Several different initiatives have been adopted to boost the link between research and innovation, or between research and industrial exploitation. Many of these are highly relevant though not specific to the bio-based economy, such as incubators and spin-offs. Making this connection work is one of the key challenges of the bioeconomy.

Box 7.6. Different strategies for the use of new research

Festel (2015) investigates the technology transfer role of spin-offs from universities and research institutions based on the analysis of 17 case studies from the industrial biotechnology sector. After the launch of the spin-off process the relevant technology is further developed leading to different technology transfer models between established companies and spin-offs. SMEs are more focused on research cooperation with spin-offs, especially by engaging them as service providers, mainly due to limited financial resources. On the contrary, multinational enterprises are very active both by acquiring spin-offs or engaging them as service providers.

Lab-to-market plans are also mentioned in the literature to support the flow of innovation, together with public support of business plans and business development through several forms that are more or less specific to the bioeconomy.

7.3.4 Infrastructure and human resources

7.3.4.1 *Research and knowledge infrastructure*

Infrastructure constitutes a wide area of different kinds of intervention. These may be intended as laboratories/facilities as well as institutions designed to keep and develop knowledge. Observing bioeconomy strategies worldwide this includes laboratories, competence centres, research networks and training specialists, as well as infrastructure aimed at the development of specific KET. This kind of support is very important in data-intensive fields of biological resources and hence also for bioeconomy sectors.

Box 7.7. Plant genetics infrastructures

For example, in 2008 the National Science Foundation funded the iPlant Collaborative cyberinfrastructures to support plant sciences in the US, including support for data-intensive activities in the field of plant genomics and phylogenetics. The EU has provided funding to the transPlant initiative (http://www.transplantdb. eu/), an e-infrastructure to support computational analysis of genomic data from crops and model plants.

Infrastructure is a key topic, especially for developing countries and countries undergoing major structural change (see Ahn *et al.* (2012) for the example of India).

Infrastructure is increasingly oriented towards cyberinfrastructure and to be less 'physical' in nature. Cyberinfrastructure is a type of infrastructure (in both technological and organizational components) connecting people, laboratories, computers and databases. It can include, for example, data storage, management and integration and other computing and information processing services.

Box 7.8. Microbial resources centres for the bioeconomy

Conservation of microbial biodiversity was not previously a priority but is now considered to be of significant relevance for the future. Overmann (2015) provides the example of the conservation of microbial strains through microbial resource centres, intended as institutions capable of safeguarding, maintaining and distributing authenticated microbial strains, their genomic DNA and the associated data. They can help taxonomy, support scientific studies and improve recognition of scientific work. This infrastructure is also acknowledged to be able to provide relevant economic benefits. Overmann (2015) estimates that a microbial strain isolated and deposited can achieve a total value of €10,754.

Suryanarayanan *et al.* (2015) present a roadmap for the creation of a national genetic resource for fungi in India, as a basis not only for education, but also for accelerating technology development and the bioeconomy.

7.3.4.2 Human resources, education and training

The issue of human resources in the bioeconomy is a key one, based on the high degree of knowledge intensity and innovation-focus of the sector. Human and social capital are usually indicated as strategic areas of investment (McHenry, 2015). Investing in innovation capital, i.e. forms of human capital such as imagination, entrepreneurial skills and relationships that support the development of an innovation economy, is also identified as a way of supporting supply-side market development.

Education is at the core of preparing society for the bioeconomy. This includes at least two levels of action. One is that of general education. The development of the bioeconomy is based on a wide tendency towards building the basis through education in new fields, for example, biology and biotechnology, and at the different levels of the education system, from primary school to PhD level. In addition, vocational training, further education life-long learning in a disciplinary perspective are promoted in order to maintain updated competences and develop professional skills.

The lower levels of education are more connected to society's basic understanding of the bioeconomy and related technologies. Professional and life-long learning are more closely connected to technology understanding and dissemination among professionals, including support for boosting technology uptake by, for example, company managers. Managerial and technology transfer/brokerage skills in curricula are also important.

Furthermore, education at the intermediate levels is more easily understandable when focused on specific fields (e.g. bioenergy), while more holistic bioeconomy focuses are linked more to the doctorate level, research initiatives or management roles requiring vision and leadership.

Clearly, the potential of education to contribute to the development of the bioeconomy is linked to other instruments, most notably infrastructure. Motivations and attractiveness are linked to market and sector development, having hence a mix of a pull-and-push relationship with overall bioeconomy development.

Ray *et al.* (2016) illustrate the role of social innovation in biotechnology education. They emphasize that this is also connected to new technologies such as virtual laboratories, while other sources advocate for the use of massive open online courses.

Certainly, the education side of the bioeconomy remains difficult in the absence of a clear profession, while a system-wide view of the bioeconomy is a clear value-added factor for professionals working in specific sub-sectors, especially in the field of innovation.

7.3.4.3 Rural development infrastructure

Rural development, as well as marine and forest-based infrastructures, is the basis for an economically viable production of biomass. They provide basic services that make it possible to carry out economic activities and live in rural areas. This includes classical rural development infrastructure such as roads, electricity, health and education services digitalization and communication in general. This is a clear priority in most remote areas and in developing countries.

7.3.5 Market support and regulation

7.3.5.1 Overview

The functioning of markets is not straightforward; not only do new markets often struggle in the beginning, but existing ones may ultimately not work well. Several instruments are used to assist market functioning, which, in the field of bioeconomy, means transmitting the right signals (consumer preferences and societal values) across markets. In this context, it is relevant to distinguish between business-to-business and business-to-consumer environments. Business-to-business is more related to highly technical instruments for chain coordination, contracting and traceability and is discussed in section 7.4. Business-to-consumer is more related to labelling and information sharing. Marketing may have a major role here when valorizing their positive contribution to consumer needs and society concerns.

7.3.5.2 Good definitions, identification and standards

Given the complexity, diversity and variability of biological resources and their derivatives, and the wide range of attributes of interest to consumers, the clear definition and identification of goods in all stages of value chains is a strong requirement for markets to work properly. This starts upstream, including at the level of research. Indeed, due to the inherent complexity of biological systems, standards are needed to represent accurately the various biological functions, together with adequate language for data description and exchange (de Lorenzo and Schmidt, 2018). In practice, product definitions and standards are pervasive for all products related, for example, to food, where very complex sets of different quality standards are identified worldwide and locally.

7.3.5.3 Traceability, certification and labelling

Besides qualifying the goods involved in transactions, policies may impose that the characteristics of goods be traced along the production chain and communicated to consumers. Traceability, certification and labelling are ways of ensuring and communicating the positive features of goods when they are not otherwise evident. Several of these features are available in the bioeconomy as a result of the quantity of attributes

that are of interest to the consumer, but not evident from the product itself.

Traceability, certification and labelling is a major issue in food that has already been taken up by a considerable body of literature, which account is far beyond the scope of this book. Some of the most widely used examples of labelling in the food sector concern organic products, non-GM products or origin of products. Labelling to identify non-GM crops may be imposed by state regulation and have an interplay with voluntary private labels. A very broad topic is that of geographical certifications widely used in the EU for agricultural and food products.

Similarly, environmental and ethical features are increasingly included in certification and labelling.

Bio-based products are now increasingly also subject to certification and labelling Golden et al. (2015) illustrate the biopreferred programme and the US Department of Agriculture (USDA)-certified bio-based product scheme in the US. Several other certifications also exist, for example, for recyclable plastic, wood characteristics, etc. Forest certification is seen as one of the primary needs for forestry to meet bioeconomy goals in Europe (Sikkema et al., 2016). According to the authors, existing legal frameworks for legal timber harvesting (EUTR) and renewable energy use (EU-RED), as well as waste treatment provide valuable principles for EU member states and market sectors.

While already widely studied, traceability is currently undergoing major changes due to digital innovations. In particular, blockchain is quickly spreading in the food industry (Pendrous, 2017) and is expected to contribute significantly to food security (Ahmed and Broek, 2017). More generally, it can smooth systems based on multiple transactions such as those implied in the bioeconomy web, including in the application of environmental policy instruments such as permit trading (Khaqqi et al., 2018).

7.3.5.4 Marketing

Product traceability and labelling systems are effective when accompanied by appropriate marketing activities. Marketing is mainly a private sector activity aimed at matching consumer needs with supply. Initiatives may be part of the daily activities of a company, but may also explicitly benefit from public support, for example, in the case of actions aimed at promoting products abroad. The marketing strategies of big players in the retail sector and the collaboration of retailers with public initiatives are important (e.g. elimination of plastic bags). The regulation of marketing and advertising is also important to make sure fair and truthful information reaches the consumers.

7.3.6 Demand-side instruments

Demand-side instruments are used to promote demand for bio-based products. In a way they are creating an artificial demand in order to boost market development or to incorporate public goods components into private demand. Based on the models of markets depicted at the beginning of the chapter, these instruments tend to increase demand for (move the demand function upwards in Fig. 7.1) for bio-based products, hence increasing the quantity demanded as well as market prices.

Here we focus on instruments providing direct incentives to the demand side. Other instruments sometimes referred to in the demand side, such as standards, labels and social dialogue, are discussed as market functioning tools in the previous section.

7.3.6.1 Mandates

Mandates usually refer to obligations to use renewable or bio-based sources for some share or percentage composition of products. A noteworthy example concerns the mandate requiring that specified amounts of renewable fuels be blended into transportation fuels, according to the Renewable Fuel Standard (RFS), in the US Energy Independence and Security Act, 2007. These mechanisms force fuel companies to buy a specified amount of bio-based fuel. The use of mandates produces a number of effects for the different players and on the overall welfare; several indirect effects also occur due to prices and trade effects. In addition, they affect the interplay between food, energy and different environmental issues.

Box 7.9. Effects of biofuel mandates in the US

Moschini *et al.* (2017) simulate the effects of different scenarios concerning the RFS policies in the US. The current RFS programme leads to overall welfare gains for the US, mostly by way of beneficial terms of trade effects. It also yields considerable benefits for the farming sector. The implementation of proposed 2022 mandates, requiring the further expansion of biodiesel production, led to a considerable welfare loss relative to 2015 mandate levels. Constrained (second-best) optimal mandates would entail more corn-based ethanol and less biodiesel than currently mandated.

7.3.6.2 Public procurement

Public procurement is the procurement of goods and services by public institutions. It may be used to start new markets by creating a targeted demand towards goods with specific characteristics, such as bio-based products, which is often seen as a part of green procurement. It is one of the most frequently used instruments through which public intervention can stimulate the market for bio-based products and one of the main demand-side instruments. Its adoption is by no means trivial. Public managers often struggle with cost management and need, from the purchased goods, a sufficiently low cost beyond the quality of the bio-based origin, which implies that public procurement may involve serious trade-offs and may not always be an easy solution. Indeed, recent works in this field show that bio-based products are not yet considered a relevant category in public procurement and that bio-based content on its own is most often not regarded as a relevant justification for inclusion in green public procurement schemes. However, the incorporation of bio-based content as a criteria in eco-labelling schemes could help promote the acceptance of bio-based products in the public sector (Peuckert and Quitzow, 2017). Interests and suitability are apparently more straightforward for the public procurement of certified wood (Brusselaers, *et al.*, 2017) or renewable bioenergy (Jansson, 2016). However, the literature also highlights the potential discrepancy between public procurement actions in the field of energy and sustainability criteria (Jansson, 2016).

7.3.6.3   Direct incentives and tax relief

Direct incentives involve one-time or regular payments to consumers to support demand. Payments on the demand side are used for bio-energy and biofuel, but rarely for other products. An indirect incentive to consumers is tax relief and consists of the tax reduction for those adopting a 'virtuous behaviour'.

7.3.6.4   Information

Based on the widely recognized role of information in affecting consumer behaviour, information campaigns on the relevance of the bioeconomy as a whole and on the characteristics of specific products is an obvious instrument to help develop markets and general awareness of the potential role of the bioeconomy. Information may be part of marketing campaigns but can also be addressed by public institutions towards consumers.

7.3.7   Supply-side instruments

Supply-side instruments are those addressing the production of raw materials, direct support to companies and cost reduction. They can also include instruments such as financial support, venture capital and support to start-ups. The primary sector has traditionally been the object of public support since ancient times, especially in connection with the need to secure stable and sufficient food production. Strong agricultural policies are still active in most countries.

7.3.7.1   Support to biomass production

The most obvious kind of support is that provided to biomass producers, which, in many countries, goes through agricultural support. It can take several different forms:

- income support and risk reduction to famers; this maintains a basic interest in being involved in primary production and reduces (distributes) the risk connected to this activity;
- cost reduction for key inputs, still used in some countries, e.g. for fertilizers and machinery;
- support to investment and innovation; this affects long-term productivity by allowing smoother changes in technologies (innovation)

and higher capital intensity compared to privately profitable levels of capital use (investment);
- support to the price paid to producers (in addition to the market price).

The result of this support is that the supply curve for raw materials would be moved downwards, allowing for cheaper procurement costs and a potentially higher market size. However, different instruments can have different effects. For example, income support could be largely detached by the cost curve of individual products.

This also includes specific investment support to bioeconomy-relevant business, for example, bio-gas plants; subsidized producer prices for bioenergy is also a form of support used by some countries.

Box 7.10. Biofuel support in the US

Gkritza *et al.* (2012) illustrates the case of biofuel in the US boosted by the US Department of Energy's target to achieve 20% of US transportation fuels from bio-based sources by 2030. The federal government strongly promoted biofuels as an answer to imported energy sources and global climate change. As a result, ethanol production in the US grew at an annual rate of 32% between 2005 and 2008.

Support does not generally benefit different bioeconomy-related producers in an indiscriminate way, but rather selected areas; for example, in some countries biofuel producers receive public support for specific crops, whereas other kinds of bioeconomy products do not.

Moreover, support to some parts of the bioeconomy can offset incentives for the others; for example, coupled support to agricultural raw materials for food production produces an opportunity cost that increases costs for biomass production for biofuels or bio-based materials.

7.3.7.2   Finance

Finance plays a unique role in supply-side policies and related action. This may involve not only public funding, but also targeted funding by private banks.

Start-up funding is a special case of funding for new businesses. The bioeconomy is characterized by a mix of large and very small companies, especially in some sub-sectors (e.g. agriculture, food, etc.), and hence it can be a lively environment for start-ups given that it is less competitive than some others. It may also be a fertile context for the development of new and specialized business models serving a large audience with horizontally relevant services.

Yet it is often claimed that it is difficult to attract new capital in the bioeconomy. This issue is addressed by Hansen and Coenen (2017), studying resource mobilization for biorefineries by investigating investment decisions of incumbent pulp and paper firms in Sweden and Finland. They identify four issues as empirically important for improving resource mobilization for biorefinery technologies: the establishment of divisions in pulp and paper firms; the creation of internal markets for new bioproducts (aimed at further technological development); the establishment of purchasing agreements with downstream actors; and investment in new appropriate managerial competencies.

The European Investment Bank (2017) highlights that existing public financial instruments are utilized by the sector, at least in Europe, but that their ability to catalyse private investment is insufficient. They argue for the need for policy action through new or modified public financial instruments that could focus more on de-risking investments in the bioeconomy industry and catalyse (crowd-in) private capital. Two practical actions proposed by the European Investment Bank (2017) are: (i) the development of a new EU risk-sharing financial instrument dedicated to the bioeconomy that could take the form of a thematic investment platform aimed at mobilizing private capital; and (ii) the creation of an EU-wide contact, information exchange and knowledge-sharing platform to facilitate relationships between bioeconomy project promoters, industry experts, public authorities and financial market participants seeking to become active in the bioeconomy.

7.3.8 Policy framework conditions and policy coherence

Several framework conditions may be relevant to improve and regulate the development of the bioeconomy. This includes, among others, plans for an industrial renaissance providing a general stimulus to renewed industrial vitality, especially with respect to innovation, e.g. industry 4.0. It can also involve green taxes on competing fossil products, putting a general burden on more polluting sectors/technologies and giving a competitive advantage to more environmental friendly solutions such as those provided for by the bioeconomy.

A general improvement of the legal system, especially with regard to improving the functioning of regulations (simplification, enforcement), would clearly benefit all other instruments. In particular, improved regulation could be sought for: (i) improved access to renewable resources by new institutional mechanisms (easier when not harmful, but at the same time regulated); (ii) easier understanding, policy consistency and common infrastructures and organizational solutions for recycling; and (iii) approval of new technologies, where in principle a more flexible and smoother legislation could simplify the development of new bioeconomy technologies, but also a better regulated mechanism (e.g. with some level of selection and lower costs) could improve competitiveness of small scale solutions and increase access to patenting by small firms.

Policy coherence is also an issue in the relationship between bioeconomy policy and existing sector policies, even within the bioeconomy itself (such as agriculture). This can be the most difficult requirement because closer fields, such as agriculture, food, energy and the environment, benefit of lasting, and hence path-dependent, legislation, sometimes already poorly coherent. For this reason, and due also to the variety of different fields of regulation that the bioeconomy comes into contact, policy coherence is a sensitive issue and a priority for action in different areas. For example, in the EU the coherence with agriculture (incentives provided by the common agricultural policy [CAP]) and water regulations have been very challenging and have emphasized several mismatches, as well as being constantly high in the agenda for future action in both the bioeconomy and the CAP.

Several papers provide examples of how lack of policy consistence affects the bioeconomy. Philippidis et al. (2016) investigate the effects of broad policy scenarios ('inward-looking' and 'outward-looking') related to bioeconomy development in the EU. The overall result is similar across scenarios and with the baseline; however, different scenarios yield a rather different effect across sectors, in particular concerning the bioenergy sector and biomass production.

Concerning Brazil and Latin America, Dias and De Carvalho (2017) argue that essential elements for the development of the bioeconomy include a wide variety of different fields of intervention, notably: (i) human capital; (ii) research; (iii) protection of IPR; (iv) adequate regulation; (v) technology transfer support and regulation; (vi) market incentives; and (vii) a legal framework ensuring legal certainty.

With regard to forest policy in Ontario (Canada), a sector that has previously experienced a reduction in size and difficulties in finding new strategies based on bioeconomy concepts, Majumdar *et al.* (2017) argue that forest policy needs to take an integrated approach to promote all aspects of the bioeconomy. In addition, to support the bioeconomy it is necessary to integrate and coordinate the actions of government with industry players and academia involved in the bioproducts sector.

A specifically important issue is that of policy coordination between agriculture and bioenergy, as biogas production chains can serve as inputs for bio-based products in a biorefinery concept. For example, Turley (2015) argues that in Europe, with the exception of the energy sector, there is no direct incentive promoting the use of bio-based products, including bio-based chemicals and materials. Rather than going with different and often contrasting policies, it is advocated that a consistent policy promoting the optimal use of biomass for multiple goals across the two sectors is needed (Pfau *et al.*, 2017).

Another area of attention is local-specific frameworks and problems, in particular in rural contexts. McHenry (2015) argues that three crucial investment drivers are receiving insufficient consideration in research and extension in the rural bioeconomy, namely human collaborative knowledge, sustainable production capability, and cross-sectoral transformational science and policy.

7.4   Contracts, Integration and Organization in the Bio-based Value Web

7.4.1   Chain and system organization

Market and chain organization is a constant concern of private players, as well as policy intervention, in the food and bioeconomy sectors. This is for several reasons, such as information asymmetries, market power along the chains, risk sharing and stabilization of supply. This is largely emphasized by the growing number of separability points in chain and cross-chain connections of the bioeconomy. Aggregation among farmers, through horizontal or vertical integration, is a very common topic in the agriculture and food economics literature. Moreover, chain integration is a growing concern in the forest industry (Korhonen *et al.*, 2016) and the bioenergy sector (Yang *et al.*, 2016).

The structure of the chain may be an important determinant of innovation ability in bioenergy. For example, Heinonen *et al.* (2017) contrast maize-based biofuel production with switchgrass, showing that the latter experiences difficulties due to the lack of a well-established value chain to promote production stabilization around a series of inter-related process innovations. In such contexts, policy alone is not sufficient, or should intervene on multiple entry points, including not only incentives to production, but also direct promotion for chain coordination.

The aggregation of consumers is also a frequently addressed issue, as well as the integration of retailers and upstream firms. New ways of interaction between consumers and the different stages of the chain is also a topic that is attracting increasing interest. For example, one approach is that of short supply chains, used in the food sector and also advocated for energy and wood. This approach could, in principle, facilitate a common vision for producers and consumers alike, increase awareness and minimize the rents paid to intermediaries, but would also involve costs such as conservation costs and an increase in the overall time societies devote to primary goods provision. Chain structuring is also a topic in relation to innovation processes.

Many different options for integration are possible depending on the different context. This goes from firm take over to simple networking, with intermediate solutions represented by co-operatives or consortia. For most cases (value chains), vertical integration and contracting (see section 7.4.2) may coexist and serve the needs of different kinds of industry firms (e.g. depending on need of supply stability) and landowners (e.g. depending on land quality and risk preference) (Korhonen *et al.*, 2016; Yang *et al.*, 2016).

7.4.2 Contracts

Contracts are either written or oral agreements among parties, especially aimed at the transaction of a good. Contracts are implicitly everywhere in the modern world. Contracts are signed when accepting conditions in online purchasing, for example. Contracting is one of the most widespread trends in the agriculture and food sector, deeply embedded in the nature of farming activity (Allen and Dean, 2002), and particularly relevant for the bioeconomy as a whole. Notable kinds of contracts for the bioeconomy are those related to inter-firm relationships along the chain and company-farmer contracting for products. Contracts are now widespread and seen as a key instrument to ensure biomass for energy and biorefinery.

Different types of contracts exist, notably ranging from purely commercial contracts to production contracts involving the definition of technical prescriptions. Different contract parameters may be set, such as price (fixed, variable, mixed), product quality (with higher or lower premium prices attached to them), length of the contract, and so on. Contract design interacts with context variables (e.g. price stability and trends; markets of competing products) and farmer preferences, to determine quantity, stability and quality of biomass production for the bioeconomy (Wamisho Hossiso *et al.*, 2017).

Contracts may also have a role in ensuring socially sustainable development or counteracting harsh ways of accessing land such as land grabbing. However, it is observed that even contracts tend to maintain or strengthen unbalanced power relationships between industry and farmers, as well as between different groups of farmers (Vicol, 2017).

A special type of contracts is that related to agri-environmental schemes or payments for ecosystem services. Such contracts may be particularly useful to promote conservation practices towards land and biological resources.

7.4.3 Land tenure

A very specific type of contract is applied to land tenure. This is a key topic as it concerns access to the very basic resources at the heart of the bioeconomy. In most of the Western world, primary access to land is guaranteed through private ownership, which is in fact a selected set of rights on land. Landowners can transfer the right of using land for cultivation through rent contracts. Rent contracts can have different features in terms of length, obligations and price determination. One key distinction is between cash rent and crop-sharing schemes in which the owner receives a share of the harvest and may contribute to a share of the costs (with risks distributed accordingly). In addition, they may involve the transfer of rights, such as those connected to public payments. A number of other cases exist, such as common land managed as club goods or open access.

One general problem is that of growing perennial crops when tenure is based on short-term contracts, as the plantation may have implications for the transfer of tenure. Another issue concerns the distribution of market benefits or policy into rents for owners. This is also connected to contracting in downstream production. For example, there is evidence that production contracts by farmers in GM soybean areas may be associated with higher rents (Choumert and Phélinas, 2017).

> **Box 7.11.** Contracts for bioenergy in the US
>
> Scott and Endres (2014) illustrate how producers of biomass need to grow dedicated, high-yielding energy crops of a perennial nature on leased property in the US. The ability to involve tenant farmers is very important due to the relevance of the tenure system in the US. Using contracts as governance schemes, landowners and tenants can address three key challenges of the bioeconomy: (i) the need for long-term access to land; (ii) managing landowner concerns regarding the potential invasiveness of some new bioenergy crops; and (iii) developing the reclamation of rhizomes as an additional revenue stream from perennial biomass crops.

7.5   Institutions, Governance, Participation and Information

To better address the challenges of the bioeconomy at the global level, global awareness and global coordination in defining bioeconomy strategies is advocated (El-Chichakli *et al.*, 2016). The general need for adapting institutional arrangements as an answer to new technology or economic issues brought about by the bioeconomy is also supported by the literature. This is at the boundary with political views of the bioeconomy and the perspectives of the political economy discussed in Chapter 8, but is also linked to managing new technologies.

Box 7.12. Restructuring power relationships in sheep breeding

---

Gibbs *et al.* (2009) illustrate how the increasing significance of genetic techniques affects beef cattle and sheep breeding in the UK. In particular, they find that the stronger involvement of international and corporate interests is restructuring the network of interests in the sector and modifying the knowledge and decision-making approach of individual breeders. The authors argue that the negotiation between breeders, traditional knowledge-practices and 'geneticized' techniques are complicating the debate and the pathways for breeding transformation, and is restructuring power relationships in the UK livestock-breeding sector.

Increasing participation and political representation in governance is considered essential to allow for better participation of society in the processes of regulation and market functioning (McCormick and Kautto, 2013). This also goes together with changes in capabilities and responsiveness.

In addition, the rules governing the interplay between different subsystems and their power relationships can determine the way in which entire systems react to stimuli. Informal interactions may in fact play a major role and are key in understanding complexities in bioeconomy governance systems (Clark and Phillips, 2013).

Inclusive project development is another way of ensuring participation from the early stages of new projects. Inclusive project development is reported for biofuel from *Jathropa* in Madagascar, where companies are countering criticisms of land grabbing with small-size actions (see Neimark, 2016). In part, the ability to support participatory approaches, in fact, goes back to the scope and conceptual view of the bioeconomy. In particular, Mukhtarov *et al.* (2017) note that industry and governments often have a too narrow vision of the bioeconomy, centred on technical issues, while proper stakeholders' involvement would require a much wider accounting of far-reaching economic and social issues.

Information has already been listed as a critical element of demand and supply management. Here, information is important as a component of governance and participation, as well as a support to the stronger connection between supply and demand. New communication technologies may have a key role in boosting these connections, for example, by improving the awareness of consumers and their ability to choose, or provide information, for example, about the quality and eco-friendliness of products. Examples include apps that help consumers search for non-GMO, organic, in-season food or environmental or ethically produced food, through simple coding of product features. The role of these technologies in shaping awareness, preferences and future demand–supply interaction is still largely unexplored.

Box 7.13. Responsiveness and capabilities in the forestry sector

---

Mustalahti (2017) discusses responsive governance in the forest-based bioeconomy and capabilities with respect to the forest sector in Finland, highlighting that climate change concerns and bioeconomy could challenge the way states, citizens and companies interact in natural resource governance. The study highlights that citizens' participation, though often advocated, is not yet addressed sufficiently. Citizens could have the capabilities and the ability to participate in decisions on the forest-based bioeconomy, but this would require a collaborative approach to empower people and institutions to debate the development of bioeconomy opportunities. In spite of the fact that citizens may be unable to find solutions and develop innovations themselves, their values, interests, know-how and environmental entitlements need to be taken into account.

7.6 International Dimensions: Cooperation and Management of Global Markets

The bioeconomy is concerned with strategic resources distributed worldwide and ultimately benefits from complementarities across countries. This involves the management of international markets, including careful management of trade barriers, international finance, access to bio-resources, as well as dealing with global issues such as climate change. A key topic is also cooperation in research, which is widely advocated in the field of the bioeconomy (Bertrand et al., 2016).

International cooperation is also very important, especially among developed and emerging economies. The example of EU–Russia is discussed in Sharova et al. (2016), while an example of reasoning on cooperation between US and China is illustrated in Snyder (2016).

International cooperation is directly related to research, but also involves financing and international specialization, as well as global and bilateral trade agreements and trade policies.

Box 7.14. Benefits of cooperation between the US and China

The example of potential cooperation between the US and China is illustrated in Snyder (2016), who highlights the potential for significant economic and environmental benefits from commercializing bio-based products and biofuels. The author proposes a pathway for developing collaborations in which the US deploys foreign capital, notably from China. Joining foreign capital (a limiting factor in biotech project development) with US technical experience could drive commercialization in the US and internationally, which could in turn be boosted by commercial cooperation between the US and China. The author argues that the US and China could possibly develop parallel supply chains without directly competing with each other, due to the potential territorial specialization and distribution of biomass supply chains.

Another topic for international collaboration is the management of emerging global bio-based markets. Trends in market globalization envisage higher trade for bio-based products and energy over time, unless regionalized scenarios prevail.

In addition, it is also possible that a greater use of biomass and a specialization towards areas with higher biomass availability and lower production costs would occur. This is already occurring in the case of wood for example.

Increased trade implies the problem of ensuring sustainability of production in areas where biomass is produced. On this topic, Sánchez et al. (2016) highlight the position of Europe, where demand for biomass is increasing year-by-year. The EU is already a net importer of biomass for bioenergy and dependency on imports could become even more relevant in the near future. Therefore, it is important to guarantee that biomass supply from outside the EU is sustainable, hence minimizing negative socioeconomic and environmental impacts.

Furthermore, stability and sufficient independence from importing goods are advocated. This started with food sovereignty issues, but is expanding to more general biomass self-sufficiency.

Markets and substitution potential can also have a regional dimension when the supply of bio-based raw materials is highly concentrated in some areas. For example, Paggiola et al. (2016) discuss this issue with respect to the case of citrus waste-derived limonene as a solvent for cleaning applications, finding that Brazil, Florida (US), Spain, India and South Africa would be able to meet their demand by deploying mainly their domestic limonene extraction potential and with important import–export dynamics. On the contrary, this is more difficult to reach on a global scale and would require major investments in infrastructure and distribution logistics.

More in general, logistical infrastructure through the network/chain, already seen on the local or regional scale, is also a focus of public intervention or market organization to support international market development. Attention to this topic is more focused on the supply side (Lamers et al., 2016).

The international dimension of the bioeconomy needs to be properly managed and the development of some regional markets could have the effect of helping other regions. For example, Mathews (2008) argues that investment in biorefinery in the south can only be undertaken if markets in the northern hemisphere are guaranteed. The authors advocate the need for a comprehensive global trade agreement in the field of biofuels in order to achieve this target smoothly and ensuring the connection between investment and market development.

7.7   Outlook

Instruments to stimulate the development of the bioeconomy, either already in place or being envisaged for the future, are myriad. Some characteristics of these instruments are presently rather clear. First, given that the bioeconomy is a sector in its inception that also has important values linked to public goods, it needs strong public intervention. Second, a mix of instruments is needed to promote consistent 'push' and 'pull' effects. Third, a strong focus is necessarily related to research and innovation. Finally, the public–private nature of collaboration gives at the same time a role to all actors (government, consumers, the private sector) and to the interplay between private-based instruments and public intervention. One clear message from the current experience in the bioeconomy is that if market forces are not enough to reach societal objectives, also public policy alone is insufficient to solve all problems.

How this can work in practice depends on different contexts and proper timing, which hints at the need for a flexible policy mix coupled with stable political conditions able at the same time to minimize uncertainties (Purkus *et al.*, 2017). In addition, the features of the bioeconomy require a re-thinking of policies to ensure that they are not excessively aimed at determining innovation direction, but rather that they support facilitation activities and the emergence of innovation initiatives (Klerkx *et al.*, 2010).

The relevance of the interplay between public and private initiatives, as well as among multiple actors, calls for the need for an improved understanding and more careful design/regulation of the role of intermediary organizations, including the need to develop alternative concepts on the performance of these organizations and institutions in spatial interactions (Kearnes, 2013).

One of the most innovative aspects of bioeconomy development is the new role of the connection between citizens/consumers and the supply side. This is due to the comprehensive nature of the bioeconomy and is leading to strong institutional innovation with respect to connecting needs to their fulfilment in the bioeconomy. This is even more important when taking into account the observation that emerged from the previous three chapters, namely that supply and demand are not a 'given', but rather evolve through interaction with each other. A systems perspective is often advocated to address these bioeconomy issues.

Box 7.15. Need for a system perspective thinking in the bioeconomy

---

Rex *et al.* (2017) explore opportunities for the use of feedstock from food waste for the production of bio-based chemicals in Sweden. They consider the following aspects: (i) feedstock feasibility; (ii) societal drivers and barriers for technology progress; and (iii) resource availability. They find that production seems possible from a technical feasibility and resource availability perspective. However, this use lacks institutional support, commitment on the part of the relevant actors and alignment. The authors argue that a holistic and interdisciplinary systems perspective contributes valuable insight when assessing prospects for these products.

A broad example is also provided by Ahn *et al.* (2012) with regard to the emerging bioeconomy in India. The study identified the critical role of government (e.g. in building infrastructure and accelerating bi-directional technology and capital flows), the need for public–private sector collaboration, as well as partnering between academia and government.

A systems view is also needed for the diagnosis of the innovation system, first understanding weaknesses and then identifying the package of required solutions. For example, the analysis by Giurca and Späth (2017) on lignocellulosic biorefinery systems in Germany highlights a number of internal and external system weaknesses (e.g. fragmented policies, underdeveloped market formation, technological immaturity and incomplete actor networks, among others).

For the future these characteristics will largely remain the same, though two variables may lead to very different results. One is the degree of market impact due to the depletion of fossil resources, which could create a strong push towards a market driven bioeconomy; the second is the prevailing nature of public goods-led pressures on the bioeconomy (e.g. due to climate change) that could lead to a bioeconomy that is increasingly led by environmental concerns and public incentive considerations, as well as variable perceptions on the part of consumers.

A potential major change for the future will be the reconsideration of basic principles of bioeconomy regulation when the technical basis and environmental effects will be better known. In this direction, it could be expected that risk perceptions will be less prominent and the use of the precautionary principle could become more pragmatic and less frequent. Yet this will only be possible in the presence of a higher degree of awareness and trust. In addition, several choices will depend on practical considerations related to the effectiveness and efficiency of different policy instruments.

Finally, this chapter has hopefully made even more evident that a sustainable and circular bioeconomy will not develop by itself, but will need the activation of a wide range of clever instruments and institutional solutions, of which accurate design will be one of the main challenges for future socioeconomic research.

References

Ahmed, S. and Broek, N.T. (2017) Food supply: Blockchain could boost food security. *Nature* 550(7674), 43. doi: 10.1038/550043e.

Ahn, M.J., Hajela, A. and Akbar, M. (2012) High technology in emerging markets. Building biotechnology clusters, capabilities and competitiveness in India. *Asia-Pacific Journal of Business Administration* 4(1), 23–41. doi: 10.1108/17574321211207953.

Allen, D.W. and Dean, L. (2002) *The Nature of the Farm.* Cambridge, Massachusetts, USA, Massachusetts Institute of Technology.

Auer, V., Zscheile, M., Engler, B., Haller, P., Hartig, J., Wehsener, J., Husmann, K., Erler, J., Thole, V., Schulz, T., Hesse, E., Rüther, N. and Himsel, A. (2016) Bioeconomy cluster: Resource efficient creation of value from beech wood to bio-based building materials. In WCTE 2016, World Conference on Timber Engineering, August 22-25, 2016, Vienna, Austria, CD-ROM Proceedings.

Beachy, R.N. (2014) Building political and financial support for science and technology for agriculture. *Philosophical Transactions of the Royal Society B: Biological Sciences* 369, 20120274. doi: 10.1098/rstb.2012.0274.

Bertrand, E., Pradel, M. and Dussap, C.-G. (2016) Economic and environmental aspects of biofuels. In Soccol, C.R. and Brar, D.K., Faulds, C, Pereira Ramos, L. (eds) *Green Energy and Technology.* New York, USA, Springer, pp. 525–555. doi: 10.1007/978-3-319-30205-8_22.

Brusselaers, J., Van Huylenbroeck, G. and Buysse, J. (2017) Green public procurement of certified wood: Spatial leverage effect and welfare implications. *Ecological Economics* 135, 91–102. doi: 10.1016/j.ecolecon.2017.01.012.

Chapotin, S.M. and Wolt, J.D. (2007) Genetically modified crops for the bioeconomy: Meeting public and regulatory expectations. *Transgenic Research* 16(6), 675–688. doi: 10.1007/s11248-007-9122-y.

Choumert, J. and Phélinas, P. (2017) Farmland rental prices in GM soybean areas of Argentina: Do contractual arrangements matter? *Journal of Development Studies* 53(8), 1286–1302. doi: 10.1080/00220388.2016.1241388.

Clark, L.F. and Phillips, P.W.B. (2013) Bioproduct approval regulation: An analysis of front-line governance complexity. *AgBioForum*, 16(2), 1–14.

Cota, M.M.G., De Paula Silva Gomes, J., Lunardi, L.M., De Andrade Gomes, C., Salles, A.M., Di Blasi, G. and Soares, E.E. (2016) Patent policies and intellectual property challenges in Brazil. *Industrial Biotechnology* 12(1), 58–61. doi: 10.1089/ind.2015.0020.

de Lorenzo, V. and Schmidt, M. (2018) Biological standards for the Knowledge-Based BioEconomy: What is at stake. *New Biotechnology* 40(Pt A) 170–180. doi: 10.1016/j.nbt.2017.05.001.

Dias, R.F. and De Carvalho, C.A.A. (2017) Bioeconomia no Brasil e no mundo: panorama atual e perspectivas. *Revista Virtual de Quimica* 9(1), 410–430. doi: 10.21577/1984-6835.20170023.

Dietrich, K., Dumont, M.-J., Del Rio, L.F. and Orsat, V. (2017) Producing PHAs in the bioeconomy — Towards a sustainable bioplastic. *Sustainable Production and Consumption*, 9. doi: 10.1016/j. spc.2016.09.001.

El-Chichakli, B., von Braun, J., Lang, C., Barben, D. and Philp, J. (2016) Five cornerstones of a global bio-economy. *Nature* 535, 221–223.

European Investment Bank (2017) *Study on Access-to-Finance. Conditions for Investments in Bio-Based Industries and the Blue Economy*. Kirchberg, Luxembourg, European Investment Bank.

Festel, G. (2015) Technology transfer models based on academic spin-offs within the industrial biotechnology sector. *International Journal of Innovation Management* 19(4), 1550031. doi: 10.1142/S1363919615500310.

Fevolden, A.M., Coenen, L., Hansen, T. and Klitkou, A. (2017) The role of trials and demonstration projects in the development of a sustainable bioeconomy. *Sustainability* 9(3), 419. doi: 10.3390/su9030419.

García, A.M., López-Moya, J.R. and Ramos, P. (2013) Key points in biotechnological patents to be exploited. *Recent Patents on Biotechnology* 7(2), 84–97.

German Bioeconomy Council (2015a) Bioeconomy policy (Part I) Synopsis and analysis of strategies in the G7. Available at biooekonomierat.de/fileadmin/international/Bioeconomy-Policy_Part-I.pdf (accessed on 16 November 2015).

German Bioeconomy Council (2015b) Bioeconomy policy (Part II) Synopsis of national strategies around the world. Available at http://gbs2018.com/fileadmin/gbs2018/Downloads/GBS_2018_Bioeconomy-Strategies-around-the_World_Part-III.pdf (accessed 26 May 2018).

Gibbs, D., Holloway, L., Gilna, B. and Morris, C. (2009) Genetic techniques for livestock breeding: Restructuring institutional relationships in agriculture. *Geoforum* 40(6), 1041–1049. doi: 10.1016/j.geoforum. 2009.07.011.

Giurca, A. and Späth, P. (2017) A forest-based bioeconomy for Germany? Strengths, weaknesses and policy options for lignocellulosic biorefineries. *Journal of Cleaner Production* 153, 51–62. doi: 10.1016/j. jclepro.2017.03.156.

Gkritza, K., Nlenanya, I. and Jiang, W. (2012) Bioeconomy and transportation infrastructure impacts: a case study of iowa's renewable energy, green energy and technology. In Gopalakrishnan, K., van Leeuwen, J. and Brown, R.C. (eds) *Sustainable Bioenergy and Bioproducts: Value Added Engineering Applications*. London, UK, Springer Ltd, pp. 173–188. doi: 10.1007/978-1-4471-2324-8_9.

Golden, J.S., Handfield, R.B., Daystar, J. and McConnell, T.E. (2015) An Economic Impact Analysis of the U.S. Biobased Products Industry: A Report to the Congress of the United States of America. Available at https://www.biopreferred.gov/BPResources/files/EconomicReport_6_12_2015.pdf (accessed on 11 March 2017).

Greene, C., Wechsler, S.J., Adalja, A. and Hanson, J. (2016) Economic Issues in the Coexistence of Organic, Genetically Engineered (GE), and Non-GE Crops. *Economic Information Bulletin No. (EIB-149)*. United States Department of Agriculture, Economic Research Service, 41 pp.

Hansen, T. and Coenen, L. (2017) Unpacking resource mobilisation by incumbents for biorefineries: the role of micro-level factors for technological innovation system weaknesses. *Technology Analysis and Strategic Management* 29(5), 500–513. doi: 10.1080/09537325.2016.1249838.

Heinonen, T., Pukkala, T., Mehtätalo, L., Asikainen, A., Kangas, J. and Peltola, H. (2017) Scenario analyses for the effects of harvesting intensity on development of forest resources, timber supply, carbon balance and biodiversity of Finnish forestry. *Forest Policy and Economics* 80, 80–98. doi: 10.1016/j. forpol.2017.03.011.

Jansson, M.S. (2016) Public procurement and biofuel sustainability criteria: Is there a link?. *Climate Law* 6(3–4), 296–313. doi: 10.1163/18786561-00603006.

Kearnes, M. (2013) Performing synthetic worlds: Situating the bioeconomy. *Science and Public Policy* 40(4), 453–465. doi: 10.1093/scipol/sct052.

Khaqqi, K.N., Sikorski, J.J., Hadinoto, K. and Kraft, M. (2018) Incorporating seller/buyer reputation-based system in blockchain-enabled emission trading application. *Applied Energy* 209, 8–19. doi: 10.1016/j. apenergy.2017.10.070.

Kircher, M., Breves, R., Taden, A. and Herzberg, D. (2016) How to capture the bioeconomy's industrial and regional potential through professional cluster management. *New Biotechnology* 40(Pt A) 119–128. doi: 10.1016/j.nbt.2017.05.007.

Klerkx, L., Aarts, N. and Leeuwis, C. (2010) Adaptive management in agricultural innovation systems: The interactions between innovation networks and their environment. *Agricultural Systems* 103(6), 390–400. doi: 10.1016/j.agsy.2010.03.012.

Korhonen, J., Zhang, Y. and Toppinen, A. (2016) Examining timberland ownership and control strategies in the global forest sector. *Forest Policy and Economics* 70, 39–46. doi: 10.1016/j.forpol.2016.05.015.

Lamers, P., Searcy, E. and Hess, J.R. (2016) Transition strategies: Resource mobilization through merchandisable feedstock intermediates. In Lamers, P., Searcy, E., Hess, J.R. and Stichnothe, H. (eds) *Developing the Global Bioeconomy: Technical, Market, and Environmental Lessons from Bioenergy.* London, UK, Academic Press, pp. 165–186. doi: 10.1016/B978-0-12-805165-8.00008-2.

Lee, D.-H. (2016) Levelized cost of energy and financial evaluation for biobutanol, algal biodiesel and biohydrogen during commercial development. *International Journal of Hydrogen Energy* 41(46), 21583–21599. doi: 10.1016/j.ijhydene.2016.07.242.

Lettner, M., Schöggl, J.-P. and Stern, T. (2017) Factors influencing the market diffusion of bio-based plastics: Results of four comparative scenario analyses. *Journal of Cleaner Production* 157, 289–298. doi: 10.1016/j.jclepro.2017.04.077.

Levin, N. and Leonelli, S. (2017) How does one 'open' science? Questions of value in biological research. *Science Technology and Human Values* 42(2), 313–323. doi: 10.1177/0162243916672071.

Lewandowski, I. (2015) Securing a sustainable biomass supply in a growing bioeconomy. *Global Food Security* 6, 34–42. doi: 10.1016/j.gfs.2015.10.001.

Lewandowski, I., Clifton-Brown, J., Trindade, L.M., Van Der Linden, G.C., Schwarz, K.-U., Müller-Sämann, K., Anisimov, A., Chen, C.-L., Dolstra, O., Donnison, I.S., Farrar, K., Fonteyne, S., Harding, G., Hastings, A., Huxley, L.M., Iqbal, Y., Khokhlov, N., Kiesel, A., Lootens, P., Meyer, H., Mos, M., Muylle, H., Nunn, C., Özgüven, M., Roldán-Ruiz, I., Schüle, H., Tarakanov, I., Der Weijde, T., Wagner, M., Xi, Q. and Kalinina, O. (2016) Progress on optimizing miscanthus biomass production for the European bioeconomy: Results of the EU FP7 project OPTIMISC. *Frontiers in Plant Science* 7(NOVEMBER20). doi: 10.3389/fpls.2016.01620.

Lochhead, K., Ghafghazi, S., Havlik, P., Forsell, N., Obersteiner, M., Bull, G. and Mabee, W. (2016) Price trends and volatility scenarios for designing forest sector transformation. *Energy Economics* 57, 184–191. doi: 10.1016/j.eneco.2016.05.001.

Lopes, M.S.G. (2015) Engineering biological systems toward a sustainable bioeconomy. *Journal of Industrial Microbiology and Biotechnology* 42(6), 813–838. doi: 10.1007/s10295-015-1606-9.

Majumdar, I., Campbell, K.A., Maure, J., Saleem, I., Halasz, J. and Mutton, J. (2017) Forest bioeconomy in Ontario-A policy discussion. *Forestry Chronicle* 93(1), 21–31. doi: 10.5558/tfc2017-007.

Małyska, A. and Jacobi, J. (2017) Plant breeding as the cornerstone of a sustainable bioeconomy. *New Biotechnology* 40(Pt A), 129–132. doi: 10.1016/j.nbt.2017.06.011.

Mathews, J.A. (2008) Biofuels, climate change and industrial development: Can the tropical South build 2000 biorefineries in the next decade? *Biofuels, Bioproducts and Biorefining* 2(2), 103–125. doi: 10.1002/bbb.63.

McCormick, K. and Kautto, N. (2013) The bioeconomy in Europe: An overview. *Sustainability* 5(6), 2589–2608. doi: 10.3390/su5062589.

McHenry, M.P. (2015) A rural bioeconomic strategy to redefine primary production systems within the Australian innovation system: Productivity, management, and impact of climate change In *Agriculture Management for Climate Change*. Hauppauge, New York, USA, Nova Science Publishers, Inc.

Mengal, P., Wubbolts, M., Zika, E., Ruiz, A., Brigitta, D., Pieniadz, A. and Black, S. (2017) Bio-based Industries Joint Undertaking: The catalyst for sustainable bio-based economic growth in Europe. *New Biotechnology* 40(Pt A), 31–39. doi: 10.1016/j.nbt.2017.06.002.

Mertens, A., Van Meensel, J., Mondelaers, K., Lauwers, L. and Buysse, J. (2016) Context matters—Using an agent-based model to investigate the influence of market context on the supply of local biomass for anaerobic digestion. *Bioenergy Research* 9(1), 132–145. doi: 10.1007/s12155-015-9668-0.

Millinger, M. and Thrän, D. (2016) Biomass price developments inhibit biofuel investments and research in Germany: The crucial future role of high yields. *Journal of Cleaner Production* 172, 1654–1663. doi: 10.1016/j.jclepro.2016.11.175.

Moschini, G. (2015) In medio stat virtus: Coexistence policies for GM and non-GM production in spatial equilibrium. *European Review of Agricultural Economics* 42(5), 851–874. doi: 10.1093/erae/jbu040.

Moschini, G., Lapan, H. and Kim, H. (2017) The renewable fuel standard in competitive equilibrium: Market and welfare effects. *American Journal of Agricultural Economics* 99(5), 1117–1142. doi: 10.1093/ajae/aax041.

Mukhtarov, F., Gerlak, A. and Pierce, R. (2017) Away from fossil-fuels and toward a bioeconomy: Knowledge versatility for public policy? *Environment and Planning C: Politics and Space* 35(6), 1010–1028. doi: 10.1177/0263774X16676273.

Mustalahti, I. (2017) The responsive bioeconomy: The need for inclusion of citizens and environmental capability in the forest based bioeconomy. *Journal of Cleaner Production* 172, 3781–3790. doi: 10.1016/j.jclepro.2017.06.132.

Myking, T., Walløe Tvedt, M. and Karlsson, B. (2017) Protection of forest genetic resources by intellectual property rights – exploring possibilities and conceivable conflicts. *Scandinavian Journal of Forest Research* 32(7), 598–606. doi: 10.1080/02827581.2017.1293151.

Neimark, B.D. (2016) Biofuel imaginaries: The emerging politics surrounding 'inclusive' private sector development in Madagascar. *Journal of Rural Studies* 45, pp. 146–156. doi: 10.1016/j.jrurstud.2016.03.012.

Nuhoff-Isakhanyan, G., Wubben, E.F.M. and Omta, S.W.F. (2016) Sustainability benefits and challenges of inter-organizational collaboration in bio-based business: A systematic literature review. *Sustainability* (Switzerland), 8(4). doi: 10.3390/su8040307.

OECD (2017) *The Next Production Revolution. Implications for Governments and Business*. Paris, France, The Organisation for Economic Co-operation and Development.

Overmann, J. (2015) Significance and future role of microbial resource centers. *Systematic and Applied Microbiology* 38(4), 258–265. doi: 10.1016/j.syapm.2015.02.008.

Paggiola, G., Van Stempvoort, S., Bustamante, J., Barbero, J.M.V., Hunt, A.J. and Clark, J.H. (2016) Can bio-based chemicals meet demand? Global and regional case-study around citrus waste-derived limonene as a solvent for cleaning applications. *Biofuels, Bioproducts and Biorefining* 10(6), 686–698. doi: 10.1002/bbb.1677.

Pendrous, R. (2017) Blockchain takes off in food and drink. *Food Manufacture* October.

Peuckert, J. and Quitzow, R. (2017) Acceptance of bio-based products in the business-to-business market and public procurement: Expert survey results. *Biofuels, Bioproducts and Biorefining* 11(1), 92–109. doi: 10.1002/bbb.1725.

Pfau, S.F., Hagens, J.E. and Dankbaar, B. (2017) Biogas between renewable energy and bio-economy policies—opportunities and constraints resulting from a dual role. *Energy, Sustainability and Society* 7(1), 17. doi: 10.1186/s13705-017-0120-5.

Philippidis, G., M'barek, R. and Ferrari, E. (2016) Is 'bio-based' activity a panacea for sustainable competitive growth? *Energies* 9(10), 806. doi: 10.3390/en9100806.

Philp, J. (2015) Balancing the bioeconomy: Supporting biofuels and bio-based materials in public policy. *Energy and Environmental Science* 8(11), 3063–3068. doi: 10.1039/c5ee01864a.

Philp, J. and Winickoff, D.E. (2017) Clusters in industrial biotechnology and bioeconomy: The roles of the public sector. *Trends in Biotechnology* 35(8), 682–686. doi: 10.1016/j.tibtech.2017.04.004.

Purkus, A., Gawel, E. and Thrän, D. (2017) Addressing uncertainty in decarbonisation policy mixes - Lessons learned from German and European bioenergy policy. *Energy Research and Social Science* 33, 82–94. doi: 10.1016/j.erss.2017.09.020.

Ray, S., Srivastava, S., Diwakar, S., Nair, B. and Özdemir, V. (2016) Delivering on the promise of bioeconomy in the developing world: Link it with social innovation and education. In Srivastava S. (ed) *Biomarker Discovery in the Developing World: Dissecting the Pipeline for Meeting the Challenges*.New Delhi, Springer, pp. 73–81. doi: 10.1007/978-81-322-2837-0_6.

Rex, E., Rosander, E., Røyne, F., Veide, A. and Ulmanen, J. (2017) A systems perspective on chemical production from mixed food waste: The case of bio-succinate in Sweden. *Resources, Conservation and Recycling* 125, 86–97. doi: 10.1016/j.resconrec.2017.05.012.

Sánchez, D., Del Campo, I., Janssen, R., Rutz, D., Fritsche, U., Iriarte, L., Fingerman, K., Diaz-Chávez, R., Junginger, M., Mai-Moulin, T., Visser, L., Elbersen, B., Nabuurs, G.J., Elbersen, W., Staritsky, I. and Pelkmans, L. (2016) Towards the development of a European bioenergy trade strategy for 2020 and beyond (Biotrade2020plus project). In *European Biomass Conference and Exhibition Proceedings*, pp. 1356-1363. doi: 10.5071/24thEUBCE2016-4CO.6.6.

Schütte, G. (2017) What kind of innovation policy does the bioeconomy need? *New Biotechnology* 40(Pt A), 82–86. doi: 10.1016/j.nbt.2017.04.003.

Scott, E. C. and Endres, A. B. (2014) Demanding supply: Re-envisioning the landlord-tenant relationship for optimized perennial energy crop production. *Duke Environmental Law and Policy Forum* 25(1), 101–129.

Sharova, I., Dzedzyulya, E., Abramycheva, I. and Lavrova, A. (2016) Instruments of international scientific cooperation in the field of bioeconomy as driver of emerging economies. The experience of the EU-Russia cooperation. *International Journal of Environmental and Science Education* 11(18), 11845–11853.

Sikkema, R., Dallemand, J.F., Matos, C.T., van der Velde, M. and San-Miguel-Ayanz, J. (2016) How can the ambitious goals for the EU's future bioeconomy be supported by sustainable and efficient wood sourcing practices? *Scandinavian Journal of Forest Research* 32, 551–558. doi: 10.1080/02827581. 2016.1240228.

Smart, R.D., Blum, M. and Wesseler, J. (2017) Trends in approval times for genetically engineered crops in the United States and the European Union. *Journal of Agricultural Economics* 68(1), 182–198. doi: 10.1111/1477-9552.12171.

Snyder, S.W. (2016) A path forward: Investment cooperation between the United States and China in a bioeconomy. In Snyder, S.W. (ed.) *Commercializing Biobased Products: Opportunities, Challenges, Benefits, and Risks*. London, UK, Royal Society of Chemistry Publishing, pp. 352–365. doi: 10.1039/9781782622444-00352.

Stadler, T. and Chauvet, J.-M. (2017) New innovative ecosystems in France to develop the bioeconomy. *New Biotechnology* 40(Pt A), 113–118. doi: 10.1016/j.nbt.2017.07.009.

Suryanarayanan, T.S., Gopalan, V., Sahal, D. and Sanyal, K. (2015) Establishing a national fungal genetic resource to enhance the bioeconomy. *Current Science* 109(6), 1033–1037.

Turley, D. (2015) Policies and strategies for delivering a sustainable bioeconomy: A European perspective. In Clark, J. and Deswarte, F. (eds) *Introduction to Chemicals from Biomass 2e*. Hoboken, New Jersey, USA, John Wiley and Sons, pp. 285–309. doi: 10.1002/9781118714478.ch8.

Twardowski, T., Aguilar, A., Puigdomenech, P., Linkiewicz, A., Sowa, S. and Zimny, T. (2017) European Union needs agro-bioeconomy. *Biotechnologia* 98(1), 73–78. doi: 10.5114/bta.2017.66619.

Vicol, M. (2017) Is contract farming an inclusive alternative to land grabbing? The case of potato contract farming in Maharashtra, India. *Geoforum* 85, 157–166. doi: 10.1016/j.geoforum.2017.07.012.

Wamisho Hossiso, K., De Laporte, A. and Ripplinger, D. (2017) The effects of contract mechanism design and risk preferences on biomass supply for ethanol production. *Agribusiness* 33(3), 339–357. doi: 10.1002/agr.21491.

Yang, X., Paulson, N.D. and Khanna, M. (2016) Optimal mix of vertical integration and contracting for energy crops: Effect of risk preferences and land quality. *Applied Economic Perspectives and Policy* 38(4), 632–654. doi: 10.1093/aepp/ppv029.

Zilberman, D., Kim, E., Kirschner, S., Kaplan, S. and Reeves, J. (2013) Technology and the future bioeconomy. *Agricultural Economics* 44(S1), 95–102. doi: 10.1111/agec.12054.

8

The Political Economy of the Bioeconomy, Regulation, Public Policy and Transition

_____

8.1 Introduction and Overview

As discussed in the previous chapters, different actors can have very different views of bioeconomy technology and strategies. Some bioeconomy technologies have shown totally different levels of acceptance in different areas of the world, whereas others have encountered strong opposition or are spreading steadily. As a result, the actual level of development of the bioeconomy is largely dependent on public acceptance and regulation, which is a result of how different interest groups interact in the political arena (Zilberman _et al._, 2015). Given the difficulties that some key bioeconomy technologies, most notably genetically modified organisms (GMOs), face in navigating their way through political legitimation and public acceptance, the political economy of the bioeconomy is emerging as one of the most focused economic research areas applied to the bioeconomy. A special issue of the _German Journal of Agricultural Economics_ was dedicated to this topic (Zilberman _et al._, 2015; Pannicke _et al._, 2015; Puttkammer and Grethe, 2015; Wesseler _et al._, 2015). The authors stress the importance of understanding and representing explicitly the interplay among and within different stakeholder groups and the usefulness of insights from behavioural economics, especially in connection with how to approach unknown futures.

The need to cast the development of the bioeconomy in a dynamic framework is also highlighted.

The use of a dynamic approach leads to transition analysis, understanding how major changes in technologies may become possible through different steps and enabling conditions (Pannicke _et al._, 2015).

The regulation of new technologies has been at the forefront of this analysis, as a way of ensuring the compatibility between profitability and economic, social and environmental objectives related to bioeconomy technologies (Wesseler _et al._, 2010).

The history of bioeconomy technologies is marked by contradictions; three among the most striking are highlighted in Herring and Paarlberg (2016):

- the contrast between the enthusiastic political acceptance of new crop technologies during the original Green Revolution of the 1960s and 1970s versus the considerable resistance to genetically modified (GM) crop technologies today;
- the contrast between today's highly precautionary regulation of biotechnology in farming versus the more permissive regulation of biotechnology in medicine; and
- the contrast between the greater political acceptance of agricultural biotechnology in some countries versus others, for some crops versus others, and for some crop traits versus others.

 © D. Viaggi 2018. _The Bioeconomy: Delivering Sustainable Green Growth_ (D. Viaggi)

These contradictions are rooted in the dynamics of interest groups in different contexts, which can be understood using a political economy approach.

This section is largely based on insights of two reviews: Zilberman *et al.* (2015) and Herring and Paarlberg (2016). Both (the former, in particular) provide access to the majority of relevant literature on the topic, in particular on agricultural biotechnologies.

8.2 The Positioning of Stakeholders

A number of interest groups are playing key roles in the bioeconomy field (Zilberman *et al.*, 2015). Zilberman *et al.* (2015) provide an analysis of actors and interest groups in the field of genetic engineering (GE) (Table 8.1).

These groups may have different positions in different countries. In addition, each group is internally heterogeneous, even within the same country. Dynamics, path dependency and regional identity may play a major role. For example, the chemical industry in agriculture is largely located in the EU and logically opposes genetically modified technology that results in reduced chemical usage, while the GM industry is more of a US-centred industry and is interested in expansion of these crops worldwide.

Moreover, each interest group may have an interest arising from a mix of different pros and cons; for example, consumers benefit from lower food prices due to GE, but may also have environmental concerns. Health issues linked to GE may also play either in favour (e.g. reduction of chemicals) or against (e.g. fears of health problems caused by GE) the diffusion of GM crops.

Farmers have advantages in terms of costs; however, these advantages may be short term, as price adaptations would partially offset them. This may also affect farmers heterogeneously depending on the relative advantages of GE technology on their farms and the timing of their adaptation with respect to competitors.

Science is also part of the game. Research institutions may support GE technologies due to potential funding opportunities; on the other hand, funding is also available for competing streams of research, such as organic production.

The diversity within different groups is a key aspect of the topic and has been widely illustrated in the previous chapters; this heterogeneity affects how different attitudes and efforts in support of, or in opposition to, GE emerge within each group. Zilberman *et al.* (2015) note that even within groups that may significantly benefit overall from GE technologies, or that are not affected, there may be subgroups that take strong opposing positions, including providing financial contributions to activism against GE technologies.

Table 8.1. Different interest groups in the field of genetic engineering (GE).

Interest group	Objectives	Attitude towards GE	Influence	Credibility
Policymakers	Being elected or promoted, gross national product, well-being of their region	Varies	High	Depends
Consumer	Low prices, health, ecosystems	Varies	Low	High
Farmers	Farm profit; work quality, health	Varies	High	Medium
Food retailers, wholesalers and distributors	Food sales, customer satisfaction, reduction of processing costs	Varies	High	High
Major biotech companies	Profits, market power	Positive	High	Low
Competing input supliers	Profits, market power	Negative	High	Low
Startup biotech companies	Sell to large companies	Positive	Low	Low
Academic institutions and scientists	Obtain funding	Varies	Medium	Varies
Environmental organizations	Power; affecting outcomes	Negative	High	Varies

Source: Modified from Zilberman *et al.* (2015)

Stakeholder involvement is at the core of several bioeconomy projects and bioeconomy discourses.

Examples of promising coordination facing different technological alternatives are available in the literature and prove the possibility of matching different views and interests (Höher *et al.*, 2016).

One case study investigating the positions with regard to GMOs in Poland found that 'scientists were rather optimistic about the use of GE in the economy, agricultural advisors were mostly against the use of GMOs in food and feed production, and farmers showed particular interest in the profitability and safety of specific GM products' (Malyska and Twardowski, 2014).

Another example of opposition comes from Argentina where Arancibia (2013) studied collective action between 1996 and 2011 in opposition to the adoption of GM crops and the growing use of glyphosate-based herbicides. The results demonstrate that diverse and innovative collective strategies as well as promoting the creation of new scientific data, can achieve some degree of influence on decisions regarding risk.

The complex interplay of resources and stakeholder positioning can be seen in examples of studies to prepare a policy agenda for countries that do not yet have a strategy. An example from Ireland is provided in Devaney and Henchion (2017).

An important issue is how visions and stakeholder positioning interact with options and opportunities given by new technologies. This topic is covered, for example, in connection with the topic of flexible crops.

Borras *et al.* (2016) highlight that in 'flex crops and commodities', the 'multiple-ness' and 'flexible-ness' are two distinct but connected features that modify the political economy of crops. In particular, they change the power relationships between landholders, agricultural labourers, crop exporters, processors and traders and so altering their patterns of production, circulation and consumption. In particular, flexible-ness intensifies market competition among producers (see Chapter 7, section 7.2.2.) and produces incentives to modify in land-tenure arrangements. The author distinguishes three interconnected types of flexing: real flexing, anticipated/speculative flexing and imagined flexing, which have many intersections and interactions. The political–

economic intersection includes technical aspects as well as 'flex narratives' and concepts by companies and public institutions to justify and promote flexible-ness. An extreme view in this direction is that of the envisaged future bioeconomy 'value web'. Kröger (2016) deals with flexible-ness for trees, observing that this feature has a large impact on power relations in the global political economy of forestry and the forest industry. Higher power is being gained by who is best able to flex or de-multiply, which allows better control of commodity webs and processing technology. While it allows diversification of process industry, in contrast it tends to push primary production towards specialization and monoculture.

8.3   Costs, Benefits and Uncertainty

An issue for understanding the political economy of the bioeconomy is the degree of uncertainty over the outcome of new technologies (see Chapters 3 to 6). To some extent, both costs and benefits are (highly?) uncertain.

In these circumstances, different interpretations of risk may proliferate and parties will engage in strategic behaviour to secure support for the framing of regulatory positions they prefer. This is, on the other hand, partially understandable due to the significant role played by risk in politics.

Noteworthy and evident benefits help to contrast perceived risks. Accordingly, cases with more evident benefits for consumers have been more easily accepted compared to those in which benefits are less straightforward and mostly for farmers or companies. One key reaction to uncertainty is the precautionary principle, which has permeated the related regulation to date (Herring and Paarlberg, 2016).

Herring and Paarlberg (2016) try to develop an extended framework for the political economy of GE, in which risk is a central component. The standard political economy model used in most reasonings is that of regulators determining policy by responding to a vector sum of diverse material interests weighted by influence. This needs modification in the field of biotechnology, since these new crops carry a special burden of proof because of the technology used from which there is a social lack of experience-based farmer or citizen opinion. This makes it difficult to judge actual costs and benefits from the material and objective features of the technologies, but also makes it difficult to identify interests and increases the complexity and difficulties in interest group formation and collective actions. As a result, risk-related opinion building may prevail while these technologies must seek case-by-case regulatory approval before any public use. In this context, even the precautionary principle is applied selectively and not in an always straightforward way, leading to opposite results in different cases with apparently similar interest settings. These effects have had the most evident case in recombinant DNA modifications used for GM crops, in which uncertainty due to lack of evidence has been recast in the discussion as a risk of negative impacts causing a slowing down of diffusion due mainly to non-market interventions (Herring and Paarlberg, 2016)

8.4 Dynamic Perspectives

Taking a dynamic perspective is key for the analysis of political economy of the bioeconomy. Zilberman *et al.* (2015) propose a dynamic framework in which the dynamics emerge at the interplay of different dynamic phenomena:

- The regulatory process of a technology is dynamic and includes multiple stages; these stages can be summarized as follows: (i) the introduction and initial assessment of the technology (and of its impacts); (ii) the policy debate, where each interest group aims to influence others; and (iii) the decision-making (which may itself be different depending on the institutional set-up).
- The different parties that participate in the development of and debate over GE have dynamic perspectives, in the sense of potentially changing positions over time, due to changes in internal or external conditions.
- The attitudes of voters towards a technology are evolving, again due to changes in the information available, external conditions, and so on. This change may be summarized as an evolution of goodwill (see Chapter 5).
- Evolution of technology is path dependent; in other words, once outcomes occur, they provide initial conditions for future choices and action. This is especially relevant in a heterogeneous world characterized by varied initial conditions in terms of institutions, regulations and past experiences (e.g. with food safety issues).

8.5 Local and Global Governance

Due to its need to consider simultaneously issues related to food, energy, resource security, environmental degradation, climate change, economic growth and development, the bioeconomy poses unique questions concerning governance and stakeholder's role. Devaney *et al.* (2017) explore the application of good governance principles to the bioeconomy in Europe, with particular attention to accountability and participation. The paper highlights that good governance requires input from a diverse range of stakeholders, besides public policies.

Different models of governance have been experienced. Two extremes are investigated by Bosman and Rotmans (2016): a rather centralized one (e.g. in the Netherlands) and a more distributed one (e.g. in Finland). The paper highlights that the latter seems to work better, but generalizations may be misleading.

One relevant issue for a new business is its 'political ecology' and what kind of motivations and strategies it has for selling it to society. A pertinent example is that of the change of focus of arguments before and after failures of several *Jathropa* projects in Kenya. In this case, the arguments shifted from a very broad approach (ability to solve many problems at the same time) to a more focused approach, in which single objectives are distinguished and the role of biofuel is better qualified (Hunsberger, 2016).

In the global arena, Zilberman *et al.* (2015) and Herring and Paarlberg (2016) highlight the existence of international areas of influence with regard to regulatory attitude. This is highlighted by the diffusion of GM, in which two main blocks can be recognized. One is driven by the US (as well as South American countries and the Philippines), which is more open to this technology and the other by the EU (and African countries), standing on the 'safe side' where regulation tends to block GM based on precautionary principles.

8.6   Vision and Imaginaries

The vision of the bioeconomy, as discussed in the introduction, is largely driven by conscious strategy building on the part of governments in many countries. In others, however, this is not the case. In these cases, the imaginary about the bioeconomy is building up in an unstructured way. An example is provided by Canada (Birch, 2016).

Hausknost et al. (2017) investigate the visions of the bioeconomy in Austria and classify them according to two main dimensions: one related to agro-ecology versus industrial bioeconomy and the other related to the pathway of economic development, contrasting sufficiency with capitalist growth. Their results highlight an association between the two dimensions, that is, the strategies and opinions tend to be classified either as combining agro-ecology and self-sufficiency or combining industrial biotechnology and capitalist growth. The authors emphasize the diverging visions of different interest groups and the existence of considerable gaps between official policy papers and visions supported by stakeholders. Besides underscoring the diverging visions of society arising from the bioeconomy view, these gaps show the highly political nature of the concept of the bioeconomy.

8.7   Transition Perspectives

According to Priefer et al. (2017), the bioeconomy can be seen as a comprehensive societal transition that includes a variety of sectors, actors and interests and that is related to far-reaching changes in today's production systems. The authors note that the objectives it pursues (reducing dependence on fossil fuels, mitigating climate change, ensuring global food security, increasing the industrial use of biogenic resources, etc.) are not generally contentious. Yet there is a strong controversy over the possible pathways for achieving these objectives. The main lines of criticism of the current bioeconomy concept are identified in the (too strong) focus on technology, 'the lack of consideration given to alternative implementation pathways, the insufficient differentiation of underlying sustainability requirements, and the inadequate participation of societal stakeholders' (Priefer et al., 2017).

As society is not able to predict the most suitable pathway, the authors suggest a transition pathway based on diversity.

Hansen and Bjørkhaug (2017) provide an analysis of visions of, and expectations from, transitions towards a bioeconomy in Norway, based on theories of transition and transition management and on an empirical survey data from biosector representatives. The results show that there are clear differences in motivation across sectors and this suggests the need for more cross-sectoral integration, heading towards a system shift to encourage a consistent transition.

Levidow (2015) illustrates the transition perspective related to the bioeconomy, in connection to the nascent 'corporate-environmental food regime', together with the 'sustainable intensification' (neoproductivism) agenda. Broadly speaking, the three concepts are going in the same neoliberal conceptual direction, that is, more resource-efficient methods are needed for increasing production to be able to meet the greater market demand for food, feed, fuel and bio-materials. Levidow (2015) contrasts these visions with the agro-ecology narrative which rather sees the problem as 'profit-driven agro-industrial monoculture systems making farmers dependent on external inputs, undermining their knowledge, and distancing consumers from agri-producers'. This narrative encourages new alliances linking social movements, farmers' knowledge, research, food sovereignty and citizens' initiatives, facilitated by civil society organizations. These contending narratives justify and represent different trajectories for an agro-food transition, and this should be made more explicit in the current debate (Levidow, 2015).

Transition perspectives and studies also point to issues such as contents, forms of possible transition, obstacles and chances. Monteleone (2015) illustrates this point by focusing on a 'paradigm shift' from industrial agriculture to agro-ecology. The author argues that new approaches should be multifunctional and diversified, local-based and self-sustained.

8.8   Political Economy and Value

Birch (2017), in an article called 'Rethinking Value in the Bio-economy: Finance, Assetization, and the Management of Value', highlights that the prevailing notion of the bioeconomy, seen as the articulation of capitalism and biotechnology, is built on central notions such as commodity production, commodification and materiality. These notions emphasize that it is possible to derive value from body parts, cellular tissues and biological processes. However, this perspective, largely built on technology and markets, lacks consideration of the role of political–economic actors, knowledge and practices that contribute to the creation of value. Birch (2017) identifies and analyses three key political–economic processes contributing to value creation: financialization, capitalization and assetization. The author concludes that value is not inherent in biological materialities, but rather managed as part of a series of valuation practices.

While more evident in the medical bioeconomy, for bioeconomy and high-technology industries in general, sociotechnical futures and new forms of promissory value are co-produced. This emphasizes the need to deepen the understanding of promissory economies and the role of expectations in building new markets (products, sectors) (Martin, 2015).

8.9   Outlook

The high degree of uncertainty and the diverse interests and political positions of stakeholders, together with the high degree of political commitment towards the bioeconomy, find in the political economy analysis a key for understanding its past and future development.

A key point of this discussion is understanding what is specific about the bioeconomy and how technology features and political interplay lead to decisions about a diverse pattern of transition pathways. Wield *et al.* (2013) argue that investment in biotechnology has produced disappointing results compared with expectations. This brings to asking: 'Does research on "life" bring different complexities and uncertainties that act as a barrier to the application of new biology in global health and agriculture?'. Though a good deal of research has been carried out on the social and economic aspects of biotechnologies, few systematic attempts to undertake interdisciplinary research and address these constraints exist. Wield *et al.* (2013) provide an analysis of contemporary and future understanding of the bioeconomy using a co-evolutionary and interactive approach to examine the extent to which it may be different from other technological transformations and can be the basis for the further understanding of this topic.

While, in the initial steps of different building blocks of the bioeconomy, the contrasting views of the sector, and of the emerging technologies, have been largely a challenge hindering the development of the bioeconomy, it is now clearer how the field of political economy can help substantially in making the potential benefits of the bioeconomy become real. Concerning the specific case of flex crops, for example, Borras *et al.* (2016) argue that 'a future research agenda should investigate questions about material bases, real-life changes, flex narratives and political mobilization'. This will help reveal in advance the potential obstacles and devise suitable steps for smooth and fast transition.

A point of attention for the future debate on the bioeconomy is the 'post-fact' nature that is currently developing as a feature of our society. This means the tendency to give less weight to factual considerations, and notions of truth, and rather to debate issues based on emotional factors and responses. This is not incidental and not marginal for the bioeconomy. On the contrary, the bioeconomy development is highly interwoven with these trends.

References

Arancibia, F. (2013) Challenging the bioeconomy: The dynamics of collective action in Argentina. *Technology in Society* 35(2), 79–92. doi: 10.1016/j.techsoc.2013.01.008.

Birch, K. (2016) Emergent imaginaries and fragmented policy frameworks in the Canadian bio-economy. *Sustainability*, 8(10), 1007. doi: 10.3390/su8101007.

Birch, K. (2017) Rethinking value in the bio-economy: Finance, assetization, and the management of value. *Science Technology and Human Values* 42(3), 325–339. doi: 10.1177/0162243916661633.

Borras, S.M., Franco, J.C., Isakson, S.R., Levidow, L. and Vervest, P. (2016) The rise of flex crops and commodities: implications for research. *Journal of Peasant Studies* 43(1), 617–628. doi: 10.1080/03066150.2015.1036417.

Bosman, R. and Rotmans, J. (2016) Transition governance towards a bioeconomy: A comparison of Finland and The Netherlands. *Sustainability* 8(10), 1017. doi: 10.3390/su8101017.

Devaney, L.A. and Henchion, M. (2017) If opportunity doesn't knock, build a door: Reflecting on a bioeconomy policy agenda for Ireland. *Economic and Social Review* 48(2), 207–229.

Devaney, L., Henchion, M. and Regan, A. (2017) Good governance in the bioeconomy. *EuroChoices* 16(2), 41–46. doi: 10.1111/1746-692X.12141.

Hansen, L. and Bjørkhaug, H. (2017) Visions and expectations for the Norwegian bioeconomy. *Sustainability* 9(3), 341. doi: 10.3390/su9030341.

Hausknost, D., Schriefl, E., Lauk, C. and Kalt, G. (2017) A transition to which bioeconomy? An exploration of diverging techno-political choices. *Sustainability* 9(4), 669. doi: 10.3390/su9040669.

Herring, R. and Paarlberg, R. (2016) The political economy of biotechnology. *Annual Review of Resource Economics* 8(1), 397–416. doi: 10.1146/annurev-resource-100815-095506.

Höher, M., Schwarzbauer, P., Menrad, K., Hedeler, B., Peer, M. and Stern, T. (2016) From wood to food: Approaching stakeholder integration in forest-based biorefinery development. *Bodenkultur* 67(3), 165–175. doi: 10.1515/boku-2016-0014.

Hunsberger, C. (2016) Explaining bioenergy: representations of jatropha in Kenya before and after disappointing results. *SpringerPlus* 5(1), 2000. doi: 10.1186/s40064-016-3687-y.

Kröger, M. (2016) The political economy of 'flex trees': a preliminary analysis. *Journal of Peasant Studies* 43(4), 886–909. doi: 10.1080/03066150.2016.1140646.

Levidow, L. (2015) European transitions towards a corporate-environmental food regime: Agroecological incorporation or contestation? *Journal of Rural Studies* 40, 76–89. doi: 10.1016/j.jrurstud.2015.06.001.

Malyska, A. and Twardowski, T. (2014) The influence of scientists, agricultural advisors, and farmers on innovative agrobiotechnology. *AgBioForum* 17(1), 84–89.

Martin, P. (2015) Commercialising neurofutures: Promissory economies, value creation and the making of a new industry. *BioSocieties* 10(4), 422–443. doi: 10.1057/biosoc.2014.40.

Monteleone, M. (2015) Reshaping agriculture toward a transition to a post-fossil bioeconomy. In Monteduro, M., Buongiorno, P., Di Benedetto, S. and Isoni, A. (eds) *Law and Agroecology: A Transdisciplinary Dialogue*. Berlin, Germany, Springer Verlag, pp. 359–376. doi: 10.1007/978-3-662-46617-9_18.

Pannicke, N., Gawel, E., Hagemann, N., Purkus, A. and Strunz, S. (2015) The political economy of fostering a wood-based bioeconomy in Germany. *German Journal of Agricultural Economics* 64(4), 224–243.

Priefer, C., Jörissen, J. and Frör, O. (2017) Pathways to shape the bioeconomy. *Resources* 6(1), 10. doi: 10.3390/resources6010010.

Puttkammer, J. and Grethe, H. (2015) The public debate on biofuels in Germany: Who drives the discourse? *German Journal of Agricultural Economics* 64(4), 263–273.

Wesseler, J., Banse, M. and Zilberman, D. (2015) Introduction special issue 'The political economy of the bioeconomy'. *German Journal of Agricultural Economics* 64(4), 209–211.

Wesseler, J., Spielman, D.J. and Demont, M. (2010) The future of governance in the global bioeconomy: Policy, regulation, and investment challenges for the biotechnology and bioenergy sectors. *AgBioForum* 13(4), 288–290.

Wield, D., Hanlin, R., Mittra, J. and Smith, J. (2013) Twenty-first century bioeconomy: Global challenges of biological knowledge for health and agriculture. *Science and Public Policy* 40(1), 17–24. doi: 10.1093/scipol/scs116.

Zilberman, D., Graff, G., Hochman, G. and Kaplan, S. (2015) The political economy of biotechnology. *German Journal of Agricultural Economics* 64(4), 212–223.

9

The Bioeconomy and Sustainable Development

9.1 Introduction and Overview

Several of the topics discussed in previous chapters are already providing insights into the potential role of the bioeconomy in society as a whole and its role in growth, development and sustainability. This chapter addresses the understanding of the net effect towards achieving society's sustainability goals, their measure and interpretation.

One way of framing this chapter is given by the United Nations Sustainable Development Goals (SDGs), which the bioeconomy may be expected to contribute towards (Schütte, 2017). This contribution is particularly focused on the SDGs related to food security and nutrition (Goal 2), healthy lives (Goal 3), water and sanitation (Goal 6), affordable and clean energy (Goal 7), sustainable consumption and production (Goal 12), climate change (Goal 13), oceans, seas and marine resources (Goal 14), and terrestrial ecosystems, forests, desertification, land degradation and biodiversity (Goal 15). It is also very relevant to sustainable economic growth (Goals 8 and 9) and sustainable cities (Goal 11) (El-Chichakli et al., 2016). The SDGs also set precise targets directly related to the bioeconomy, such as the production of 30% of materials from bio-based sources by 2030.

The association between sustainability and the bioeconomy has been well-established in the literature for several years (e.g. Smyth et al. (2011)), highlighting not only the potential positive contribution often emphasized in policy, but also the questions raised by new bioeconomy technologies in terms of environmental sustainability, human survival, social justice and human rights (Bryden et al., 2017). A strong reason for interest in sustainability assessment is that the bioeconomy is often depicted as a new development model focused on green knowledge-based solutions, especially promoted for areas affected by economic crisis or lagging behind in terms of development (Koukios et al., 2016). Clearly, the future role of the bioeconomy is linked to its ability to actually confront the major challenges of our world, in particular the finite nature of key resources and climate change (Roy, 2016).

Indeed, a specific area of concern is that of the natural resources needed for biomass production and hence for the sustainability of the bioeconomy itself. A widely explored field focuses on the ability of resources to meet biomass demand, and the innovation needed to match such demand. Among others, land is the primary resource under pressure, but increasing attention is being paid to alternative sources of biomass (e.g. seas). There is also an increasing focus on the need for water and fertilizer (especially non-easily renewable ones, such as phosphorus), which are the main other factors needed for vegetal production (Hertel et al., 2013; Rosegrant et al. 2013). This chapter discusses the specificities of each of these resources and their relationship with the bioeconomy. Though keeping in mind that all of them are somehow interconnected.

One problem with sustainability is that it implies taking a holistic view, which is very appealing, but which also makes it difficult to disentangle individual issues. It is to be highlighted that several of the sustainability dimensions listed above are connected to each other and some parts of the discussion can overlap. However, in the following discussion we examine each one separately to emphasize the individual specificities. Before doing so, however, section 9.2 discusses the overall concept of sustainability as applied to the bioeconomy, in order to set the scene for the sections that follow.

9.2   Bioeconomy and Sustainability

As mentioned, sustainability entails ensuring a satisfactory level of welfare today, without compromising the future welfare or that of future generations. This broad definition involves three main dimensions of sustainability (economic, environmental and social) and implicitly involves some understanding or assumptions about the future (sometimes even the long-term future).

The bioeconomy is being presented as a solution to various ecological and social challenges, covering all dimensions of sustainability and notably including, among others: climate change mitigation, environmental sustainability through cleaner production processes, economic growth, and the creation of new employment opportunities (Bennich and Belyazid, 2017).

Pfau *et al.* (2014) identify four main visions of the relationship between bioeconomy and sustainability: the first assumes sustainability as an inherent characteristic of the bioeconomy; the second is the expectation of prevailing benefits, but under certain conditions; the third is mainly critical, taking into account potential failures; and the fourth sees the bioeconomy as impacting negatively on sustainability.

This clearly hints at the potential opposite, or at least ambiguous, understanding of (or expectations concerning) the effects of the bioeconomy on sustainability. In addition, the literature highlights that bioeconomy strategies are often contradictory, resulting in different views on the actions needed to implement its potential (Bennich and Belyazid, 2017).

This may be also connected to the fact that the bioeconomy integrates different sectors and technologies with totally different perceptions from the point of view of society. In addition, it can be attributed to the role of location and context-specific variables in effecting both perception and the actual contribution to sustainability. A wide review by Nuhoff-Isakhanyan *et al.* (2016) shows that sustainability benefits (e.g. reduced emissions, reduced waste, economic synergies and socioeconomic development) depend on geographical proximity and complementarities, and that several of them (energy availability, lower emissions, improved socioeconomic life and poverty reduction) are essential, in particular in emerging economies.

A key difficulty with the topic of sustainability is that it is to a large extent related to socially constructed notions. The literature advocates the need for the political process to better define sustainability (Schepers, 2015). But sustainability is also recognized as being a continuous social learning path needing a deep transformation in society. In addition, problems encountered when dealing with sustainability are not only conceptual, but also related to the practical measurement of sustainability indicators.

As sustainability is a key selling point of the bioeconomy (Ramcilovic-Suominen and Pülzl, 2016), a relevant issue is to determine what kind of sustainability and sustainable development is envisaged in bioeconomy discourse. This is important because the idea of sustainability determines the direction of bioeconomy policies. It is even more relevant as the variety of interpretations and the (apparent) self-explanatory nature of the concept of sustainability encourage its wider use without necessarily taking strong action beyond the status quo. Ramcilovic-Suominen and Pülzl (2016) analyse the approaches to sustainability adopted by the bioeconomy policy debate in the EU, showing that they tend towards prudentially conservationist, utilitarian and instrumental approaches to sustainable development, as well as towards weak sustainability (Fig. 9.1).

The authors foresee an urgent need to define and broaden the scope of the bioeconomy, as well as to include environmental and social concerns more directly in order to prevent possible negative consequences from biomass production. The authors also argue for a broader definition

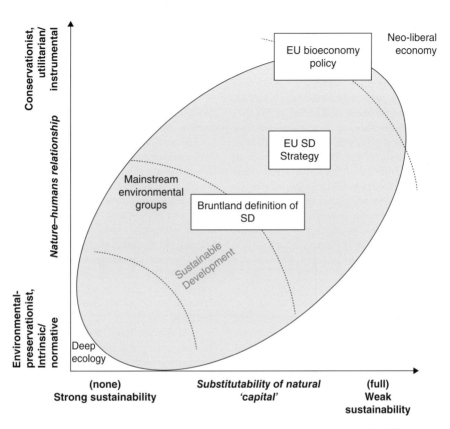

Fig. 9.1. Mapping the European Union's bioeconomy approach with regards to sustainability and sustainable development. Source: Ramcilovic-Suominen and Pülzl (2016)

of sustainability within the EU bioeconomy policy framework that includes issues such as social justice, fairness, equity, social and environmental safeguards, and local traditional knowledge. Bennich and Belyazid (2017) also confirm the view that current developments of the bioeconomy are oriented towards a weak sustainability paradigm and argue that a move towards the notion of 'strong sustainability' would require a better acknowledgment of limits to growth.

9.3 Economic Sustainability

9.3.1 Economic growth

As an economic activity, the bioeconomy is expected to contribute to the value-added of the economy and hence to the economy's vitality and consumers' welfare through economic growth.

In the short term, it is plausible that investment effort and bioeconomy development based on public funds and regulation will impose more costs to society compared to alternative economic strategies. On the contrary, in the long term, due to efficient management of key resources (biomass), valorization of underutilized resources (wastes, by-products) and the substitution of non-renewable resources (e.g. fossil fuels), the bioeconomy is expected to improve economic growth. This also means that, in the longer term, the ability to maintain low prices for goods by substitution of (increasingly costly) non-renewable resources will yield a better consumer surplus.

While this is true in general, some counterbalancing arguments may come from the fact that bioeconomy markets are built by displacement of other industries or goods, so that the net economic benefit is not as big as the total size of the bioeconomy as envisaged by naïve forecasts.

Economic studies are usually positive about the potential contribution to growth of the bioeconomy. In the literature, an economy-wide quantitative assessment covering the full diversity of the bioeconomy is lacking due to poor data availability for disaggregated bio-based activities. However, examples are available. Philippidis *et al.* (2016) use an EU-wide input–output model. In a baseline scenario, they project GDP growth from bio-based sectors in the order of +13% from 2013 to 2020 and +28.9% until 2030. Most studies are more narrower in geographical scope and tend to assess additional effects from bio-based development. For example, the benefits of establishing a tropical biorefinery in Queensland (Australia) is reported by O'Hara and Glenn (2017) estimating the benefit to be US$1.4 billion and more than 6 600 jobs.

Of interest is also the disaggregation of the different bioeconomy components. Orlović *et al.* (2014) employs social accounting matrixes (SAMs) for each EU27 member with a highly disaggregated treatment of 'bio-based' activities. The results identify six clusters of EU member countries with homogeneous bioeconomy structures. Milk and dairy are mostly 'key sectors' for the whole bioeconomy.

9.3.2 Development

Bioeconomy and economic development are linked not only to potential value creation by the bioeconomy, but also to the way it affects income distribution among regions and countries, and may have asymmetric implications between developed and developing economies or areas needing development. This is connected not only to single-state policies, but also international relationships and global economic integration. Several bioeconomy products, such as food and agricultural products, have seen a huge increase in trade and import–export in recent years. This is increasingly evident in explicit strategies towards building global value chains for key bio-based products and ensuring access to biomass sources. A discussion of trends and issues in this direction is provided by Kircher (2014).

Internal bioeconomy dichotomies are also emerging in relation to different pathways of development of the industry. In particular,

industrial bioeconomy is seen by some authors as potentially negative for development, in contrast to a small-scale inclusive bioeconomy. This is also linked to how the bioeconomy affects economic competitiveness.

Several examples in the literature emphasize the potential of the bioeconomy for less competitive areas. For example, Koukios *et al.* (2016) provide a manifesto, co-authored by scientists and engineers from Portugal, Spain, Italy, Greece and Cyprus, which envisages the bioeconomy as a new development model for the Southern European and Mediterranean areas, also able to boost development in other sectors of the economy. However, the ability of this pathway to materialize will depend on several other factors including research models, the broader policy environment, the (lack of) required skills and wider societal factors, largely linked to participation and consensus building (Koukios, 2015).

On the one hand, it may be expected that the bioeconomy will entail a redistribution of income and growth potential towards those countries/areas with a better endowment of biological resources, and background resources (land and water). Having a biomass production potential distributed on the land and seas, the bioeconomy may ensure a wider distribution of growth potential worldwide compared with the rather concentrated fossil resources.

On the other hand, the bioeconomy relies heavily on new technologies, the ownership of which may encourage a redistribution of economic benefits across countries and players in a much less equitable way, that is, allowing for a better positioning of those with stronger biotechnology research and bioscience know-how, enabling technologies, and leading companies in the field.

In several cases, the bioeconomy has been presented as a promising technology to turn the tables and allow developing economies to recover or surpass more developed ones (Li *et al.*, 2006). In the same way, it may be expected to modify the geography of relevant natural resources, and hence modify the balance among countries. Indeed, according to some authors, the bioeconomy has improved the economic potential of the developing world and modified social perspectives, with the relevant, yet not exclusive, contribution of medical applications. Developments in science and technology related to the bioeconomy have contributed to economic growth and the enhancement of the social status of emerging economies, such as Brazil, Russia, India, China, South Africa and Mexico (Kumar, 2015).

Other subject areas yield more contrasting results, notably in connection to the biofuel–food nexus. Mathews (2008) provides an optimistic view of this issue by arguing that biofuel production in the South can be increased through global market liberalization and that 'biofuels have the potential to bring together North and South in a new Biopact' benefiting both the North and South. Until now, biomass demand, driven in particular by bioenergy production, has mixed interpretations. Neimark (2016) report that biofuels are seen as prime drivers in global 'land grabbing', rainforest clearance and the dispossession of farmers. This is partly true also for food need increases and, in perspective, for bio-based products, especially when they compete with food production. In addition to visible land-grabbing issues, effects on markets may negatively affect the livelihoods of the poor. This is not simply an industrial-level issue, but can rather be traced back to consumer preferences and consumption patterns. Herring and Paarlberg (2016) emphasize the link between consumption and regulatory preferences of rich countries and the development opportunities of developing countries. A relevant topic is that of genetically modified organism (GMO) dissemination, and the way in which concerns by high-income consumers have hindered the growth of genetically modified (GM) technologies of potential interest for farmers and consumers in developing countries.

Wield (2013) presents research evidence focused on the relationships between changes in the bioeconomy and changes in the global economy. The author argues that the complex interactions in life science innovation between technology, markets, regulation and civil society could significantly impact on the global distribution of the industry. The paper also supports the interpretation that new forms of governance and regulation are necessary to strengthen industrial policies needed for emerging and developing countries. A key emerging issue is the ability of different countries or contexts to benefit from bioeconomy innovations. Several works highlight the link between development, biotechnology education and social innovation to support the contribution of the bioeconomy to development in developing countries (Ray et al., 2016).

Another issue is that of the bioeconomy and rural (and/or marginal area) development. In many areas this has been seen once again through the lens of the effects of bioenergy production. Johnson and Altman (2014) investigate how this affects the economies of rural regions. The authors identify three main determinants of the net effect of bioenergy development in rural areas: the costs of bioenergy and especially transportation costs, the regional energy balance and the utilization of waste streams. Low and Isserman (2009) describe the ethanol industry in the US and investigate local effects of ethanol plants. They note that an ethanol plant provides benefits comparable with those of a manufacturing plant, with 35 to 40 jobs. However, several characteristics of the industry may entail uncertainties and require a careful analysis to support local economic development decisions.

Rossi and Hinrichs (2011) argue that US government policies and programmes promoting agricultural bioenergy have focused on national goals rather than on local needs and experiences. Based on a survey of farming and non-farming participants in two switchgrass bioenergy projects, they discuss the locally perceived benefits of these projects. The results highlight that local and regional revitalization are somehow 'the benefit most desired and also least expected'. Scepticism about the bioeconomy is largely led by the observation of corporate control. The work highlights the need to go beyond the instrumental views of farmers and rural communities as technical providers of feedstocks. Rather, it is necessary to look at the local social and cultural context that determines responsiveness to stimuli towards bioeconomy development in rural areas.

This issue can be also interpreted in the context of the relationship between global technologies and local social structures, which emphasizes the role of local networks (Chen, 2015). Indigenous communities and their interplay with the bioeconomy and natural resource management is also emerging as an issue (Davidson-Hunt and Turner, 2012).

Box 9.1. The role of local communities

Davidson-Hunt *et al.* (2012) investigate the role and position of Inuit culture, commons, indigenous entrepreneurship and opportunity recognition for commercial caribou harvests in the bioeconomy in Canada. The authors found that Inuit people identified traditional uses of caribou for health care and explored its potential uses for biomedicines, nutraceuticals and functional foods. They discussed partnerships for development highlighting preferred partnering with Inuit and northern businesses and government. Moreover, Inuit people hold the opinion that the right to develop and sell caribou products and decisions about products and processes should be in the hands of the community ensuring food security was perceived as critical.

Neimark (2016) reports a case study in Madagascar that found that access to biofuel land and labour, and ultimately the success of *Jatropha* development, is dependent upon the inclusion of Malagasy in rural development projects.

9.4   Social Sustainability

9.4.1   Employment

Another expectation of the bioeconomy that is often put forward is that of job creation. This may apply especially in rural and poor areas connected to the distribution of biomass production and in relation to the occupation potential of the research and innovation industry and of new technologies. Although this is at the core of the discourse related to the bioeconomy and concerns in most countries with respect to employment in times of crisis, the literature provides different views about the realization of this potential.

Philippidis *et al.* (2016) show the forecasted change in contributions of the bio-based economy to employment in Europe. The main trend is increasing employment in non-agriculture bio-based sectors, in contrast to an overall reduction of employment in the bioeconomy due to the ongoing reduction in agricultural and food employment. Using a SAM approach, Orlović *et al.* (2014) found that, with the exception of forestry, fishing and wood activities, bio-based employment generation prospects are below non-bioeconomy activities. On the other hand, Heinonen *et al.* (2017) question the overly optimistic statements regarding growth potential in rural areas and employment potential in general. One major issue is that of displacement of jobs from one sector to the other. This would normally relax the estimates of the additional effects of the bioeconomy. However, this would not question the contribution of the bioeconomy as a whole as a guarantee of employment in more sustainable and lasting activities.

9.4.2   Poverty reduction

Poverty reduction relates to several of the issues discussed in the previous sections, coupled with additional issues that impact on income distribution and access to jobs. This topic is strictly connected to the distribution of benefits from the bioeconomy across countries and areas, as poverty reduction may have much higher relevance in developing economies. This may be true by combining the bioeconomy with its potential effects in terms of biomass needs and use of basic resources such as land and water, with implications for access to employment, income and food. On the consumer side, as long as connections with food prices are relevant, access to food for the poor is also a major topic. A specific issue relates to the distribution of bioeconomy value-added among the different players of the chain, especially smallholders and poor farmers in general, but also workers.

Box 9.2. Smallholders and genetically modified crops

A specific but telling issue concerns GM crops and farmer benefits. Smyth *et al.* (2011) report different ways in which benefits accrue to smallholders in developing economies, in particular from GM crops, including: (i) reduced use of pesticides, which yields health improvement for those directly involved in cultivation (farmers) as well as reduced costs for the public health system; (ii) higher yields from crops, which is especially relevant for self-consumption and farm profitability; and (iii) the possibility of accessing better finance and technical support when introducing new crops. At the same time, the (decreasing) net long-term benefit to farmers' income from GM crops and the alleged dependency on seed producers are reported as potential drawbacks of this technology.

9.4.3 Health

Health is an issue often mentioned in the bioeconomy documents, though the connection between the bioeconomy and health is not always straightforward, except for biotechnology applications related to medicine. For most of the non-medical bioeconomy, health issues may be related to positive or negative impacts of the new bioeconomy technologies.

It is well documented that biotechnology innovations aimed at reducing the use of chemicals, pesticides or air pollution contribute to a reduction in health problems.

In contrast, other technologies have, in principle, the potential to create health issues. GM technology has for a long time been alleged to pose potential negative health (as well as environmental) effects (Bennett *et al.*, 2013). New crops are still sometimes considered to potentially cause uncontrolled allergens and toxicity-associated impacts on human well-being (Sheppard *et al.* 2011a). Even the newest technologies, such as synthetic biology, are claimed to have potential occupational risks, especially in moving from the lab scale to the industry scale (Howard *et al.*, 2017).

In general, due to the strong attention to this issue, new technologies are characterized by improvements in terms of health at different stages of the chain. On the other hand, being a field in which uncertain negative effects may have a very high social importance, risk management actions linked to health are often high in the policy and political agenda.

9.4.4 Other issues

The topics above are so wide, complex and partially unexplored that their further understanding can bring in different directions and involve a number of complementary issues. One issue is that of social responsibility that is mentioned in various contributions (e.g. Shortall *et al.*, 2015). Social awareness may also be included, as there is a lot of emphasis on this in connection to responsibility. The same applies for a number of development-related issues, such as democracy, empowerment, and so on, which can be supported by a 'small size bioeconomy'

but are also key to ensuring the balanced development of the bioeconomy itself.

A problem common to all of these issues is the extent to which they are inherent to the bioeconomy or are instruments to ensure the development of the bioeconomy or if the bioeconomy itself can contribute to them as an ancillary objective of its development (e.g. to democracy). Even more, the literature on the bioeconomy brings into discussion human survival, social justice and human rights (Bryden *et al.*, 2017).

Box 9.3. Inclusive innovation issues

Cavicchi *et al.* (2017) analyse the case of sustainable bioenergy development in Emilia-Romagna (Italy) using an analytical framework called grounded innovation platform and the local governance approach. The authors find that biogas has not fostered inclusiveness and triple bottom line sustainability benefits, while forest-based bioenergy has. Their interpretation of results hints at the fact that sustainability of bio-based industries may be hindered by the assignment of an excessive role to industrial and national interests, at the expense of local actors, particularly municipalities. The results also highlight that policies focused on financial incentives may only lead to a land-acquisition rush, unforeseen local environmental effects and exacerbated conflicts.

Though answering these issues is beyond the scope of this book, it should also be acknowledged that the bioeconomy messages are non-neutral with respect to these concerns. For example, the bioeconomy makes it clear that technology cannot be detached from social change and this is somehow more evident for a wide-ranging view of the bioeconomy as a technology–society transition. Certainly, the bioeconomy encourages a comprehensive understanding of the management of biological resources that is pushing strongly for much greater individual awareness and embedding in social structures. For this reason, connections with politics are also explicitly addressed in this book (see Chapters 8 and 11).

9.5 Environmental and Resource Sustainability

9.5.1 Biological resources and ecosystems

The bioeconomy requires biomass and hence puts pressure on biological resources in terms of existing anthropic biomass production, but also on new sources of biomass from land (Mola-Yudego *et al.*, 2017) and seas (Børresen, 2017). In a way there is a 'qualitative pressure', linked to the availability and variety of genetic material, and a 'quantitative pressure', linked to the need for biomass.

In a more articulated way, different levels of concern can be identified. One is the pressure on populations of certain species, seen as biomass stocks. A second concerns the impact on biodiversity and agrobiodiversity, and related property right issues associated with genetic resources. The third is the pressure on complex ecosystems. Though, as discussed in Chapters 3 and 4, the bioeconomy can work in improving these resources, concerns arise when there are risks of depletion.

Concerning the first topic, the main management issue is related to the preservation of sufficient stocks and the potential for biomass harvesting without altering such stocks. Indeed, Heinonen *et al.* (2017) studied the case of Finnish forestry and found that a high harvesting intensity (potentially connected to high demand from the bioeconomy), would reduce stock volumes. On the other hand, Mola-Yudego *et al.* (2017) provide an example of potential from increased cultivation by investigating the production of wood biomass for energy in Europe, both from forest land and agriculture (fast growing plantations). The paper reviews the estimated potential for wood biomass production in 25 countries in Europe, based on previous models and estimations. Notably compared to the current 76 million m³ of potential wood biomass from the forests, 90 million m³ can be obtained by increasing the utilization of forest lands, while an additional 98 million m³ can be obtained from using 5% of the current agricultural land for fast growing plantations. Different countries would do better to specialize in one solution or another.

Improved biomass production and better use of land is possible through technology change at the agronomic level. For example, Jordan *et al.* (2016) argue that temperate-zone agricultural systems, currently organized for the production of summer-annual crops, can be diversified using fallow-season and perennial crops. This would also improve the sustainability of these agro-ecological systems through diversification. However, several of the candidate crops are not yet well adpated and suffer from poor marketability. The authors outline that these problems require a suitable approach, which they term as 'sustainable commercialization'. This approach, benefiting from a multi-stakeholder governance approach, would address innovation at least in three critical steps: germoplasm development, multifunctional agroecosystem management and development of end uses (together with linked supply chains and markets).

A useful concept for interpretation is that of human appropriation of net primary production (HANPP), an indicator used by system ecologists that quantifies human-induced changes on the productivity and harvest of biomass flows (Temper, 2016).

With regard to the second topic, a first area of concern is that the bioeconomy stimulates incentives that lead to a reduction in agrobiodiversity. This may happen when there are strong incentives towards a specific crop (variety) (e.g. for bioenergy). To a certain extent this issue also applies in the context of potential trade-offs between economic development and agrobiodiversity arising from the use of abandoned land for biomass production (Miyake *et al.*, 2016). The conservation of genetic diversity and of gene pools for native tree species is very relevant for forestry and is affected by the selection of practices encouraged by the development of the bioeconomy (Orlović *et al.*, 2014).

This also applies to different properties of GM plants. In a study by Krishna *et al.* (2016) on cotton in India, the authors investigate the hypothesis that GM cotton reduces incentives for farmers to diversify crops at the farm level, due to higher profitability and yield stabilization. In fact, they find little evidence of this as agrobiodiversity has basically remained stable in spite of the fact that more than 90% of cotton is now transgenic in India.

The potential of bio-invasions by energy crops is highlighted by Barney and DiTomaso (2011), in particular based on the broad climatic tolerance of such plants, which facilitates their adaptiveness and ability to thrive in a wide variety of climate contexts, similar to past cases of bio-invasion by plants introduced for agronomic purposes. The risk of bio-invasion, driven by the production of bioenergy, is reported and discussed in the US context by Barney (2014), in relation to the vast acreage of mainly exotic perennial grasses used for bioenergy. Risk assessment studies usually highlight a high risk of invasion, while other studies, based on empirical evidence on, for example, giant miscanthus, suggest a relatively low risk of invasion. However, the most important bioenergy crops are still poorly studied in the perspective of bio-invasion. Bio-invasion risk may also be connected to harvesting, transportation and feedstock storage.

The bioeconomy also interacts with biodiversity conservation and hence with the international provisions for the protection of biodiversity, including the distribution of the benefits arising from its use. The literature reports problems in applying protection protocols, especially to microorganisms, and there are claims that it will in fact hinder the development of the bioeconomy (Overmann and Scholz, 2017).

Impacts on ecosystems are even more difficult to assess and to manage due to the complexity of the dimensions involved, and to the claimed scale of this change, which leads some authors to state that 'the emerging bioeconomy is likely to result in the single largest reconfiguration of the agricultural landscape since the advent of industrial agriculture' (Raghu et al., 2011).

On the other hand, Marchetti et al. (2014) highlight that the bioeconomy is an important opportunity to stop the loss of biodiversity and the reduction of the provision of ecosystem services. With reference to forests, they provide an overview of bioeconomy-based natural resources management. They conclude by highlighting the following needs: (i) a deeper understanding of natural capital and its changes; (ii) the effective integration of ecological, socio-cultural, and economic dimensions into the management of natural resources; and (ii) an improvement of public participation in decision-making processes at the landscape scale. Other authors also highlight the inadequacy of seeking simplistic solutions.

Raghu et al. (2011) propose to approach the topic using the framework of 'biocomplexity', which would enable a multidimensional and cross-disciplinary consideration of biodiversity issues (linked in particular to biofuel production). They also advocate for a more inclusive public engagement process.

The increasingly global issue of biosecurity has emerged as a topic in the bioeconomy; a broad view of this issue is provided by Sheppard et al. (2011a) and more in detail by Sheppard et al. (2011b). Sheppard et al. (2011a), in the abstract of the introductory paper on a special issue on this topic, summarize these concerns, highlighting that novel crops can pose significant threats to human health, agriculture, biodiversity and natural ecosystem services. The ways in which this may happen include: (i) uncontrolled allergen and toxicity impacts on humans; (ii) abandoned trial plantings of uneconomic varieties; (iii) feral individuals or invasive species from economically viable plantations invading agricultural and natural landscapes; and (iv) pests, weeds and diseases carried by novel crops and potentially impacting on pest management systems of neighbouring crops.

9.5.2 CO_2 and climate change-related emissions

A clear focus of bioeconomy strategies is on combating climate change through a reduction in CO_2 and other green house gas (GHG) emissions. This occurs broadly by the substitution of fossil fuels. In more detail, it is dependent on the net effect of bioeconomy technologies on CO_2 emissions. This issue has taken on momentum with the increasing awareness of climate change, the consideration that a stronger climate policy is required and the awareness of its link with fluctuating fossil fuel prices and with the bioeconomy development (Tsiropoulos et al., 2017)

While some political discussion always seems to assume net benefits from the bioeconomy in the fight against climate change, the issue is controversial and depends on specific technology choices. Smyth et al. (2011) report on the results of several studies on bioenergy production, highlighting their ambiguity with respect to the issue of carbon emissions. Indeed, according to several papers, the assumption of carbon-neutral bioenergy production is not corroborated by evidence. Among the potential causes, the following needs to be highlighted:

- A first issue is the link with agricultural practices, as bioenergy from dedicated agricultural crops can raise concerns about the net fossil carbon balance of the required crop cultivation.
- There is also a potential link with other intensification pathways, such as higher use of water and fertilizers, with, in turn, effects on carbon balance or other GHG emissions through energy use.
- There are leakage effects on fossil fuel markets; so, while bioethanol can reduce CO_2 emissions directly, its expansion would reduce the price for fossil fuel, hence increasing consumption which partly counterbalances emission reduction.

A general issue is that carbon neutrality is very time sensitive, as it depends on the decomposition time of the biomass involved and hence on the time frame considered for calculation.

Another issue is the cost of reducing GHG emissions, in the sense that while bioenergy or bioeconomy technologies may indeed provide benefits, they are not necessarily the least costly solutions. This is also an issue internal to different bioenergy production (e.g. biogas versus biofuel).

Some of these issues apply to aspects other than bioenergy. In particular, indirect effects of GM crops on the reduction of fertilizers, pesticides and water needs contribute to a reduction in GHG emissions (Smyth *et al.*, 2011).

Projections for the future show the dependency of this issue on technology assumptions and development.

In addition, an integrated perspective of the different stages of the life of a product (using a life cycle assessment approach) is very relevant to achieving a true account of the impact of the bioeconomy on carbon emissions. For example, Glew *et al.* (2017) show that the disposal stages could double carbon footprint of a biomaterial, or reduce it to the point that it could claim to be zero carbon.

9.5.3 Land

Land is the basic resource on which the majority of primary production of biomass is performed. Land supports plant and animal growth as well as ecosystems. Land quality may vary a lot depending on soil features (e.g. steepness, soil depth and structure, etc.). Different quality means different productivity and production costs per unit of production. Land is also distributed in space; distance from key locations (e.g. selling points or processing plants) is a key feature of land.

Land is currently used for different activities, of which agriculture is only one. Moreover, due to the heterogeneity, while there is pressure to cultivate new land in some areas, there is land abandonment in others. The transformation of land into land usable for cultivation is often an arduous one, in part due to the conversion costs.

Access to land has historically be one of the key areas of conflict and legislation. Access to land is determined by state boundaries and by internal property rights on land. Private property is one of the options, but common properties also exist. Land grabbing is a recent phenomenon. In light of the predicted food needs and potential land scarcity, countries most likely to

---

Box 9.4. Technology and contribution to CO_2 emissions reduction

Tsiropoulos *et al.* (2017) assess pathways to 2030 for the energy system of the Netherlands using different scenarios combined with a cost-minimization model. The biomass use and CO_2 balance are highly influenced by technology development, fossil fuel prices and climate change targets. The study finds a clear trade-off between biomass consumption and reductions in CO_2 emissions. A high degree of technology development causes an increase in biomass consumption by 100–270 PJ and emission reductions of 8–20 Mt CO_2 compared to a low degree of technology development. In high technology development scenarios, additional emission reduction is primarily achieved by bioenergy and carbon capture and storage. The authors conclude that 'high technology development is a no-regrets option to realize deep emission reduction as it also ensures stable growth for the bioeconomy even under unfavourable conditions'.

have problems with land availability have started appropriating land in countries with higher availability of land and lower populations (or income), mostly through long-term contracts. Also, the globalization of the economy pushes for companies to produce biomass using lands from countries different than their origin. For these reasons, the land market is increasingly an international one.

Besides existing land rights, the development of the bioeconomy is bringing additional issues and segmentations in the land market:

- first, management of natural biological resources is creating areas protected by virtue of their ecosystems and subtracted to use for, for example, construction, and constrained use for agriculture (protected areas, natura 2000, etc.);
- segmentation of areas based on districts with specific agricultural features, e.g. based on diffuse organic farming (e.g. the concept of organic district), no fossil districts or, for example, countries that are GMO free; and
- segmentation of specific plots of land based on bio-features; for example, previous organic farming that allows the land to be ready to use for organic production without conversion time; plots organized to avoid contamination with GMO, etc.

The bioeconomy will affect land in different ways. The most straightforward effect is that, in order to meet increasing food, energy and material needs and in order to substitute fossil-derived energy and materials with bio-based materials, more land is required for primary production, other things being equal. Land can be regarded as a finite resource and the most direct limit to growth of the bioeconomy. On the other hand, the bioeconomy trends point in the direction of tackling these issues, through specifically:

- plant biotechnology, that can increase yields and hence reduce the amount of land needed for cultivation;
- the use of marginal land through appropriate crops, especially for non-food uses (e.g. energy) in order to minimize competition with food production;
- the reduction of losses and wastes during the process, in order to produce more final product with the same land;

- circular approaches to the re-use of wastes and by-products; and
- production processes not using land or using small amounts of land (e.g. vertical farming or sea resources).

Given pressure for land due to the development of the bioeconomy, Miyake et al. (2016) highlight the importance of the use of 'underutilized agricultural land', including 'abandoned agricultural land'. Land abandonment is a major issue in several parts of the EU, in particular in the Eastern EU. A focus of the bioeconomy on these areas would allow for agricultural reactivation on these lands for biomass production and create economic opportunities for rural regions; however, trade-offs with environmental effects, especially for biodiversity, are highly controversial.

Juerges and Hansjürgens (2016) discuss how relevant soil is for the sustainability of the bioeconomy and highlight that soil governance is an underdeveloped research field that deserves more attention in the context of environmental and natural resource governance. Indeed, farming can impact negatively on land. The most common effects are soil erosion and soil pollution/contamination and/or degradation of soil quality (e.g. through the reduction of organic matter). On the other hand, bioeconomy innovation can improve soil management. Indeed, Smyth et al. (2011) report evidence of reduction in land use needs thanks to GM crops with higher yields. In addition, they emphasize the connection between GM crops and minimum tillage from several US studies, which makes it possible to derive a link between biotechnologies and land quality conservation, including conservation agriculture.

O'Brien et al. (2015) show that in the EU-27, between 2000 and 2011, per capita cropland footprints have dropped by around 1% annually. However, the authors argue for the need to promote a further decrease in per capita cropland requirements (of around 2% annually) to make sure that consumption is kept within the safe operating space of planetary boundaries by 2030.

A major topic related to land use is the competition for land, in particular between food and energy production. This is by the way a multidimensional issue involving several aspects of sustainability and different scales.

Box 9.5. Trade-offs on land use at different scales

Trade-offs on land use are discussed at different scales in the literature. Using the agricultural sector model ESIM (European Simulation Model), Choi *et al.* (2016) provide a simulation of land available for energy crops in 2030 and 2050 in different scenarios. They find that the land available would be 5 (+4.6/–6.8) million ha in 2030 and 1.5 (+8.3/–5.1) million ha in 2050; overall, the released area is unlikely to exceed 10 million hectares in EU28 compared to the base year, but these figures are highly uncertain and depend primarily on the degree of market liberalization. The authors advocate that bioeconomy policies should enhance land productivity so that it reduces its land competition with food, fodder and energy biomass.

Mathews (2009) highlights the conflict over land use brought about by biofuels as opposed to food, feed and fibre production. At the time of the study, biofuels represented only 1% of agricultural production, but caused significant concerns including an alleged contribution to spikes in food prices in 2008. Contrary to the prevailing focus on negative impacts, the author focuses on options aimed at achieving benefits from land use dedicated to biofuels, including a shift from wasteful annual crops to low-input high-diversity perennial crops, increased sequestration of carbon in soils and improvements to conservative water management practices.

Jha *et al.* (2009) investigate the costs and environmental effects of switching a substantial amount of cultivated land to switchgrass in Iowa. Farm level economic results favour traditional crops, especially corn for ethanol, which means that farmers would convert to switchgrass only if substantial subsidies are provided. However, the shift would reduce sediment, nitrate and phosphorus loads substantially.

9.5.4   Seas and oceans

Seas and oceans are where the majority of fishing activity is carried out. Similar to land, due to competition for fish resources, seas have been the object of establishing of property rights on accessing them as well as on the use of fish stocks, such as quotas, or time for fishing. In addition, some species are protected from fishing.

This demonstrates that seas also have limits to exploitation that can be put under pressure by the development of the bioeconomy. However, unlike land a good deal of the potential of the seas is still unexplored and seas are at the forefront of bioeconomy development. For example, in the EU, there is now a specific sub-programme referred to as 'blue growth'.

9.5.5   Water

Water is a primary need for plants and animals. Most primary production is obtained in an open environment where plants and animals use water. In these conditions, production is largely affected by water availability. For this reason, a lack of water means reduced production. Besides average availability, agriculture is also strongly affected by uncertainty in water availability. For example, weather is not fully predictable and droughts may occur that affect crop production.

Water can also be a medium for production. This concerns, in particular, fisheries and aquaculture, including algae production. This may have implications for water quality and is affected by water availability, quality and quantity.

There are several links between water resources and the bioeconomy. A broad distinction involves three different issues:

● water abstraction (quantity);
● water pollution (quality); and
● other effects on water cycles.

From the quantity point of view, irrigation is a common practice to bring water to cultivated plants. Worldwide, irrigated areas have, on average, a productivity of twice that of non-irrigated areas. Clearly, bioeconomy development relying on increased biomass production will put more pressure on water resources, due to the likely expansion of production by enlarging irrigated areas.

Agriculture alone uses about 70% of water resources worldwide, but several processes in the downstream chain also use water (e.g. for washing, processing, etc.). The issue with water abstraction is that it has an effect on the environment by damaging rivers and related ecosystems, as well as subsidence due to groundwater abstraction, and the subtraction of water for alternative uses.

A second issue is that of water quality. The quality of water is affected by the release of polluting susbtances emitted from human activities. In relation to the bioeconomy, the most noteworthy negative effects on water are non-point pollution emissions, which, in the context of the bioeconomy, are mainly related to agriculture. The main topics concern fertilizers and pesticides, together with nitrogen from livestock production. At present, innovation is seeking to reduce the impact of these activities on water: reduction of pesticide toxicity and minimization of fertilizer use. The bioeconomy concept reinforces, in particular, the technology in this direction. For example, pest-resistant crops would help avoid or reduce the spraying of pesticides.

Also in the field of bioeconomy, a growing role is taken by intelligent technologies (e.g. digital technologies) minimizing the use of pesticides and fertilizers, such as precision agriculture. Finally, alternative technology pathways are emerging with low or non-use of chemical pesticides, such as organic production.

Water quality is also affected by downstream activities (food processing, etc.). However, these are usually point sources (i.e. pollution is emitted in one precise location and not widespread such as in the case of agriculture), so in most cases the potential pollution can be controlled and treated.

As for the third point, effects on the water cycle may come from activities linked to the bioeconomy without specific water-using practices. Simple changes in land use may affect water abstraction and regulation. An example is reforestation or forest management, which affect ecosystems and water cycles. This may be also related to soil erosion, flood management and provision of drinking water.

At the same time, efforts are being made to reduce water use (e.g. through drought-resistant crops). In addition, technologies to reduce water use, such as precision irrigation, drought detection and meteorological forecasts, are in the scope of current research. Smyth *et al.* (2011) report several studies related to GM crops with better efficiency in the use of fertilizers and with drought-resistant characteristics, which allow the potential to reduce the impact on water resources in terms of both quantity and quality and to allow better resilience when confronted with variability of water availability. In most

cases this yields a reduction of water use per unit of product. In addition there is evidence that switching from annual to perennial crops in connection to bioenergy crops would improve the water cycle (Mathews, 2009).

Water is one of the primary topics in resource re-use, and water re-use is a strong focus for current research, by way of connecting the quality and quantity aspects of water. It also directly affects different connecting stages of the bioeconomy itself, for example, with the use of food industry wastewater in agriculture.

9.5.6 Fertilizers

Plants also need mineral elements to grow. These are partly provided by soil and water, and partly artificially introduced to fields through fertilizers. Agriculture productivity is maintained and boosted through the use of fertilizers. Increased biomass production will put more pressure on fertilizer use.

The three main components of fertilizers are:

*   those derived from non-renewable resources (e.g. mineral phosphorus);
*   those derived from renewable resources (e.g. nitrogen); and
*   those derived from recycling or the re-use of resources (e.g. phosphorus from wastewater).

The first example is the one that is most urgently constraining biomass production. The second implies a problem of higher pressure on the supply of the production factor and hence on upstream resource use (e.g. energy). This is especially an issue for nitrogen fertilizers. The third case implies a balance between recycled and extracted resources, similar to those referred to in Chapter 7.

9.6 Outlook

Due to the complexity of the interlinkages between the bioeconomy, society and ecosystems, sustainable development is the 'natural' context in which the bioeconomy concept should be cast. This will be even more important in the future. The current literature emphasizes the potential of the bioeconomy in this direction, but, at the same time, also clarifies that this contribution is not necessarily positive or un-contentious, but largely depends on the ability of the socioeconomic systems to guide the bio-economy in suitable directions. In addition, the solutions leading in this direction are to be as-sessed case-by-case in the different country and regional contexts, especially in relation to re-source availability (i.e. water, biomass), compet-ing sectors (food) and needs for economic development. In fact, this observation requires a distinction between the political need to empha-size the potential of the bioeconomy, which is in fact a 'must' in the launch stage of a new vi-sion of technology and development, and the need to support the bioeconomy with a careful monitoring and evaluation process to allow for the exploitation of potential while avoiding failures and minimizing damages in the face of trade-offs.

The overall resource and environmental sustainability of the bioeconomy turns around a triple set of issues:

- its higher demand for biomass, with related pressures on resources and the environment;
- its ability to increase efficiency of resource use, with an expected reduction of such input;
- its substitution of fossil fuels and chemical sources.

The balance of these effects is not straightfor-ward and may vary from sector to sector, and de-pends on specific technology pathways. On the other hand, this is largely dependent on policy incentives, especially on the side of appropriate regulatory constraints to the use of environ-mental resources and policy coherence.

This is also emphasized if economic and so-cial sustainability are concerned, as this depends

not only on the role of technology in economic development, but also on its actual impact on employment and wealth distribution and access to opportunities (which is largely driven by insti-tutional settings and policy).

The way in which different aspects of sustainability entail trade-offs is visible at dif-ferent scales. An example is the positive reac-tion of stock exchanges to the decision by the US president to exit the Paris Agreement on climate change. Trade-offs are still there, and easy win–win solutions are not the norm, though are often claimed in the field of the bioeconomy. For example, in the biofuel ver-sus food debate, Mathews (2009) provides a discussion of trade-offs and their manage-ment, as well as how a lack of information on specific topics can induce excessive reactions. The author also emphasizes the role of in-novation in solving, at least partially, these conflicts.

Indeed, one key ways of managing the potential conflicts and benefits related to the bioeconomy is through proactive innovation management. Bryden *et al.* (2017) emphasize the roles that institutions play regarding innov-ation in the bioeconomy as well as in the adop-tion of inclusive innovation processes such as those 'improving the lives of the most needy'.

A weakness in making sustainable develop-ment real is the absence of quality data and ap-propriate monitoring systems. This is a common topic in any field related to sustainability and is hence not specific to the bioeconomy. However, there are bioeconomy specificities. O'Brien *et al.* (2017) argue that a systemic monitoring system, capable of connecting human–environment interactions and multiple scales of analysis in a dynamic way, is needed to ensure that the bio-economy transition meets its goals, like the SDGs. More details on this topic are provided in Chapter 10.

References

Barney, J.N. (2014) Bioenergy and invasive plants: Quantifying and mitigating future risks. *Invasive Plant Science and Management* 7(2), 199–209. doi: 10.1614/IPSM-D-13-00060.1.
Barney, J.N. and DiTomaso, J.M. (2011) Global climate niche estimates for bioenergy crops and invasive species of agronomic origin: Potential problems and opportunities. *PLoS ONE* 6(3), e17222. doi: 10.1371/journal.pone.0017222.

Bennett, A.B., Chi-Ham, C., Barrows, G., Sexton, S. and Zilberman, D. (2013) Agricultural biotechnology: Economics, environment, ethics, and the future. *Annual Review of Environment and Resources* 38, 249–279. doi: 10.1146/annurev-environ-050912-124612.

Bennich, T. and Belyazid, S. (2017) The route to sustainability-prospects and challenges of the bio-based economy. *Sustainability* 9(6), 887. doi: 10.3390/su9060887.

Børresen, T. (2017) Blue bioeconomy. *Journal of Aquatic Food Product Technology* 26(2), 139. doi: 10.1080/10498850.2017.1287477.

Bryden, J., Gezelius, S.S., Refsgaard, K. and Sutz, J. (2017) Inclusive innovation in the bioeconomy: Concepts and directions for research. *Innovation and Development* 7(1), 1–16. doi: 10.1080/2157930X.2017.1281209.

Cavicchi, B., Palmieri, S. and Odaldi, M. (2017) The influence of local governance: Effects on the sustainability of bioenergy innovation. *Sustainability* 9(3), 406. doi: 10.3390/su9030406.

Chen, T.-W. (2015) Global technology and local society: Developing a Taiwanese and Korean bioeconomy through the vaccine industry. *East Asian Science Technology and Society* 9(2), 167–186. doi: 10.1215/18752160-2876770.

Choi, H.S., Entenmann, S. and Grethe, H. (2016) Sensitivity analysis on land availability for energy crops in the EU by 2050. In *European Biomass Conference and Exhibition Proceedings*, pp. 277–282. doi: 10.5071/24thEUBCE2016-1BV.4.99

Davidson-Hunt, I.J. and Turner, K.L. (2012) Indigenous communities, the bioeconomy and natural resource development. *Journal of Enterprising Communities: People and Places in the Global Economy* 6(3). doi: 10.1108/jec.2012.32906caa.001.

Davidson-Hunt, I.J., Turner, K.L., Meis Mason, A.H., Anderson, R.B. and Dana, L.-P. (2012) Inuit culture and opportunity recognition for commercial caribou harvests in the bio economy. *Journal of Enterprising Communities: People and Places in the Global Economy* 6(3), 194–212. doi: 10.1108/17506201211258388.

El-Chichakli, B., von Braun, J., Lang, C., Barben, D. and Philp, J. (2016) Five cornerstones of a global bioeconomy. *Nature* 535, 221–223.

Glew, D., Stringer, L.C., Acquaye, A. and McQueen-Mason, S. (2017) Evaluating the potential for harmonized prediction and comparison of disposal-stage greenhouse gas emissions for biomaterial products. *Journal of Industrial Ecology* 21(1), 101–115. doi: 10.1111/jiec.12421.

Heinonen, T., Pukkala, T., Mehtätalo, L., Asikainen, A., Kangas, J. and Peltola, H. (2017) Scenario analyses for the effects of harvesting intensity on development of forest resources, timber supply, carbon balance and biodiversity of Finnish forestry. *Forest Policy and Economics* 80, 80–98. doi: 10.1016/j.forpol.2017.03.011.

Herring, R. and Paarlberg, R. (2016) The political economy of biotechnology. *Annual Review of Resource Economics* 8(1), 397–416. doi: 10.1146/annurev-resource-100815-095506.

Hertel, T., Steinbuks, J. and Baldos, U. (2013) Competition for land in the global bioeconomy. *Agricultural Economics* 44(S1), 129–138. doi: 10.1111/agec.12057.

Howard, J., Murashov, V. and Schulte, P. (2017) Synthetic biology and occupational risk. *Journal of Occupational and Environmental Hygiene* 14(3), 224–236. doi: 10.1080/15459624.2016.1237031.

Jha, M., Babcock, B.A., Gassman, P.W. and Kling, C.L. (2009) Economic and environmental impacts of alternative energy crops. *International Agricultural Engineering Journal* 18(3–4), 15–23.

Johnson, T.G. and Altman, I. (2014) Rural development opportunities in the bioeconomy. *Biomass and Bioenergy* 63, 341–344. doi: 10.1016/j.biombioe.2014.01.028.

Jordan, N.R., Dorn, K., Runck, B., Ewing, P., Williams, A., Anderson, K.A., Felice, L., Haralson, K., Goplen, J., Altendorf, K., Fernandez, A., Phippen, W., Sedbrook, J., Marks, M., Wolf, K., Wyse, D. and Johnson, G. (2016) Sustainable commercialization of new crops for the agricultural bioeconomy. *Elementa* 4, 000081. doi: 10.12952/journal.elementa.000081.

Juerges, N. and Hansjürgens, B. (2016) Soil governance in the transition towards a sustainable bioeconomy – A review. *Journal of Cleaner Production* 170. doi: 10.1016/j.jclepro.2016.10.143.

Kircher, M. (2014) The emerging bioeconomy: Industrial drivers, global impact, and international strategies. *Industrial Biotechnology* 10(1), 11–18. doi: 10.1089/ind.2014.1500.

Koukios, E.G. (2015) Knowledge-based greening as a new bioeconomy strategy for development: Agroecological Utopia or revolution? In Monteduro, M., Buongiorno, P., Di Benedetto, S. and Isoni, A. (eds) *Law and Agroecology: A Transdisciplinary Dialogue*. Berlin, Germany, Springer Verlag, pp. 439–450. doi: 10.1007/978-3-662-46617-9_23.

Koukios, E., Monteleone, M., Texeira Carrondo, M.J., Charalambous, A., Girio, F., Hernández, E.L., Mannelli, S., Parajó, J.C., Polycarpou, P. and Zabaniotou, A. (2016) Targeting sustainable bioeconomy: A new development strategy for Southern European countries. The Manifesto of the European Mezzogiorno. *Journal of Cleaner Production* doi: 10.1016/j.jclepro.2017.05.020.

Krishna, V., Qaim, M. and Zilberman, D. (2016) Transgenic crops, production risk and agrobiodiversity. *European Review of Agricultural Economics* 43(1), 137–164. doi: 10.1093/erae/jbv012.

Kumar, D. (2015) Socioeconomic outcomes of genomics in the developing world. In Kumar, D. and Chadwick, R. (eds) *Genomics and Society: Ethical, Legal, Cultural and Socioeconomic Implications*. London, Elsevier, pp. 239–258. doi: 10.1016/B978-0-12-420195-8.00012-4.

Li, Q., Zhao, Q., Hu, Y. and Wang, H. (2006) Biotechnology and bioeconomy in China. *Biotechnology Journal* 1(11), 1189–1194. doi: 10.1002/biot.200600133.

Low, S.A. and Isserman, A.M. (2009) Ethanol and the local economy: Industry trends, location factors, economic impacts, and risks. *Economic Development Quarterly* 23(1), 71–88. doi: 10.1177/0891242408329485.

Marchetti, M., Vizzarri, M., Lasserre, B., Sallustio, L. and Tavone, A. (2014) Natural capital and bioeconomy: Challenges and opportunities for forestry. *Annals of Silvicultural Research* 38(2), 62–73. doi: 10.12899/ASR-1013.

Mathews, J.A. (2008) Biofuels, climate change and industrial development: Can the tropical South build 2000 biorefineries in the next decade? *Biofuels, Bioproducts and Biorefining* 2(2), 103–125. doi: 10.1002/bbb.63.

Mathews, J.A. (2009) From the petroeconomy to the bioeconomy: Integrating bioenergy production with agricultural demands. *Biofuels, Bioproducts and Biorefining* 3(6), 613–632. doi: 10.1002/bbb.181.

Miyake, S., Mizgajski, J.T., Bargiel, D., Wowra, K. and Schebek, L. (2016) Biodiversity and socio-economic implications of the use of abandoned agricultural land for future biomass production in Central and Eastern Europe (CEE). In *European Biomass Conference and Exhibition Proceedings,* pp. 1422–1430. doi: 10.5071/24thEUBCE2016-4DO.8.4.

Mola-Yudego, B., Arevalo, J., Díaz-Yáñez, O., Dimitriou, I., Freshwater, E., Haapala, A., Khanam, T. and Selkimäki, M. (2017) Reviewing wood biomass potentials for energy in Europe: the role of forests and fast-growing plantations. *Biofuels* 8(4), 401–410. doi: 10.1080/17597269.2016.1271627.

Neimark, B.D. (2016) Biofuel imaginaries: The emerging politics surrounding 'inclusive' private sector development in Madagascar. *Journal of Rural Studies* 45, 146–156. doi: 10.1016/j.jrurstud.2016.03.012.

Nuhoff-Isakhanyan, G., Wubben, E.F.M. and Omta, S.W.F. (2016) Sustainability benefits and challenges of inter-organizational collaboration in bio-based business: A systematic literature review. *Sustainability* 8(4), 307. doi: 10.3390/su8040307.

O'Brien, M., Schütz, H. and Bringezu, S. (2015) The land footprint of the EU bioeconomy: Monitoring tools, gaps and needs. *Land Use Policy* 47, 235–246. doi: 10.1016/j.landusepol.2015.04.012.

O'Brien, M., Wechsler, D., Bringezu, S. and Schaldach, R. (2017) Toward a systemic monitoring of the European bioeconomy: Gaps, needs and the integration of sustainability indicators and targets for global land use. *Land Use Policy* 66, 162–171. doi: 10.1016/j.landusepol.2017.04.047.

O'Hara, I.M. and Glenn, D. (2017) The economic case for bioeconomy development in Australia. *Industrial Biotechnology* 13(2), 65–68. doi: 10.1089/ind.2016.29046.imo.

Orlović, S., Ivanković, M., Andonoski, V., Stojnić, S. and Isajev, V. (2014) Forest genetic resources to support global bioeconomy. *Annals of Silvicultural Research* 38(2), 51–61. doi: 10.12899/ASR-942.

Overmann, J. and Scholz, A.H. (2017) Microbiological research under the Nagoya Protocol: facts and fiction. *Trends in Microbiology* 25(2), 85–88. doi: 10.1016/j.tim.2016.11.001.

Pfau, S., Hagens, J., Dankbaar, B. and Smits, A. (2014) Visions of sustainability in bioeconomy research. *Sustainability* 6(3), 1222–1249. doi: 10.3390/su6031222.

Philippidis, G., M'barek, R. and Ferrari, E. (2016) Is 'bio-based' activity a panacea for sustainable competitive growth? *Energies* 9(10), 806. doi: 10.3390/en9100806.

Raghu, S., Spencer, J.L., Davis, A.S. and Wiedenmann, R.N. (2011) Ecological considerations in the sustainable development of terrestrial biofuel crops. *Current Opinion in Environmental Sustainability* 3(1–2), 15–23. doi: 10.1016/j.cosust.2010.11.005.

Ramcilovic-Suominen, S. and Pülzl, H. (2016) Sustainable development – A 'selling point' of the emerging EU bioeconomy policy framework? *Journal of Cleaner Production* 172. doi: 10.1016/j.jclepro.2016.12.157.

Ray, S., Srivastava, S., Diwakar, S., Nair, B. and Özdemir, V. (2016) Delivering on the promise of bioeconomy in the developing world: Link it with social innovation and education. In Srivastava S. (ed)

Biomarker Discovery in the Developing World: Dissecting the Pipeline for Meeting the Challenges. New Delhi, Springer, pp. 73–81. doi: 10.1007/978-81-322-2837-0_6.

Rosegrant, M.W., Ringler, C., Zhu, T., Tokgoz, S. and Bhandary, P. (2013) Water and food in the bioeconomy: challenges and opportunities for development. *Agricultural Economics* 44(S1), 139–150. doi: 10.1111/agec.12058.

Rossi, A.M. and Hinrichs, C.C. (2011) Hope and skepticism: Farmer and local community views on the socio-economic benefits of agricultural bioenergy. *Biomass and Bioenergy* 35(4), 1418–1428. doi: 10.1016/j.biombioe.2010.08.036.

Roy, C. (2016) Les potentiels de la bioéconomie: De la photosynthèse à l'industrie, de l'innovation aux Marchés. *Futuribles: Analyse et Prospective* No. 410.

Schepers, S. (2015) Managing the politics of innovation and sustainability. *Journal of Public Affairs* 15(1), 91–100. doi: 10.1002/pa.1521.

Schütte, G. (2017) What kind of innovation policy does the bioeconomy need? *New Biotechnology* 40(Pt A), 82–86. doi: 10.1016/j.nbt.2017.04.003.

Sheppard, A.W., Gillespie, I., Hirsch, M. and Begley, C. (2011a) Biosecurity and sustainability within the growing global bioeconomy. *Current Opinion in Environmental Sustainability* 3(1–2), 4–10. doi: 10.1016/j.cosust.2010.12.011.

Sheppard, A.W., Raghu, S., Begley, C., Genovesi, P., De Barro, P., Tasker, A. and Roberts, B. (2011b) Biosecurity as an integral part of the new bioeconomy: a path to a more sustainable future. *Current Opinion in Environmental Sustainability* 3(1–2), 105–111. doi: 10.1016/j.cosust.2010.12.012.

Shortall, O.K., Raman, S. and Millar, K. (2015) Are plants the new oil? Responsible innovation, biorefining and multipurpose agriculture. *Energy Policy* 86, 360–368. doi: 10.1016/j.enpol.2015.07.011.

Smyth, S.J., Aerni, P., Castle, D., Demont, M., Falck-Zepeda, J.B., Paarlberg, R., Phillips, P.W., Pray, C.E., Savastano, S., Wesseler, J.H.H. and Zilberman, D. (2011) Sustainability and the bioeconomy: Synthesis of Key themes from the 15th ICABR conference. *AgBioForum* 14(3), 108–186.

Temper, L. (2016) Who gets the HANPP (Human Appropriation of Net Primary Production)? Biomass distribution and the bio-economy in the Tana Delta, Kenya. *Journal of Political Ecology* 23(1), 328–491.

Tsiropoulos, I., Hoefnagels, R., van den Broek, M., Patel, M.K. and Faaij, A.P.C. (2017) The role of bioenergy and biochemicals in CO_2 mitigation through the energy system – a scenario analysis for the Netherlands. *Global Change Biology Bioenergy* 9(9), 1489–1509. doi: 10.1111/gcbb.12447.

Wield, D. (2013) Bioeconomy and the global economy: Industrial policies and bio-innovation. *Technology Analysis and Strategic Management* 25(10), 1209–1221. doi: 10.1080/09537325.2013.843664.

10

Impact Evaluation and Management Tools

———————————

10.1 Introduction and Overview

The final part of the 20th century has been characterized by the growing role of evaluation tools in policy making and private decisions, in all fields of activity. To some extent, this has been connected to the growing number of concerns in the area of environment and health. In addition, the push for participation in public decision making by a large number of stakeholders, and also the growing need for companies to speak to myriad interested parties, including consumers and environmental organizations, has made these instruments increasingly popular. Some of them have become the basis for legally established administrative procedures (e.g. environmental impact assessment [EIA]); others have become, among other roles, a support to marketing (like the life cycle assessment [LCA]).

Most of them are not specific to the bioeconomy. However, the types of technology development and sustainability approaches of the bioeconomy highlight and encourage the use of some of these instruments and stimulate bioeconomy-specific adaptations of them.

This chapter first reviews the issues and challenges in assessing the impact of decisions as well as of research and innovation in the bioeconomy, and then illustrates some of the most common methodologies used, with a specific focus on LCA, highlighting the connections with the previous chapters. Plenty of manuals and easily accessible web-based information provide introductions and illustrations of these tools. Benefiting from this, this chapter highlights only selected points in connection with the bioeconomy.

10.2 Challenges in Tracing the Impacts of Bioeconomy Technologies

Evaluating choices and supporting decisions means, first of all, being able to trace the effects of decisions and of new technology adoption. Unfortunately, a number of aspects of bioeconomy technologies go in the direction of making it more relevant, but also more difficult to evaluate the effects and impact of new technologies (Viaggi, 2015).

In some cases, the difficulty comes from the fact that new technologies improve the range of opportunities, but do not provide clarity on which ones will be realized. This is the case for technologies such as flexible platforms, i.e. basic intermediate compounds with multiple potential uses, which makes it difficult to provide clear expectations on future outlets and assign impacts between downstream drivers and upstream sources. This is amplified by the knowledge-based nature of the bioeconomy, in which the impact of research critically depends on the stock of available knowledge and how it is mobilized to exploit newly produced knowledge.

© D. Viaggi 2018. *The Bioeconomy: Delivering Sustainable Green Growth* (D. Viaggi)

In many cases, the actual impact is also uncertain because the process addressed is 'far' from the final products, the demand for which is more easily detected or new to the market (or new substitutes of existing ones). In this case, there is also higher probability that new technologies may yield a displacement of effects from one step to the other along the product chain rather than producing a net change in performances.

Bioeconomy research can also lead to modified economic relationships, for example, new products may have the potential to transform by-products into main products, modifying the way market forces justify production, with implications for the allocation of effects. This effect is potentially more important as the number of potential outputs from the same raw material grows. The increase in spatial complexity of biomass flows also makes it more difficult to trace the geographical distribution of impacts.

Another issue that makes it difficult to evaluate impacts is that new technologies already incorporate a number of (sometimes conflicting) objectives, notably those linked to sustainability. Moreover, far from traditional end-of-pipe approaches, this occurs through strategies that directly address process and product design, value chain organization or even producer habits (Viaggi, 2015). This implies that most upstream solutions are expected to impact directly on a number of ramified branches of the system and on a number of environmental and social dimensions.

Working with biological materials tends to maintain a high degree of uncertainty about the characteristics of inputs and outputs, as well as of the outcome of research processes and their applications. Moreover, reactions by living organisms are much less predictable than those of simpler systems.

Finally, in a holistic view of the bioeconomy system it is difficult to account for all potential synergic (exponential) effects, countermeasures or offset/compensatory effects in the system, so that net effects are potentially substantially different from what can be expected from a simplified analysis. In addition, worldwide trends, including the opening of economies and climate change, have given additional weight to uncertainty and dynamics, with a special emphasis on non-linear and 'non-trend' dynamics and fundamental uncertainty in the outcome and relevance of any technology change and policy action.

Accounting for higher and growing complexity also requires adequate approaches for research and evaluation. A pertinent example is that of the impact pathways evaluation approach (Douthwaite et al., 2003). In another example, Borge and Bröring (2017) build on the contingent effectiveness model of technology transfer and revise it in the direction of better accounting for interdisciplinary collaborations in the bioeconomy.

10.3 Indicators

Indicators are the first way of assessing technologies and decisions in the bioeconomy and a support to all other instruments. However, the assessment of bioeconomy value chains is still incipient and limited to a handful of indicators, with the exception of a few value chains such as liquid biofuels, food crops and some biopolymers (Cristóbal et al., 2015). Most work carried out to date derives from environmental indicators and from the well-developed field of agricultural indicators (Bockstaller et al., 2008). The challenge is to provide a set of indicator and evaluation criteria able to account for trade-offs between different sustainability goals (Hildebrandt et al., 2014). An additional aspect of bioeconomy indicators is that they need to allow for a spatially explicit assessment (Höltinger et al., 2014), which may be linked to available sources of information at the right scale as well as a functional understanding of spatial interactions.

Criteria can be distinguished from indicators and indexes. However, the term 'indicator' is often used in a generic way to cover the three meanings. Some indicators relate to a single parameter of impact (e.g. nitrogen pollution). Others aggregate impacts by different dimensions, such as (land, energy, water) footprint indicators. Some indicators also connect physical and economic dimensions in investment analysis (see Chapter 10, section 10.5, concerning investment assessment techniques).

A (non-exhaustive) list of indicators suggested by the literature is available in Table (10.1); a selection of them is further discussed in this section.

Table 10.1. Selected indicators for a systemic monitoring of the bioeconomy.

Policy objectives	Indicators and targets
Environment and resources	
General	Resource efficiency
Land resources	Land use
	Land footprint of the bioeconomy
	Global land use related to the safe operating space
	Land use intensity of crops and product groups
	'Hot spots' across the life cycle of specific crops or products/product groups
	Soil quality
Water resources	Water footprint of the bioeconomy
	Whether imported crops originate from water scarce regions
	Water emissions
Mitigating climate change	Primary energy demand
	GHG savings
	GHG footprint (including upstream and downstream effects)
	Carbon footprint (especially for timber-based energy sources)
Biodiversity	Biodiversity protection
Air	Air emissions
Economy	
Raising competitiveness	Production costs
	Turnover of bioeconomy sectors
	Turnover of new and innovative bioeconomy markets
	Trade balance
Strengthening innovation	Input-related indicators (e.g. R&D expenditure and investments)
	Innovation activity indicators (e.g. patents)
	Indicators for monitoring new and emerging key bioeconomy innovations
Contributing to GDP	GDP of 'traditional' bioeconomy sectors (e.g. primary production, paper and pulp, food processing, etc.)
	GDP of innovative bioeconomy sectors and lead markets (e.g. bio-plastics)
	Share of other sectors that use biomass as their feedstock (e.g. chemical industry, construction sector, etc.)
	GDP share of bio-based and fossil
	GDP of recycling-based production and consumption
Social	
Creating jobs	Employment in bio-based sectors
	Shifts of employment between bio-based and fossil
	Share of employment in recycling-based production and consumption
Labour conditions	Labour health conditions
Food security	Food price
	Food price volatility
	Trends in household and retail food waste
Quality of life	Life quality indicators such as life satisfaction
	Healthy livelihoods
	Local community conflicts

GDP, gross domestic product; GHG, greenhouse gas; R&D, research and development.
Source: Modified from Fritsche and Iriarte (2014); Jungmeier *et al.* (2016); O'Brien *et al.* (2017).

A 'land footprint' is a description that accounts for land use, as discussed in O'Brien *et al.* (2015) who provide an application for EU-27.

Another indicator is the carbon footprint. The systematic use of this indicator may be key for the proper comparison of different technologies, especially with respect to climate change (Glew *et al.*, 2017). However, this requires harmonization of the precise measurement approaches in order to ensure comparability.

Cristóbal *et al.* (2015) have developed a harmonized procedure – the product environmental footprint (PEF) – that includes fourteen impact categories to produce an aggregate impact indicator.

The human appropriation of net primary production (HANPP) framework is also relevant as it illustrates the appropriation and distribution of primary production. It also allows for an examination of the interrelations between land availability, land use, ecosystem services and policies from a dynamic (change-related) and spatial perspective (Krausmann *et al.* 2013; Plutzar *et al.* 2016).

Various indicators are under development to address the topic of the circularity of the bioeconomy. This implies considering the degree of closeness of an economy. At the same time, the focus on circularity questions traditional productivity measures if they do not qualify the origin of raw materials (recycled or newly extracted). Hildebrandt *et al.* (2017) argue that this is needed, in particular, for characterizing progress towards more recycling-friendly bio-based

polymers. Examples of sustainability rules, indicators and indicator calculation in cascading processes are reported below as identified in Hildebrandt *et al.*, (2017) (Table 10.2).

Huysveld *et al.* (2015) developed a cumulative overall resource efficiency assessment (COREA) index, taking into account: (i) bioproductive land resources; and (ii) the distinction between the non-renewable character of fossil resources. Concerning the first point, the methodological issue of the choice of system boundaries is considered as two aspects: (i) the definition of the system boundary of solar energy input to the primary biomass production system; and (ii) the definition of the temporal boundary of this system. An empirical application shows that these choices make a difference as the bio-based products have higher resource efficiency than their comparable fossil-derived products only if fossil resources are explicitly considered as ancient consumers of solar energy.

As critical resources may change over time due to rapid technological changes in the

Table 10.2. Sustainability rules for the cascading use of bio-based polymers within a circular economy.

Material cycles	Type of bioplastic	Business as usual	Extended cascade use	Indicators
Technical cycles	PET, PE, PP	Multi-stage cascade use	Increasing recovery quota for multi-stage cascade use	Thresholds for deteriorating MFI and viscosity for extrusion cycles Average number of extrusion cycles Thresholds for blending grades Cumulative energy demand for recycling phases Energy recovery Hazardous ingredient content
Technical and natural cycles	Expanded PLA/PLA	Energetic recovery from solid recovery fuel	Mechanical recycling for multi-stage cascade use	Cumulative energy demand for recycling phase Energy recovery Biogas yield Biodegradability Thresholds for extrusion cycles Thresholds for blending grades
Natural cycles	PHA and starch based	Mineralization on field or in industrial composting	Single stage cascade use	Energy recovery Biodegradability Compost yield Biogas yield Threshold for hazardous ingredient

MFI, melt flow index; PET, polyethylene terephthalate; PE, polyethylene; PP, polypropylene; PLA, polylactic acid; PHA, polyhydroxyalkanoat.
Source: modified from Hildebrandt *et al.* (2017)

bioeconomy, attention and indicators may require the flexibility to change over time to account for critical issues. O'Brien *et al.* (2017) summarize, with reference to the European Union (EU), the key gaps and needs for a systematic monitoring of the bioeconomy, and provide a useful checklist of issues to be addressed by future research.

The indicators discussed above may be used in isolation, but, most often, are used in connection with the methodology discussed below, in particular with the LCA.

10.4   Life Cycle Assessment and Life Cycle Costing

LCA is an assessment method focusing on impacts generated by each functional unit (a unit of a product or of its functions that enable comparisons across the relevant product chain alternatives considered) along its life cycle, from 'cradle to grave'. The basis and a key phase of the method is a compilation of the inventory of inputs and outputs, usually with reference to key resources (e.g. energy, water) or pollutants (e.g. greenhouse gasses, nitrogen). A further step is that of the evaluation of differences among technology alternatives.

LCA has rapidly developed as a key tool for technology analysis, to answer the growing need to account for a product chain view of the requirement for new technology assessments in many fields. It has now been used for more than two decades as an environmental assessment tool and is increasingly used to support marketing messages. It is increasingly used, in particular, as the basis for the selection of products in 'green procurement' and the inclusion of products in various national and regional eco-labelling schemes (e.g. European Flower, German Blue Angel, Nordic Swan eco-label, etc.). It is now widely promoted for early evaluation of research and innovation processes. In particular, LCA has already been applied to the Seventh Framework programme of the EU and is now regularly required in a number of programme calls by the EU Commission.

LCA is presently one of the cornerstones of the environmental and sustainability analysis of bioeconomy products and processes. LCA addresses the need to better account for the broad impact of technologies by considering a potentially wide range of effects on complex systems. In addition, it seeks to account for environmental effects along the value chain of a given product (broken down in different key phases), hence being able to explicitly account for displacement or compensatory effects at different stages of the chain. It also takes into consideration by-products and recycling. Furthermore, LCA makes it possible to support prescriptions regarding the steps in the process where intervention/research is more urgent due to higher criticalities in terms of impacts.

In a way, LCA responds to the basic idea of productivity though it is expressed in a reverse manner, that is, aiming at minimizing the unit of input and emission per unit of product. Notably, however, it tends to maintain a multidimensional and rather broad (and diverse) view of such a ratio, depending on the environmental indicators measured for input and output.

Several papers already report applications referring to the bioeconomy. A quick search on Scopus in April 2017 yielded 547 papers for the keywords 'life cycle assessment + biorefinery' and 2571 articles for 'life cycle assessment + bioenergy'. The number of applications goes hand-in-hand with the level of interest of the different subsectors, that is, bioenergy/biofuels are dominant, followed by biorefinery, while fewer (though rapidly growing) studies currently involve bio-based materials.

Besides case studies, the literature is currently developing methodological variants that incorporate specific bioeconomy issues, such as the allocation of impact across multiple products and by-products, as well as circular flows (Ness *et al.*, 2007; Tilche and Galatola, 2008; Sandin *et al.*, 2015).

One pathway concerns specific adaptations of LCA to the biorefinery concept. In this field, the choice of allocation method is a key determinant of results, particularly in consequential studies and in studies focused on co-products representing relatively small flows (Sandin *et al.*, 2015). Some of the works related to biorefinery also show the need to consider adaptations of whole approaches, and especially adapted environmental indicators, in accommodating the consideration of new feedstock. An example of the LCA for microalgae in biorefinery is provided by Seghetta *et al.* (2016).

Another pathway involves the consideration of expansions/adaptations of LCA to complex and circular systems, beyond the 'standard' chain vision. For example, Sommerhuber *et al.* (2017) examined wood plastic using a system LCA approach, where systems with equal functions were generated to secure a comparison of end-of-life treatment systems. In another example, Mattila *et al.* (2012) discuss the methodological aspects of applying LCA to industrial symbioses and, more generally, to circular economies. Indeed, circularity has been a key topic in LCA since its inception, but it is gaining importance as it becomes an independent objective rather than a feature of the system considered to be accounted for.

An issue of specific interest, and a limitation in the light of the previous discussions about socially constructed visions of technologies, is that LCA tends to focus attention on environmental/resource use, whereas economic and social impacts remain more difficult to account for. Notably, recent attention has been devoted to the use of life cycle costing (LCC), which is an economic assessment tool considering all projected significant and relevant cost flows over a period of analysis expressed in monetary value. It can be applied to a physical asset life or to the life cycle of a product/services in analogy to LCA. This is also gaining attention for use in public procurement.

Another area of development with a specific interest for the bioeconomy is that of integrated use of multi-criteria analysis and LCA. A review of combined applications of LCA, multi-criteria decision-making (MCDM) and participatory techniques is available in De Luca *et al.* (2017), showing that MCDM methods can help LCA in considering different actors' values and to take into account trade-offs among the different dimensions of sustainability. Reeb *et al.* (2016) use stochastic multi-attribute analysis alongside LCA to develop an environmental preference single-score probability distribution function for the selection among alternative bio-based sugar feedstock sources.

Yet LCA applications have also experienced difficulties and the above-referenced literature emphasizes a number of open issues both concerning data availability and conceptual problem structuring, especially for not yet well-structured production chains and for more and more complex interactions among chains in the bioeconomy web (Viaggi, 2015).

Moreover, comparability and coherence among LCA studies remains an open issue not only for the choice of the functional unit, but also for the other methodological choices and assumptions needed. Cristóbal *et al.* (2015) use a harmonized procedure, the PEF, to analyse three case studies using sugar as feedstock: sugar (food and feed), bio-based ethanol (bioenergy) and polyhydroxyalkanoates (bio-based product). The results underline the need for methodological harmonization for LCA of bioeconomy value chains.

From an economic point of view, the impact in terms of changes in flows needs to be given a value (even if not necessarily a monetary one). This implies assumptions about, for example, the location of impacts. Taking the case of the use of water resources, abstraction can calculate unit costs of a different order of magnitude depending on the source used. It should also be acknowledged that the distribution of impacts across sectors can be non-neutral. Third, the use of LCA and location must be considered in the light of the different types of value chains addressed (i.e. short, long and global). Concepts connecting the measurement of impacts and trade relationships are also emerging, such as virtual water and water footprint and are key to the consideration of territorial and public goods type issues in bioeconomy decision making.

Accounting for these challenges remains difficult but attempts are under way using increasingly holistic approaches. An example is provided by Siebert *et al.* (2016), who expand LCA in both the territorial dimension and the social dimension (which are largely related by the way), by implementing regional specific contextualized social life cycle assessment (RESPONSA) framework to assess a product's social performance from a regional perspective.

10.5   Investment Analysis

Investment analysis has been used widely in every field of economic activity as a supporting tool for investment feasibility decisions. This can be used to evaluate any action involving an expenditure aimed at obtaining future revenue, or, in the field of public investment analysis, a cost to achieve future benefits. The basic scheme of investment analysis can be applied to physical projects, financial investments, new businesses, technology adoption, investment in research, etc., each with the necessary adaptations. There are applications to private, as well as public, decision making, which fall under the name of cost–benefit analysis.

The basic logic of this analysis is the comparison of the discounted net cash flows (the difference between positive and negative cash flows or costs and benefits in a public evaluation framework) in order to calculate a net present value or other parameters such as the internal rate of return.

Variations are available in the literature to account for uncertainty. This is accomplished by way of the real options model, in which the option to wait (gaining a better knowledge for the future) is balanced against the giving up of some return.

Uncertainty is more often dealt with in standard cost–benefit analysis through scenario analysis or sensitivity analysis. Adaptation of these topics to the bioeconomy involves potentially considering any of the issues highlighted in the previous chapters that contribute to uncertainties in tracing impacts and in the value of resources as well as of outputs, such as the opportunity cost of resources, market scenarios for output and the value of externalities and public goods.

Given the specific focus of the bioeconomy, special adaptations are available from the literature to adjust the cost–benefit analysis to the objective of resource savings or production of some environmental effect. An example is that of levelized costs. This concept is utilized for energy through the levelized cost of energy (LCOE) and is widely used to obtain discounted and normalized unit costs of energy, taking into account that different energy technologies may have a different operating period and a different distribution of cash flows over time. For an example of the use of the levelized cost of energy in a bioeconomy context, see Lee (2016), who uses the conventional formulation of LCOE to compare biobutanol, algal biodiesel and biohydrogen.

Another approach, which makes use of the reverse criteria and most suitable for investments aimed at reducing environmental damages, is to consider environmental effects per unit of investment. An example indicator, linked to climate change, is that of carbon returns on investment (Oldfield *et al.*, 2016).

A similar example is provided by the notion of energy return on investment (EROI). Atlason *et al.* (2015) provide an example of this calculation in case studies carried out by comparing conventional and organic farms in Austria, producing sugarbeet, which is transformed into ethanol. They found that organic sugarbeet production provided an overall EROI of 11.3 whereas the conventional farming practice showed an EROI of 14.1 and 15, respectively.

10.6   Other Instruments

Many other instruments are relevant for the bioeconomy. Again, though non-specific, they may entail adaptations to be useful in the bioeconomy context.

Given the importance of uncertain outcomes, risk assessment is an area of interest for the bioeconomy. Risk assessment techniques are now widely used in different fields and usually involve providing an estimate of the magnitude of potential damaging effects and the probability of their occurence.

The risk of bio-invasion is discussed in Barney (2014). The author proposes a nested-feedback risk assessment approach that considers the entire bioenergy supply chain and includes the broad components of weed risk assessment, species distribution models and quantitative empirical studies. The method yields a relative invasion risk measurement to be used to design robust mitigation plans that include record keeping, regular scouting and reporting, prudent harvest and transport practices.

Productivity (generally defined as the ability of production factors to produce output) measurement in the context of the bioeconomy is also of relevance (Viaggi, 2015). The literature on productivity measurement includes simple measures of partial productivity (e.g. relating output to individual inputs) as well as measures accounting for more than one input (e.g. multiple factor productivity). A more comprehensive measure of productivity is total factor productivity (TFP), which is a ratio of the (monetary) aggregation of all outputs and the aggregation of all inputs. Fuglie (2015) shows that TFP has a major role in accounting for the growth of agricultural production over time, though with a differing weight in different time periods and in different regions. Several multi-output, resource-saving or dynamic specifications are now attempting to tackle multiple research objectives (e.g. production, environmental improvement). This has led, among others, to proposing extensions such as environmentally adjusted TFP or even green TFP.

Basic cost accounting or profitability measures are largely used in practice to assess the economic sustainability of new technologies. While these techniques do not require apparent significant adaptations, the interpretation of costs and revenue variables in relation to future market changes do require attention. In addition, sensitivity or scenario analysis are key components of any meaningful economic analysis related to the bioeconomy.

The coupling of economic parameters with technical parameters can be applied here and can be seen as another version of the resource-oriented indicators previously mentioned in the context of investment analysis. For example, dos Santos *et al.* (2017) provide a techno-economic analysis of microalgal heterotrophic bioreactors applied to the treatment of poultry and swine slaughterhouse wastewater. They calculate a cost per unit of treated industrial wastewater as well as a cost per unit of dehydrated biomass.

Business plans are widely used for new activities, and are well-known to bioeconomy actors as a useful instrument for the planning and running of businesses and as a means of seeking necessary financial support.

Another widely used tool is the SWOT (strengths, weaknesses, opportunities, threats) analysis, a well-known yet non-specific tool that is being adapted to the specificities of the bioeconomy. For example, Kangas *et al.* (2016) develop hybrid SWOT methodologies to provide tangible suggestions to strategic choices related to new

bioeconomy products. The method is based on the use of SWOT, expanded by the inclusion of goals and actions, social choice theory and robust portfolio modelling.

Environmental management tools are pervasive and are increasingly seen as key instruments for the sustainable management of bioeconomy sectors (Straczewska, 2013). The logic of most environmental management tools is that of an improvement process, based on a cycle of target setting, planning, implementation and evaluation.

Modelling for decision support is also increasingly relevant for each of the disciplines and the processes involved in the bioeconomy and includes complex routes to account for the interconnected nature of bioeconomy components.

10.7 Integration of Methods

Another issue is that of the integration of methods and data sources. While several of the tools discussed above can be used in isolation, they can also benefit from combined use or even the integration of multiple tools.

First, the integration of databases and monitoring tools with assessment methods may be essential to ensure good quality data are available for assessment. Second, different assessment methods may help triangulation and management of uncertainty. Third, integration between classification, conceptualization and measurement tools is needed. Finally, consistency between methods and procedures for their use in decision-making is essential, as also discussed in the next two sections.

A key overall issue is the integration of hard numerical methods with stakeholder involvement procedures. With respect to timing, integration between the design and assessment perspective is also important (*ex-ante* and *ex-post*). Integration between chain and territorial perspectives is also of particular importance. Several examples of these exercises are emerging in the literature.

Box 10.1. Integration approaches

Vale *et al.* (2017) provide a hybrid economic input output – life cycle assessment (EIO–LCA) coupled with qualitative methods for consensus creation and participative policy building. This is used to construct a robust regional bioeconomic profile able to inform policy makers and other stakeholders about socioeconomic and ecological impacts.

Continued

Box 10.1. Continued.

In the evaluation of biorefinery development, Nitzsche *et al.* (2016) provide an empirical example using a set of tools including: (i) conceptual approaches for new biorefinery products and processes, with the integration of economic and environmental optimization; (ii) social life cycle assessment (SLCA) methods; (iii) an assessment of eco-efficiency; and (iv) indicator-based monitoring of the level of sustainability of the studied products. Jungmeier *et al.* (2016) integrate different aspects of the assessment of a biorefinery in a common framework, including: (i) a biorefinery classification; (ii) an assessment of the technologies and processes, including their technology readiness level, integrated in a 'Biorefinery Complexity Index (BCI)'; (iii) an economic assessment based on LCC; (iv) an assessment of environmental effects based on LCA; (v) accounting of social issues based on SLCA; (vi) an overall life cycle sustainability assessment (LCSA); (vii) the identification of the most attractive industry sectors ('hot spots'); and (viii) an identification of research and development needs.

An example of modelling integration for the evaluation of research investment in a new bio-based chemical industry is provided in Zhuang and Herrgård (2015). The authors develop a comprehensive multi-scale framework for modelling sustainable industrial chemicals production (MuSIC), integrating modelling approaches for cellular metabolism, bioreactor design, upstream/downstream processes and economic impact assessment. The model allows for an improved understanding of trade-offs among the different components and an economic assessment to guide design decisions, especially accounting for non-intuitive dependencies within the process and across scales.

10.8 Evaluation Tools and Stakeholder Engagement

A key to the best usage of indicators and evaluation tools is in the interaction of the tools with stakeholders and decision makers. This may follow different ways and needs to take into account consistency with decision-making procedures and institutions. It is particularly important for assessment tools to incorporate the views and knowledge of different stakeholders. This may be addressed in more or less ambitious ways, from stakeholder-built indicators to simple stakeholder engagement in testing or providing opinions about values to be incorporated in the analysis. An important point is the timing of involvement, especially in connection with early product design or research orientation.

For example, Hildebrandt *et al.* (2017) provide recommendations on how LCA indicators can be used to support the dialogue between designers and recyclers to promote design for recycling principles in order to meet circular economy goals.

Early stage assessments of new products can help improve processes, but are also limited by data availability. Broeren *et al.* (2017) review and compare 27 early stage assessment methods and evaluate the extent to which they are suitable for assessing bio-based chemicals in early stage

development. Among good-practice principles for early stage assessments of bio-based chemicals they identify: low data requirements; the inclusion of climate change and energy indicators; and the inclusion of environmental impacts from biomass feedstock production.

Diaz-Chavez *et al.* (2016) highlight that sustainability assessments are not 'one-size-fits-all'. Stakeholders should be engaged in determining goals and objectives for the assessment, taking into account the specific context. Moreover, transparency in the approach and the assumptions taken is very important. Finally, sustainability assessments do not work well if fixed targets are imposed and should hence be seen as continuous improvement processes.

10.9 Outlook

This chapter has illustrated the paths to be followed and the obstacles to be overcome in finding satisfactory tools to undertake early evaluations of bioeconomy technologies. Indeed, the discrepancy between the need for more accurate technical measures of impact (or at least pressures) and the limited ability to evaluate their effects in economic terms is still very large. It is likely that the further evolution of bioeconomy technologies will further contribute to these difficulties.

These problems are emphasized when expanded beyond technical difficulties and matched with problems inherent in linking instruments to actual decision-making processes. Indeed, this is common in all fields and the literature on decision-support tools contains myriad references to the lack of attention being paid to instruments on the part of decision makers and, on the other hand, the limited ability of instruments to solve real-life decision-making problems.

This is also evident from the bioeconomy literature, including for the better-established and recognized tools (i.e. LCA). Sandin *et al* (2014), for example, note that the role of LCAs with respect to decision making tends to be unclear and arbitrary. As a result, LCA work is not adequately designed for the needs of the projects considered. This highlights the need to design appropriate roles for LCA and to plan LCA work accordingly.

The issues discussed in this chapter may also vary depending on the specific field of the bioeconomy considered. While assessment issues linked to bioenergy production and the need of biomass are presently rather well known, studies on bio-based products are less developed.

In most cases, poor data availability remains the clearest limitation and a major criterion for decisions about the individual indicators and tools to be used. This was noted in Chapter 2 with regard to bioeconomy-related statistics. Bockstaller *et al.* (2008) conclude that, in light of the limited data available at the regional level, several simple indicators should be used, at least at this level. Only when more detailed information is available can indicators based on operational models be useful. In experimental studies, when possible, it is suggested to use both measured indicators and model-based indicators.

The awareness of limitations of the current knowledge and evaluation means has contributed to the emphasis on communication and on early stage evaluation, permissions and regulation procedures (see Chapter 7), in order to ensure that choices in terms of the promotion of new technologies add value to society and avoid unexpected negative effects.

Undoubtedly the issue of impact evaluation tools will continue to evolve, supported not only by perceived governance needs, but also by the emergence of new products and increasing data availability. In particular, the interface between evaluation with limited information and public participation represents a major focus for research to support the future development of the bioeconomy.

References

Atlason, R.S., Lehtinen, T., Davídsdóttir, B., Gísladóttir, G., Brocza, F., Unnthorsson, R. and Ragnarsdóttir, K.V. (2015) Energy return on investment of Austrian sugar beet: A small-scale comparison between organic and conventional production. *Biomass and Bioenergy* 75, 267–271. doi: 10.1016/j.biombioe.2015.02.032.

Barney, J.N. (2014) Bioenergy and invasive plants: Quantifying and mitigating future risks. *Invasive Plant Science and Management* 7(2), 199–209. doi: 10.1614/IPSM-D-13-00060.1.

Bockstaller, C., Guichard, L., Makowski, D., Aveline, A., Girardin, P. and Plantureux, S. (2008) Agri-environmental indicators to assess cropping and farming systems. A review. *Agronomy for Sustainable Development* 28(1), 139–149. doi: 10.1051/agro:2007052.

Borge, L. and Bröring, S. (2017) Exploring effectiveness of technology transfer in interdisciplinary settings: The case of the bioeconomy. *Creativity and Innovation Management* 26(3), 311–322. doi: 10.1111/caim.12222.

Broeren, M.L., Zijp, M.C., Waaijers-van der Loop, S.L., Heugens, E.H., Posthuma, L., Worrell, E. and Shen, L. (2017) Environmental assessment of bio-based chemicals in early-stage development: A review of methods and indicators. *Biofuels, Bioproducts and Biorefining* 11(4), 701–718. doi: 10.1002/bbb.1772.

Cristóbal, J., Matos, C.T., Aurambout, J.-P., Manfredi, S. and Kavalov, B. (2015) Environmental sustainability assessment of bioeconomy value chains. *Biomass and Bioenergy* 89, 159–171. doi: 10.1016/j.biombioe.2016.02.002.

De Luca, A.I., Iofrida, N., Leskinen, P., Stillitano, T., Falcone, G., Strano, A. and Gulisano, G. (2017) Life cycle tools combined with multi-criteria and participatory methods for agricultural sustainability: Insights from a systematic and critical review. *Science of the Total Environment* 595, 353–370. doi: 10.1016/j.scitotenv.2017.03.284.

Diaz-Chavez, R., Stichnothe, H. and Johnson, K. (2016) Sustainability considerations for the future bioec-
    onomy. In Lamers, P., Searcy, E., Hess, J.R. and Stichnothe, H. (eds) *Developing the Global Bioec-
    onomy: Technical, Market, and Environmental Lessons from Bioenergy.* London, UK, Academic Press,
    pp. 69–90. doi: 10.1016/B978-0-12-805165-8.00004-5.
dos Santos, A.M., Roso, G.R., de Menezes, C.R., Queiroz, M.I., Zepka, L.Q. and Jacob-Lopes, E. (2017)
    The bioeconomy of microalgal heterotrophic bioreactors applied to agroindustrial wastewater treatment.
    Desalination and Water Treatment 64, 12–20. doi: 10.5004/dwt.2017.20279.
Douthwaite, B., Kuby, T., van de Fliert, E. and Schulz, S. (2003) Impact pathway evaluation: an approach
    for achieving and attributing impact in complex systems. *Agricultural Systems* 78(2), 243–265. doi:
    10.1016/S0308-521X(03)00128-8.
Fritsche, U.R. and Iriarte, L. (2014) Sustainability criteria and indicators for the bio-based economy in
    Europe: State of discussion and way forward. *Energies* 7(11), 6825–6836. doi: 10.3390/en7116825.
Fuglie, K. (2015) Accounting for growth in global agriculture. *Bio-based and Applied Economics* 4(3),
    201–234.
Glew, D., Stringer, L.C., Acquaye, A. and McQueen-Mason, S. (2017) Evaluating the potential for harmon-
    ized prediction and comparison of disposal-stage greenhouse gas emissions for biomaterial products.
    Journal of Industrial Ecology 21(1), 101–115. doi: 10.1111/jiec.12421.
Hildebrandt, J., Bezama, A. and Thran, D. (2014) Establishing a robust sustainability index for the assess-
    ment of bioeconomy regions. In Proceedings of the 2014 International Conference and Utility Exhibition
    on Green Energy for Sustainable Development, ICUE 2014.
Hildebrandt, J., Bezama, A. and Thrän, D. (2017) Cascade use indicators for selected biopolymers: Are we
    aiming for the right solutions in the design for recycling of bio-based polymers? *Waste Management
    and Research* 35(4), 367–378. doi: 10.1177/0734242X16683445.
Höltinger, S., Schmidt, J., Schönhart, M. and Schmid, E. (2014) A spatially explicit techno-economic as-
    sessment of green biorefinery concepts. *Biofuels, Bioproducts and Biorefining* 8(3), 325–341. doi:
    10.1002/bbb.1461.
Huysveld, S., De Meester, S., Van Linden, V., Muylle, H., Peiren, N., Lauwers, L. and Dewulf, J. (2015) Cu-
    mulative Overall Resource Efficiency Assessment (COREA) for comparing bio-based products with
    their fossil-derived counterparts. *Resources, Conservation and Recycling* 102, 113–127. doi:
    10.1016/j.resconrec.2015.06.007.
Jungmeier, G., Hingsamer, M., Steiner, D., Kaltenegger, I., Kleinegris, D., van Ree, R. and de Jong, E.
    (2016) The approach of life cycle sustainability assessment of biorefineries. In *European Biomass
    Conference and Exhibition Proceedings*, pp. 1660–1665. doi: 10.5071/24thEUBCE2016-IBO.8.1
Kangas, J., Tikkanen, J., Leskinen, P., Kurttila, M. and Kajanus, M. (2016) Developing hybrid SWOT meth-
    odologies for choosing joint bioeconomy co-operation priorities by three Finnish universities. *Biofuels*
    8(4), 459–471. doi: 10.1080/17597269.2016.1271625.
Krausmann, F., Erb, K.-H., Gingrich, S., Haberl, H., Bondeau, A., Gaube, V., Lauk, C., Plutzar, C. and
    Searchinger, T.D. (2013) Global human appropriation of net primary production doubled in the 20th
    century. *Proceedings of the National Academy of Sciences* 110(25), 10324–10329. doi: 10.1073/
    pnas.1211349110.
Lee, D.-H. (2016) Levelized cost of energy and financial evaluation for biobutanol, algal biodiesel and
    biohydrogen during commercial development. *International Journal of Hydrogen Energy* 41(46),
    21583–21599. doi: 10.1016/j.ijhydene.2016.07.242.
Mattila, T., Lehtoranta, S., Sokka, L., Melanen, M. and Nissinen, A. (2012) Methodological aspects of apply-
    ing life cycle assessment to industrial symbioses. *Journal of Industrial Ecology* 16(1), 51–60. doi:
    10.1111/j.1530-9290.2011.00443.x.
Ness, B., Urbel-Piirsalu, E., Anderberg, S. and Olsson, L. (2007) Categorising tools for sustainability
    assessment. *Ecological Economics* 60(3), 498–508. doi: 10.1016/j.ecolecon.2006.07.023.
Nitzsche, R., Budzinski, M., Gröngröft, A., Majer, S., Müller-Langer, F. and Thrän, D. (2016) Process
    simulation and sustainability assessment during conceptual design of new bioeconomy value
    chains. In *European Biomass Conference and Exhibition Proceedings*, pp. 1723–1726. doi: 10.5071/
    24thEUBCE2016-ICO.16.5.
O'Brien, M., Schütz, H. and Bringezu, S. (2015) The land footprint of the EU bioeconomy: Monitoring tools,
    gaps and needs. *Land Use Policy* 47, 235–246. doi: 10.1016/j.landusepol.2015.04.012.
O'Brien, M., Wechsler, D., Bringezu, S. and Schaldach, R. (2017) Toward a systemic monitoring of the
    European bioeconomy: Gaps, needs and the integration of sustainability indicators and targets for
    global land use. *Land Use Policy* 66, 162–171. doi: 10.1016/j.landusepol.2017.04.047.

Oldfield, T.L., White, E. and Holden, N.M. (2016) An environmental analysis of options for utilising wasted food and food residue. *Journal of Environmental Management* 183, 826–835. doi: 10.1016/j.jenvman.2016.09.035.

Plutzar, C., Kroisleitner, C., Haberl, H., Fetzel, T., Bulgheroni, C., Beringer, T., Hostert, P., Kastner, T., Kuemmerle, T., Lauk, C., Levers, C., Lindner, M., Moser, D., Müller, D., Niedertscheider, M., Paracchini, M.L., Schaphoff, S., Verburg, P.H., Verkerk, P.J. and Erb, K.-H. (2016) Changes in the spatial patterns of human appropriation of net primary production (HANPP) in Europe 1990–2006. *Regional Environmental Change* 16(5), 1225–1238. doi: 10.1007/s10113-015-0820-3.

Reeb, C.W., Venditti, R., Gonzalez, R. and Kelley, S. (2016) Environmental LCA and financial analysis to evaluate the feasibility of bio-based sugar feedstock biomass supply globally: Part 2. Application of multi-criteria decision-making analysis as a method for biomass feedstock comparisons. *BioResources* 11(3), 6062–6084. doi: 10.15376/biores.11.3.6062-6084.

Sandin, G., Clancy, G., Heimersson, S., Peters, G.M., Svanström, M. and ten Hoeve, M. (2014) Making the most of LCA in technical inter-organisational R&D projects. *Journal of Cleaner Production* 70, 97–104. doi: 10.1016/j.jclepro.2014.01.094.

Sandin, G., Røyne, F., Berlin, J., Peters, G. M. and Svanström, M. (2015) Allocation in LCAs of biorefinery products: implications for results and decision-making. *Journal of Cleaner Production* 93, 213–221. doi: 10.1016/j.jclepro.2015.01.013.

Seghetta, M., Hou, X., Bastianoni, S., Bjerre, A.-B. and Thomsen, M. (2016) Life cycle assessment of macroalgal biorefinery for the production of ethanol, proteins and fertilizers – A step towards a regenerative bioeconomy. *Journal of Cleaner Production* 137, 1158–1169. doi: 10.1016/j.jclepro.2016.07.195.

Siebert, A., Bezama, A., O'Keeffe, S. and Thrän, D. (2016) Social life cycle assessment indices and indicators to monitor the social implications of wood-based products. *Journal of Cleaner Production* 172, 4074–4084. doi: 10.1016/j.jclepro.2017.02.146.

Sommerhuber, P.F., Wenker, J.L., Rüter, S. and Krause, A. (2017) Life cycle assessment of wood-plastic composites: Analysing alternative materials and identifying an environmental sound end-of-life option. *Resources, Conservation and Recycling* 117, 235–248. doi: 10.1016/j.resconrec.2016.10.012.

Straczewska, I. (2013) System of environmental management as an element of bioeconomy development. *Journal of International Studies* 6(2), 155–163. doi: 10.14254/2071-8330.2013/6-2/14.

Tilche, A. and Galatola, M. (2008) Corner 'EU Life Cycle Policy and Support'. *The International Journal of Life Cycle Assessment* 13(2), 166–167. doi: 10.1065/lca2008.02.378.

Vale, M., Pantalone, M. and Bragagnolo, M. (2017) Collaborative perspective in bio-economy development: A mixed method approach. *Collaboration in a Data-Rich World: 18th IFIP WG 5.5 Working Conference on Virtual Enterprises, PRO-VE 2017, Vicenza, Italy, September 18–20, 2017, Proceedings*. New York, USA, Springer, pp. 553–563. doi: 10.1007/978-3-319-65151-4_49.

Viaggi, D. (2015) Research and innovation in agriculture: beyond productivity? *Bio-based and Applied Economics* 4(3), 279–300.

Zhuang, K.H. and Herrgård, M.J. (2015) Multi-scale exploration of the technical, economic, and environmental dimensions of bio-based chemical production. *Metabolic Engineering* 31, 1–12. doi: 10.1016/j.ymben.2015.05.007.

11

At the Boundary of Economics

11.1 Introduction and Overview

While this book takes an explicitly economic perspective on the bioeconomy, it also highlights relevant topics that could be better examined by resorting to other fields of analysis that touch or overlap with economics. This is evident from the number of potential points of contact between economics and politics, sociology, law, governance and communications (just to mention a few) highlighted in the previous chapters. Several of these fields are complementary in explaining behaviour and supporting economic considerations.

The range of possible relevant fields of science is extremely wide and this chapter only touches upon a selection of the most urgent and evident connections, focusing on the broad area of social science and humanities and providing only a flavour of potential topics. The importance of the technological foundations of the economic analysis of the bioeconomy was treated at the very beginning of the book and informs large part of it, so it will not be addressed here. It is, however, very important to highlight the growing connection between technology and behaviour and the increasingly interconnected means of interpretation, especially through information and communication technology and data management.

This section starts with topics that are closer to technical aspects, such as logistics, and then moves to geography, politics and law, sociology, psychology and communication sciences, and closes with philosophy and ethical aspects.

11.2 Logistics

Biomass is comparatively costly to transport and the spatial distribution of raw materials and the move towards a global web of biomass flows puts logistics at the heart of the bioeconomy's economic performance. Logistical issues are also central to the assessment and design of more-or-less decentralized processing or combinations of different biomass flows and the understanding of trade-offs between concentrated large-scale plants and a web of small processing plants (see, for example, Lamers *et al.* (2015a, 2015b)).

Logistics relates to the costs and benefits of the use of biomass and hence the economic feasibility of the bioeconomy as a whole. It also affects the geographical distribution of the bioeconomy and hence the environmental and social effects related to concentrated production or the use of marginal areas, rural and territorial development, and the interplay with infrastructure developments.

11.3 Geography

Space and geography considerations emerge in a number of studies related to the bioeconomy, in

© D. Viaggi 2018. *The Bioeconomy: Delivering Sustainable Green Growth* (D. Viaggi)

connection either to the sources of biomass, to territorial specialization and segmentation, or to the distribution of values and economic benefits. A broad overview of the interplay between bio-economy and geography in connection with technology and strategic management studies is provided by Calvert *et al.* (2017). Notably, the geographical dimension of the bioeconomy is linked not only to resource endowment and path dependency in technology, but also to its inter-play with societal change and how it affects the way humans construct space.

Geographical considerations are also related to the distribution of property rights and owner-ship of resources, such as land and biodiversity. In some cases, the geography of property rights is beginning to take shape, such as for rights to biological resources, while others are fading, as is notably the case for land in the context of land grabbing (not only due to the bioeconomy, but connected to it). In addition, the effects of new bioeconomy-related concepts, such as sustain-able intensification, may affect rural areas differ-ently than urban areas and form the basis for location-specific policies, such as rural policies (McDonagh, 2015).

Birch (2009) also noted that the distribu-tion of innovation activities in the field of the bioeconomy is rather uneven, with concentra-tions in certain countries and in particular re-gions of those countries. The author notes, for example, that the UK shows four major concen-trations of the bioeconomy, each one having dis-tinct patterns of knowledge and spatial inputs in the innovation process (Birch, 2009). This un-even distribution is particularly strong for know-ledge distribution as embedded in intangible and immaterial resources and labour, though the value generated is then 'abstracted in global standards, regulations and exchanges' (Birch, 2012). Spatial aspects in the development of in-novation clusters and interplay among actors are also highlighted by Kearnes (2013).

11.4   Politics and Law

Political sciences are a key focus of research be-cause political processes affect the development of the bioeconomy through power relationships, consensus building, regulation and policy making.

The main contact point with economics is polit-ical economy, but it should be noted that quali-fications from politics, culture and social psychology are essential to extend the classical political economy based on interest groups, as the bioeconomy is much more affected by ideas about the world than is usually the case in a nar-row approach to political economy (Herring and Paarlberg, 2016).

The political relevance of the bioeconomy is due to the fact that the bioeconomy is positioned at the crossroads of major development issues from which conflicts may arise. As stated by Bertrand *et al.* (2016):

> Within the framework of sustainable development, a quadruple transition (demographic, food, energetic and environmental) is necessary. The integrated biomass valorization, including the production of biofuels via the biorefinery concept lies at the cross section of these global challenges. Biofuels production competes for water and land use with respect to feed and food production. Securing these resources in an uncertain geopolitical environment is a strategic issue for many countries. It also has favourable (climate change mitigation) and adverse implications (fertilizers consumption, local pollutions).

Even when an overall positive view of the bioec-onomy prevails, conflicts may still be very rele-vant. Within Europe, for example, Richardson (2012) argues that the 'win–win' rhetoric of the knowledge-based bioeconomy (KBBE) narrative is misleading as in its background several con-flicts may be identified, notably between farmers and agribusiness, between those convinced and those sceptical of technological solutions, as well as between 'pro-corporate' and 'anti-corporate' non-governmental organizations.

Other authors point to a rather different as-pect of the bioeconomy concept, namely the na-ture of the bioeconomy as a 'master narrative' having an implicit ability to reach consensus by accommodating different views and positions (Levidow *et al.*, 2013).

Political science issues may be also viewed in the light of 'political ecology'. A link between biomass distribution and social issues is provided by Temper (2016). The paper uses human appro-priation of net primary production (HANPP) to study the distribution of biomass between different social groups (and not only between humans and non-humans).

> **Box 11.1.** Distribution of HANPP
>
> Temper (2016) studies the sugar cane plantation economy in a region of Kenya. In this case study, the sugar cane plantation economy would increase the biomass production and the proportion appropriated by humans. However, the Orma pastoralists and the Pokomo farmers would see their access to biomass reduced and less biomass would be available for local 'wild' biodiversity. A much larger proportion of the net primary production would be exported as sugar or ethanol.

A good deal of works have been devoted to how political action translates into the development of institutions to deal with reducing risks, including the topic politics of science and science-society connections, in which GM and biotechnology in general have been at the forefront. Based on an analysis of China, India and the United Kingdom, Salter *et al.* (2016) argue that there is a variable capacity of different states and political systems to work with science and in benefiting the potential of life sciences innovation.

> **Box 11.2.** Legitimacy in conflicts in genetically modified crops
>
> Gupta (2011) analyses two highly controversial developments in Indian agriculture: (i) the approval (or not) of the very first genetically modified food crop in India; and (ii) proposals to re-launch the existing biosafety regulatory system. The author contrasts legitimacy derived from innovative participatory democracy with legitimacy claims based upon 'objective' science. The study shows that, despite an acknowledged need for participatory legitimization, science-based arguments are, in practice, being prioritized.

The bioeconomy concepts linked to governance tend to question political processes, especially when connected to participation and bottom-up versus top-down approaches, making the bioeconomy a dynamic construct. Delvenne and Hendrickx (2013) present a special issue about bioeconomy development in Latin American countries, taking the perspective of scientific knowledge production, public participation,

regulation and governance. The special issue demonstrates the existence of regional differences in terms of public policy spaces, and scientific knowledge production into which bioeconomy innovations are introduced. Local bottom-up dynamics complement, challenge or undermine (depending on the case) the top-down policy narratives on the bioeconomy. On the other hand, grassroot dynamics are influenced and conditioned by macro-sociological, political and economic factors. These dynamics also interact with globalized markets in which states and national actors actively contribute.

In some cases, bioeconomy-related processes can be viewed as a restatement of the contrast between a productivist view of the use of biological resources versus a more ecological, or conservationist, view.

> **Box 11.3.** Productivist versus conservationism in forest policy
>
> Kröger and Raitio (2017) notice that the recent period has seen a political shift back towards more hierarchical policymaking that promotes a productivist forest policy under the guise of a 'forest bioeconomy'. This has co-aligned with the global bioeconomy meta-discourse that has contributed to the re-legitimization of policy goals (including an emphasis on timber production) from previous industrial forestry approaches. At the same time, this process has found obstacles in a more ecologically aware environment, leading to trade-offs with tourism uses. It could be argued that this productivist return under the umbrella of the bioeconomy, in a much more 'aware' context than in the past, will likely produce a more balanced consideration of different sustainability needs.

Other authors point to the focus on bioeconomy as a mainly political process aimed at ensuring the survival of current capitalist structures (Goven and Pavone, 2014). In this context, political concepts linked to the bioeconomy are also emerging, such as biocapitalism and biopolitics. McMichael (2012) uses these concepts to analyse the issue of land grabbing and related events, such as the 2008 'food crisis', and the emergent bioeconomy. The author interprets land grabbing as an attempt to solve the crisis of the current neoliberalism in a context in which geopolitical relationships are re-arranging towards the south, backed by the emergent bioeconomy.

International political and legal issues are also relevant, not only for international relationships, but more in general to the politics of property rights distribution. In this field, Kotsakis (2014) calls for attention to micropolitics, incorporated in apparatuses that operate outside the formal structures of the international legal spheres. The paper discusses, for example, the notion of 'genetic gold', highlighting that it is somehow an outdated conception of environmental value, now challenged by the interplay of biodiversity, biotechnology and neoliberalism.

11.5 Psychological and Sociological Aspects

Economic behaviour has its basis in a number of aspects addressed by other sciences. In a way, all of the fields of research related to individual and collective behaviour are relevant for the bioeconomy. This is increasingly more important the more unknown and less familiar the solutions that are proposed by the bioeconomy, thus increasing the level of uncertainty of the consequences of decisions taken. As a result, the spectrum of potential psychological and sociological issues touched upon by the bioeconomy is very broad. On the one hand, we have a 'simple' understanding of feelings that are needed for effective communication. On the other hand, the bioeconomy comes into contact with living matters and, to some extent, human beings. In particular, when the bioeconomy touches the health sector, it begins to address the understanding of living and the human body. Examples are given in the discussion of the 'socioecological of body' in Guthman (2015).

One interesting area of study is how scientific claims can be used to support very different and contrasting policy positions by different groups; this also implies understanding how these groups engage different components of the scientific world to support their views.

> **Box 11.4.** Scientific claims over biofuels in Denmark
>
> Biofuels can generate trade-offs between economic, social and environmental objectives. In Denmark two distinct scientific perspectives about biofuels are mapped into two competing coalitions (Hansen, 2014). The first is a reductionist biorefinery perspective, originating in biochemistry and neighbouring disciplines. It tends to envision positive synergies in the use of biomass. The second is a holistic bioscarcity perspective originating in life-cycle analysis and ecology. It envisages negative consequences from an increased reliance on biomass.

Sleenhoff *et al.* (2015a) note that a transition to the bioeconomy will require collective actions by different stakeholders, including the public, which is largely unaware of this transition process. While action will require emotional involvement, the role of emotion (what the public feels and how this could be taken into account for public engagement) in relation to the bioeconomy has received scarce attention in the literature. The authors identify four emotional viewpoints that also provide a basis to understand how different individuals emotionally connect to the transition to the bioeconomy: (i) compassionate environmentalist; (ii) principled optimist; (iii) hopeful motorist; and (iv) cynical environmentalist. Based on these considerations, Sleenhoff *et al.* (2015b) advocate for a mentality change with regard to the role of citizens in the bioeconomy. This requires the feeling of a 'state of engagement' by a public that needs to feel they have a stake and a voice. This can only be built through an understanding of the value of emotions as a basis for more effective participation processes.

One emerging topic is that new forms of organization, such as public–private partnerships, clusters and networks, largely based on relationships, are bringing about new forms of anthropological concerns to explain their development (Hodges, 2012).

The relevance of interaction between new technology and society is more relevant in the medical field, from which the whole bioeconomy can actually learn, in spite of the clear differences. Here, the term 'biosociety' is currently being used and the effects of genomics on a number of societal issues are more evident (e.g. social, cultural and legal, including rights, patenting, regulation) (Kumar and Chadwick, 2015).

11.6   Perceptions, Discourses and Communication

Communication is at the core of actions to build supply and demand, as well as to coordinate them through markets or other instruments in the bioeconomy. Though connected to social sciences and political sciences in general, communication can also be seen as a specific field of investigation. The topic is particularly relevant as different groups tend to use different frameworks and different arguments in communication about the bioeconomy, though 'innovation' and 'sustainability' are the most common keywords (Kleinschmit *et al.*, 2014).

'Bioeconomy' and related terms are increasingly used by scientists and politicians and are becoming part of an influential global discourse (Pülzl *et al.*, 2014). Discourses, arguments and resulting ideas have 'performative power', meaning that they shape actors' views, beliefs and interests and ultimately influence their behaviour (Pülzl *et al.*, 2014). They also lead to institutional change. This is even more evident in the biomedical sector. Petersen and Krisjansen (2015) argue that the bioeconomy discourse plays a crucial performative role in contemporary biopolitics and acts by mobilizing actor networks, attracting venture capital and research funding, as well as shaping the possibilities for action. The authors, however, also note the 'performativity' of the bioeconomy and highlight the need for a better consideration of the implications of promissory discourses for governance.

In a communication context, the 'bioeconomy' can be qualified as a 'boundary subject' or a 'bridging concept', that is, matching specific interests of different stakeholders under a generally accepted conceptual umbrella (Hodge *et al.*, 2017). Another side of the issue is the global view in which the bioeconomy is cast. Experience shows how important the broad view that societies have may affect a number of individual issues in the bioeconomy, ranging from acceptance and action on individual technologies, to participation in collective decision making (De Witt *et al.*, 2017).

> **Box 11.5.** Bridging concept and world views
>
> Looking at the Swedish forest sector, Hodge *et al.* (2017) found strong evidence that the bioeconomy takes the role of a bridging concept, which helps to bring together actors with different views. De Witt *et al.* (2017) use the integrative worldview framework (IWF) to satisfactorily address the variety of positions and arguments emerging when analysing perceptions of biotechnology. The IWF distinguishes between traditional, modern and postmodern worldviews, among others, and makes it possible to generate insight into the paradigmatic gaps in social science. It can also support reflexive and inclusive policy making.

11.7   Philosophy and Ethics

The development of the bioeconomy also has relevant philosophical implications. Among these, three appear to be the most evident: the vision of nature; the concept of value; and the ethical issues embedded in bioeconomy technologies.

The vision and interpretation of non-human nature changes with the bioeconomy towards a sort of 'biomimicry' (Goldstein and Johnson, 2015). The bioeconomy shift is undoubtedly leaving behind crude approaches to nature such as 'the domination of nature-as-machine' and the use of chemistry-based processes that have generated the environmental and climate crises. Indeed, the new approaches to nature are less violent and exploitative. However, the new industrial paradigm brought about by the bioeconomy is nonetheless still affecting nature, pointing to two major shifts in how non-human life is connected to production processes: (i) the production of nature as intellectual property, as opposed to the old view of producer of raw materials; and (ii) the production of nature as an active subject, as opposed to a passive receptacle or vehicle as in the traditional view of environmental concerns.

The use of standards is an example of how practical economic issues (discussed in Chapter 7) relate to conceptual avenues of research concerning the way humans manage and modify living organisms and biological resources. Mackenzie *et al.* (2013) provide an intriguing view of this topic, highlighting how imposing fixed properties to the biological actually multiplies dependencies between values, materials and human and non-human agents.

The concept of value is also discussed in its philosophical dimension, besides its purely economic dimension, with the emergence of concepts such as bio-value, which see the generation of value as being derived from (underexploited) biological resources.

Ethical and moral aspects have developed side-by-side with the bioeconomy (indeed, since even before the bioeconomy became a widespread concept) and are partly connected to environmental ethics. Notably, some of the seminal works mentioned as a basis for the bioeconomy, such as Georgescu-Roegen (1975), already highlight that a different ethical commitment towards future generations, incorporated in ways of life and consumption, is at the basis of human survival in the future.

The development of biotechnologies has given rise to bioethics as a discipline dedicated to the ethical issues emerging from advances in biology and medicine and, more in general, addressing scientific evolution due to the convergence of nanotechnology, biotechnology, information technology and cognitive science. This is now an extensively studied field with at least one dedicated international journal.

Nezhmetdinova (2013) illustrates the evolution of bioethics as a scientific discipline and social practice. The paper focuses on the role of the ethical management of potential risks from scientific research and argues for the need of bioethical social control in the development of a global bioeconomy. Székács (2017) also sees bioethics as a solution to the difficulties with purely technological or economic backgrounds to decision making. This supports the position that environmental, ecological and social aspects need to be included in the assessment of bioeconomy technologies in order to achieve balanced decisions.

A strong concept in environmental ethics, largely pertinent for the bioeconomy, is that no technological intervention can be imposed on nature beyond its receptive capacity. However, ethical and moral aspects have taken even stronger roles in the field of genetic modifications (Frewer *et al.*, 2013), largely based on the idea that it is unethical to manipulate life (Bennett

et al., 2013). A noteworthy development in the field of bioethics has been the 'precautionary principle', as a means of addressing harmful uncertain effects of new technologies. In spite of the variety of meanings given to this principle (Wesseler and Von Braun, 2017), it is now widely incorporated into regulatory processes concerning different stages of bioeconomy technologies.

New emerging fields of bioeconomy technology continuously raise new questions. Dicks (2017) discusses the evolution of the bioethics needed to provide an ethical framework to underpin the transition to a circular bioeconomy. The author contrasts in practice the view of nature as a model and the view of nature as a measure, that is, as a provider of ecological standards, against which human practices can be judged (a way of thinking that the authors call 'biomimetic ethics'). This notion is compared with that of environmental ethics, which mainly sees human ethical relations with nature as involving duties to protect, preserve or conserve various values in nature. This distinction allows to move from focusing mainly on preservation to questioning ways of producing, using and consuming goods, which is much more in line with the bioeconomy.

Blasco (2013) discusses the ethical, philosophical and political challenges of synthetic biology, seeking to clarify its implicit understanding of 'the living' and the possible consequences for the relations between science, bioeconomy and society. Most recent innovations in biotechnology have already attracted attention and revitalized the use of the precautionary principle; see, for example, the article by Peters (2017) meaningfully entitled 'Should CRISPR scientists play God?' on the use of the most recent gene editing technologies. Clearly, this topic will continue to evolve rapidly together with bioeconomy technologies.

11.8 Outlook

Looking at the disciplines connected to economic analysis, even in a very quick way as in this chapter, provides a clear insight into the complexity of the topic. Clearly, a better understanding of human behaviour facing the bioeconomy is needed to meet current societal challenges and to grant a constructive management of bioeconomy opportunities.

The bioeconomy is much more than a mere new technology; indeed, it touches the way human beings think of themselves and the constructs used in their relationships with nature and technology. It questions the philosophical and ethical background of the way individual positions may be affected by others in order to align individual preferences and behaviours with wider societal challenges.

Needless to say, inter- and trans-disciplinarity is needed to address these issues. In this context, collaboration across different social sciences is still far from achieving a satisfactory level. However, its importance is now well recognized and will be a pillar of future research in the field of the bioeconomy.

References

Bennett, A.B., Chi-Ham, C., Barrows, G., Sexton, S. and Zilberman, D. (2013) Agricultural biotechnology: Economics, environment, ethics, and the future. *Annual Review of Environment and Resources* 38, 249–279. doi: 10.1146/annurev-environ-050912-124612.

Bertrand, E., Pradel, M. and Dussap, C.-G. (2016) Economic and environmental aspects of biofuels. In Soccol, C.R. and Brar, D.K., Faulds, C, Pereira Ramos, L. (eds) *Green Energy and Technology*. New York, USA, Springer, pp. 525–555. doi: 10.1007/978-3-319-30205-8_22.

Birch, K. (2009) The knowledge-space dynamic in the UK bioeconomy. *Area* 41(3), 273–284. doi: 10.1111/j.1475-4762.2008.00864.x.

Birch, K. (2012) Knowledge, place, and power: Geographies of value in the bioeconomy. *New Genetics and Society* 31(2), 183–201. doi: 10.1080/14636778.2012.662051.

Blasco, J.M. (2013) Diseñar la biología: Retos éticos filosóficos y políticos de la biología sintética. *Contrastes* 18, 18–21.

Calvert, K.E., Kedron, P., Baka, J. and Birch, K. (2017) Geographical perspectives on sociotechnical transitions and emerging bio-economies: introduction to a special issue. *Technology Analysis and Strategic Management* 29(5), 1477–1485. doi: 10.1080/09537325.2017.1300643.

Delvenne, P. and Hendrickx, K. (2013) The multifaceted struggle for power in the bioeconomy: Introduction to the special issue. *Technology in Society* 35(2), 75–78. doi: 10.1016/j.techsoc.2013.01.001.

De Witt, A., Osseweijer, P. and Pierce, R. (2017) Understanding public perceptions of biotechnology through the 'Integrative Worldview Framework'. *Public Understanding of Science* 26(1), 70–88. doi: 10.1177/0963662515592364.

Dicks, H. (2017) Environmental ethics and biomimetic ethics: Nature as object of ethics and nature as source of ethics. *Journal of Agricultural and Environmental Ethics* 30(2), 1–20. doi: 10.1007/s10806-017-9667-6.

Frewer, L.J., van der Lans, I.A., Fischer, A.R.H., Reinders, M.J., Menozzi, D., Zhang, X., van den Berg, I. and Zimmermann, K.L. (2013) Public perceptions of agri-food applications of genetic modification – A systematic review and meta-analysis. *Trends in Food Science & Technology* 30(2), 142–152. doi: 10.1016/j.tifs.2013.01.003.

Georgescu-Roegen, N. (1975) Energy and economic myth. *Southern Economic Journal* XLI, 347–381.

Goldstein, J. and Johnson, E. (2015) Biomimicry: New natures, new enclosures. *Theory, Culture and Society* 32(1), 61–81. doi: 10.1177/0263276414551032.

Goven, J. and Pavone, V. (2014) The bioeconomy as political project: A Polanyian analysis. *Science Technology & Human Values* 40(3), 302–337. doi: 10.1177/0162243914552133.

Gupta, A. (2011) An evolving science-society contract in India: The search for legitimacy in anticipatory risk governance. *Food Policy* 36(6), 736–741. doi: 10.1016/j.foodpol.2011.07.011.

Guthman, J. (2015) Binging and purging: agrofood capitalism and the body as socioecological fix. *Environment and Planning A* 47(12), 2522–2536. doi: 10.1068/a140005p.

Hansen, J. (2014) The Danish biofuel debate: Coupling scientific and politico-economic claims. *Science as Culture* 23(1), 73–97. doi: 10.1080/09505431.2013.808619.

Herring, R. and Paarlberg, R. (2016) The political economy of biotechnology. *Annual Review of Resource Economics* 8(1), 397–416. doi: 10.1146/annurev-resource-100815-095506.

Hodge, D., Brukas, V. and Giurca, A. (2017) Forests in a bioeconomy: bridge, boundary or divide? *Scandinavian Journal of Forest Research* 32(7), 582–587 doi: 10.1080/02827581.2017.1315833.

Hodges, M. (2012) The politics of emergence: Public–private partnerships and the conflictive timescapes of apomixis technology development. *BioSocieties* 7(1), 23–49. doi: 10.1057/biosoc.2011.30.

Kearnes, M. (2013) Performing synthetic worlds: Situating the bioeconomy. *Science and Public Policy* 40(4), 453–465. doi: 10.1093/scipol/sct052.

Kleinschmit, D., Lindstad, B.H., Thorsen, B.J., Toppinen, A., Roos, A. and Baardsen, S. (2014) Shades of green: A social scientific view on bioeconomy in the forest sector. *Scandinavian Journal of Forest Research* 29(4), 402–410. doi: 10.1080/02827581.2014.921722.

Kotsakis, A. (2014) Change and subjectivity in international environmental law: The micro-politics of the transformation of biodiversity into genetic gold. *Transnational Environmental Law* 3(1), 127–147. doi: 10.1017/S204710251300054X.

Kröger, M. and Raitio, K. (2017) Finnish forest policy in the era of bioeconomy: A pathway to sustainability? *Forest Policy and Economics* 77, 6–15. doi: 10.1016/j.forpol.2016.12.003.

Kumar, D. and Chadwick, R. (eds) (2015) *Genomics and Society: Ethical, Legal, Cultural and Socioeconomic Implications*. London, Elsevier. doi: 10.1016/C2013-0-12992-3.

Lamers, P., Roni, M.S., Tumuluru, J.S., Jacobson, J.J., Cafferty, K.G., Hansen, J.K., Kenney, K., Teymouri, F. and Bals, B. (2015a) Techno-economic analysis of decentralized biomass processing depots. *Bioresource Technology* 194, 205–213. doi: 10.1016/j.biortech.2015.07.009.

Lamers, P., Tan, E.C.D., Searcy, E.M., Scarlata, C.J., Cafferty, K.G. and Jacobson, J.J. (2015b) Strategic supply system design – a holistic evaluation of operational and production cost for a biorefinery supply chain. *Biofuels, Bioproducts and Biorefining* 9(6), 648–660. doi: 10.1002/bbb.1575.

Levidow, L., Birch, K. and Papaioannou, T. (2013) Divergent paradigms of European agro-food innovation: The Knowledge-based bio-economy (KBBE) as an R&D agenda. *Science Technology and Human Values* 38(1), 94–125. doi: 10.1177/0162243912438143.

Mackenzie, A., Waterton, C., Ellis, R., Frow, E.K., McNally, R., Busch, L. and Wynne, B. (2013) Classifying, constructing, and identifying life: Standards as transformations of 'the biological'. *Science Technology and Human Values* 38(5), 701–722. doi: 10.1177/0162243912474324.

McDonagh, J. (2015) Rural geography III: Do we really have a choice? The bioeconomy and future rural pathways. *Progress in Human Geography* 39(5), 658–665. doi: 10.1177/0309132514563449.

McMichael, P. (2012) The land grab and corporate food regime restructuring. *Journal of Peasant Studies* 39(3–4), 681–701. doi: 10.1080/03066150.2012.661369.

Nezhmetdinova, F. (2013) Global challenges and globalization of bioethics. *Croatian Medical Journal* 54(1), 83–85. doi: 10.3325/cmj.2013.54.83.

Peters, T. (2017) Should CRISPR scientists play god? *Religions* 8(4), 61. doi: 10.3390/rel8040061.

Petersen, A. and Krisjansen, I. (2015) Assembling 'the bioeconomy': Exploiting the power of the promissory life sciences. *Journal of Sociology* 51(1), 28–46. doi: 10.1177/1440783314562314.

Pülzl, H., Kleinschmit, D. and Arts, B. (2014) Bioeconomy – an emerging meta-discourse affecting forest discourses? *Scandinavian Journal of Forest Research* 29(4), 386–393. doi: 10.1080/02827581.2014.920044.

Richardson, B. (2012) From a fossil-fuel to a biobased economy: The politics of industrial biotechnology. *Environment and Planning C: Government and Policy* 30(2), 282–296. doi: 10.1068/c10209.

Salter, B., Zhou, Y., Datta, S. and Salter, C. (2016) Bioinformatics and the politics of innovation in the life sciences: Science and the state in the United Kingdom, China, and India. *Science Technology and Human Values* 41(5), 793–826. doi: 10.1177/0162243916631022.

Sleenhoff, S., Cuppen, E. and Osseweijer, P. (2015a) Unravelling emotional viewpoints on a bio-based economy using Q methodology. *Public Understanding of Science* 24(7), 858–877. doi: 10.1177/0963662513517071.

Sleenhoff, S., Landeweerd, L. and Osseweijer, P. (2015b) Bio-basing society by including emotions. *Ecological Economics* 116, 78–83. doi: 10.1016/j.ecolecon.2015.04.011.

Székács, A. (2017) Environmental and ecological aspects in the overall assessment of bioeconomy. *Journal of Agricultural and Environmental Ethics* 30(1), 153–170. doi: 10.1007/s10806-017-9651-1.

Temper, L. (2016) Who gets the HANPP (human appropriation of net primary production)? Biomass distribution and the bio-economy in the Tana Delta, Kenya. *Journal of Political Ecology* 23(1), 410–433.

Wesseler, J. and Von Braun, J. (2017) Measuring the bioeconomy: economics and policies. *Annual Review of Resource Economics* 9, 275–298. doi: 10.1146/annurev-resource-100516-053701.

12

Final Thoughts and Outlook

_____

12.1 Introduction

What can we expect for the future of the bioeconomy? The bioeconomy is now growing steadily and in a way that constitutes a silent revolution that is already radically changing our everyday lives. On the other hand, this change is still fragile, characterized by uncertainties that are at times emphasized by unjustified fears, and sometimes boosted by unjustified 'industrial legends'.

Attention to uncertainty and controversies about bioeconomy technologies are motivated by the varied stories of successes and failures, and sometimes mixed results in comparison to the objectives set for their implementation (Bertrand *et al.*, 2016). For example, Withers *et al.* (2017) highlight that half of the advanced biofuel projects that started in the US in 2005 had ended by 2015. The internal barriers identified were related to technology itself, while funding and renewable fuel standards were perceived as the main external barriers. The unsatisfactory effects of bioeconomy projects may be partly explained by unfavourable conditions on international markets (Bertrand *et al.*, 2016) but also by the fact that, in many areas of the world, framework conditions such as appropriate legal frameworks and intellectual property rights are still lacking (Dias and De Carvalho, 2017).

On the other hand, the list of success stories is increasing steadily. See, for example, nine successful innovative case studies for forest resources reported by Graichen *et al.* (2016) (with a rallying

title 'Yes, we can make money out of lignin and other bio-based resources').

This chapter summarizes the main lessons learned from this book and provides an outlook for the future of the bioeconomy and related economic studies. More in detail, it seeks to provide an understanding of the future pathways of development of the bioeconomy and of the contribution that economics may provide to this process. This also implies some reflections on the situation and the potential of bioeconomy economics and policy, as research fields.

The chapter first recaps and summarizes the picture of the bioeconomy in its current state, and then discusses the future of the bioeconomy and the challenges for political action. In this context, it touches upon the role of science, which is a key and specific aspect of the bioeconomy. Finally, it turns to economics, summarizing research gaps and challenges for future economic research.

12.2 A Biodiverse and Uncertain Bioeconomy Technology

The future of the bioeconomy depends on the potential of bioeconomy technologies to solve future challenges and to compete over time with other technology options. As depicted in Figure 4.4, much of the future of the bioeconomy will depend on the balance between residual use of non-renewable resources, bio-based resource

 © D. Viaggi 2018. *The Bioeconomy: Delivering Sustainable Green Growth* (D. Viaggi)

development and future development of renewable non-bio-based resources and related technologies. The most optimistic vision in this respect is that we are currently in transition towards a bioeconomy era, which will last forever. A more realistic option is that the bioeconomy will be 'just' a step in the evolution of technologies and world visions that will last for a relevant period of human history. An outstanding question is, however, whether other technologies will 'jump over' the currently developing bio-based technologies and make them obsolete before they actually become dominant technologies. The probability of this occurring will depend on the state of play of individual bioeconomy sectors. For example, this is clearly difficult for food, though there are symptoms of emerging technologies that can partly by-pass agriculture. It is more likely that energy could find a future in non-fossil, non-bio-based solutions and this is partly happening with the growth of non-bio-based renewable resources such as wind and solar energy. According to some authors, biofuels may be a transitory solution that will phase out after two or three decades and will be substituted by new electric-powered transport systems (Mathews, 2008).

The most likely option is that this fascinating naïve model of dominant technologies and transitions will rather be diluted in a world of hybrid technology solutions with a wide knowledge base and several technology components that will interact over time in a 'techno-ecology' fashion.

This perspective makes it very difficult to predict the future, but certainly the role of the bioeconomy will depend to a significant extent on the degree of efficiency achievable in the use of biomass. Currently, we use biomass in a rather inefficient way with different sectors working in isolation; the bioeconomy vision is a unique opportunity to make biomass use more cost effective compared with all alternative (groups of) technologies and result in a substantive reduction in fossil fuels.

Within the bioeconomy itself, technologies are replacing each other over time. New technologies and solutions are sought for individual areas of research or even single organisms; some will take off while others will undoubtedly disappear before they materialize.

As mentioned in the chapter on technologies (Chapter 3), a number of potential improvements

are already foreseen for the future, in particular, making the bioeconomy technologies more efficient and more suitable for a design involving the participation of society. Notably, the development of synthetic biology is also progressively eroding the distinction between bio- and non-bio, which would make the bioeconomy even more prominent yet at the same time less identifiable as a concept.

Accordingly, it is very unlikely that there will be one single dominant technology in the future of the bioeconomy. The treatment of the bioeconomy in this book emphasizes diversity. Biodiversity is a key topic; there is a diversity of technologies, countries, stakeholder groups and visions of the bioeconomy, and this diversity is a key feature of a working bioeconomy.

Diversity may be hard to manage. It is difficult to investigate and to interpret. However, it also shows the comprehensiveness and flexibility of the bioeconomy concept and the relevance of having the ability to match different conditions and needs. These pathways can also be an asset as long as they contribute to resilience in an uncertain environment.

However, bioeconomy solutions can sometimes contribute to uncertainty themselves. Uncertainty may involve process outcomes, acceptability, effects of new technologies, context scenarios and perspectives, including the future need and role of the bioeconomy. Uncertainty is even more difficult to manage, but it is at the core of the bioeconomy, not only in relation to potential negative outcomes, but also opportunities. The bigger problem with uncertainty witnessed up until now is an uncontrolled prevalence of overestimated risk perceptions, which is linked to imaginaries, information and fair practices in politics, trade and marketing, and can actually preclude new opportunities.

The discussion above, and the trends in bioeconomy research, clearly show that solely technological solutions do not work. Solutions need to come from a societal decision-making process in which social components and political forces may play their role, together with business and science, in the design and acceptability of solutions. This is even more so when uncertainty is high and there is a diversity of solutions. In particular, diversity of options implies political efforts to achieve converging positions, but also legitimates different pathways and allows for the adaptability of the concept of bio-economy to different understandings and needs.

12.3 The Future of the Bioeconomy as a Political Choice

Philp (2017) qualifies the bioeconomy as 'the challenge of the century for policy makers'. The abstract of the paper is a perfect summary of the issues at stake and the challenges ahead:

> During the Industrial Revolution, it became clear that wood was unsuited as an energy source for industrial production, especially iron smelting. However, the transition to coal was the effort of decades. Similarly, the transition from coal to oil was neither a smooth nor rapid process. The transition to an energy and materials production regime based on renewable resources can similarly be expected to be fraught with many setbacks and obstacles, technically and politically. Those earlier transitions, however, were not complicated by the so-called grand challenges faced today. Above energy security and food and water security lurks climate change. Some events of 2015 have politically legitimized climate change and its mitigation, and 2016 saw the world finally sworn to action. The Bioeconomy holds some of the answers to the economic challenges thrown up by mitigating climate change while maintaining growth and societal wellbeing. For Bioeconomy policy makers, the future is complex and multi-faceted. The issues start in regions and extend to global reach. It is hard to quantify what is going to be the most difficult of challenges.

While several projections and political commitments seem to converge towards the rapid development and a key role for the bioeconomy in the future world economy, there are also signals in opposite directions. A circular and sustainable bioeconomy has become an objective for many governments around the globe (Egelyng *et al.*, 2016). In contrast to the recent reluctance of some countries, others are presently seeing a sustainable bioeconomy as a new and promising paradigm; an example is provided by the recent manifesto of Mediterranean countries discussed in Koukios *et al.* (2016). Beyond other considerations, at the very least, the motivation provided by climate change is expected to continue and is buttressed by a large consensus around the fact

that it is, and will continue to be, a significant threat (Hodgson *et al.*, 2016).

The bioeconomy is addressing some of the main problems faced by humanity, such as climate change, environmental degradation and food needs. The concept of 'bio' is somehow reassuring and can have a strong selling point, as it is a step forward and, in a way, a step backward to our biological nature and traditional sources of feed and food. However, the bioeconomy is clearly not a panacea, but rather a complex bundle of opportunities, that could be realized in many different ways and many papers emphasize the highly political nature of the bioeconomy, especially at its current stage of development.

Signals from researchers are now rather optimistic and characterized by a strong will to make the bioeconomy a success story (Aguilar *et al.*, 2017). Dialogue and collaboration will be key, considering that the bioeconomy is, at the same time, adapting to a number of regional conditions and developing stronger and stronger international links in a globalized economy. However, difficulties due to the uncertainty of results and policy have proven to be a clear limit to building shared views and boosting the development of the bioeconomy. Green (2016) envisages the potential of the bioeconomy. Yet he admits that the current mix of political challenges and economic uncertainties makes the transition to the bioeconomy difficult. This is reinforced by 'capricious oil markets, immature bio-based technology, and competing environmental policy priorities'. In this context, it appears that the political commitment in North America slowed down in 2017. The decision by the US President to exit the Paris Agreement on climate change and the resulting positive reaction of stock exchanges demonstrates that decoupling of economic growth from environmental concerns and resources is still far from being achieved.

Indeed, the literature recognizes the potential difficulties still to be tackled in the future of the bioeconomy. Through revising different bioeconomy strategies, Meyer (2017) identifies the following areas of attention for the future development of the bioeconomy:

> First, there is the risk of disappointment because far-reaching promises of the strategies are difficult to achieve. Second, the bioeconomy is not the only way to a low carbon economy

so alternatives could impede the desired development. Third, persistent conflicts between the different uses of biomass for food, material and energy production could lead to unstable policy support with short-term shifts. Fourth, a broader success of new bioeconomy value chains could trigger new societal conflicts over bioeconomy if efficiency gains, cascading use, residue use and sustainability certification are not sufficient to ensure a sustainable supply of biomass. Fifth, the acceptance of bioeconomy could be compromised if bioeconomy policies continue to ignore the on-going societal debates on agriculture and food.

The realization of the potential of the bioeconomy remains closely linked to political will. Ideally, in order to provide a smooth development of the bioeconomy and to allow for the expression of its potential to meet grand challenges and the sustainable development goals, different conditions at the global and local level need to converge in consistent incentives from the supply, demand and policy side, supported by an adequate political coalition in favour of the bioeconomy and in contrast to a fossil-based economy (Hagemann *et al.*, 2016). In more detail, among the needs most frequently identified, the current literature (Green, 2016; Hodgson *et al.*, 2016; Aguilar *et al.*, 2017; Dupont-Inglis and Borg, 2017) highlights the following:

- excellence in science-based concepts coupled with long-term support of mission-oriented research, in an equilibrium between science push and market and social pull;
- improvement of opportunities for biomass producers and for using novel biomass;
- coherent and science-based policies encompassing the whole bioeconomy, consistent with each other and lasting over time; in particular, investment and market policies for biofuel and bio-based products together with adequate pricing of carbon;
- developing a workforce and human resources in general;
- improved access to financial support;
- increasing investor confidence through the demonstration of bio-based technologies and greater clarity concerning best conversion routes for specific feedstocks;
- building consensus and inclusion of bioeconomy topics in the international and national agendas on sustainable development; and

- improving public perception and awareness of industrial biotechnology and bio-based products, as well as promoting a culture of industrial symbiosis.

To highlight the very promising, but also difficult state of play, Mabee (2017) uses the term 'reawakening of the bioeconomy' and identifies two main issues. First, the full range of environmental, economic and social benefits of the bioeconomy need to be better understood by benchmarking these effects against competing products, such as fossil-based products. Second, the bioeconomy needs to deliver specific products of high value for society that meet intrinsic societal needs, such as renewable diesel essential to meet future greenhouse gas emission reduction targets or renewable chemicals.

A recent comprehensive overview of potential opportunities and trends in the bioeconomy, based on a worldwide expert survey, is provided in German Bioeconomy Council (2018).

12.4 The Role of Science and Economics

The role of science in the bioeconomy is clearly paramount and at the same time difficult to gather in its precise configuration. The future seems to depend on a balanced mix of advances in basic science and the capacity to use them in applied sciences and innovation systems (Spiertz, 2014). On the one hand, science is confronted with unprecedented advances and the potential to play a major role in future economic and social development. On the other hand, the now long-standing experience with bioeconomy technologies has shown that the expression of the potential of these technologies and of the bioeconomy as a whole comes as the result not only of science, but also of political consensus. In this process, the bioeconomy is often confronted with high expectations and at the same time with restrictions or difficulties in exploiting the results of science (Zilberman *et al.*, 2015).

The context is indeed often contradictory with regard to the role of science. Science is advocated for support to public and private decision making. An example is the science-based and evidence-based focus in the EU, notably highlighted in the title 'Science for policy report' in the *Joint Research Centre Bioeconomy Report 2016* (Ronzon *et al.*, 2017). At the same time, the current context shows a trend towards a post-fact era, in which emotions and relationships are more valued than strong scientific results. On the other hand, scientific advances are also less and less comprehensible, even by scientists, and research itself tends to take a variety of positions on the same topic.

Economics and behavioural sciences clearly have a role in contributing in more aware decisions, partly through a better understanding of economic costs and benefits, and partly through a better understanding of expected reactions and potential coalitions of interests, for or against individual bioeconomy technologies. As the literature shows, however, consolidated trust and positive attitudes in terms of goodwill are an important asset, built over time and highly path-dependent.

A huge topic for the social sciences and humanities is the fact that the bioeconomy is revising the relationships between humans, living beings and technologies. The bioeconomy is (part of) a process that goes beyond the dichotomy between humans and the external world, and technologies as mediators. Some of the technologies at the basis of the bioeconomy are also resulting in new medical discoveries that are having a direct impact on human beings. At the same time, information technologies are shaping opinion and behaviour. In a way the separation between human beings and the 'object' of their actions is fading. Indeed, the bioeconomy is directly questioning the boundary between human beings and nature. In addition, more and more flexible and powerful technologies are putting the burden of choice on political systems, but also push towards the embedding of technology development and use into governance systems.

Confronted with these challenges, the field of 'bioeconomy economics and policy' is still far from being a well-established and consistent discipline. However, the current literature seems to recognize scope for a specific development with a focus on the bioeconomy, together with (and contributing to) the development of the bioeconomy as a whole. In a way the main task economics can contribute is to help finding incentive mechanisms able to make the bioeconomy economically viable (i.e. competitive with fossil resources) and doing so in a sustainable way.

Looking at current trends in the economy, and in economic research concerning the bioeconomy, the picture is that of a subject under construction, based on both very traditional areas of very basic agriculture and food economics research and, at the same time, some of the more novel approaches in economics and management.

This is also inherent to the subject, as the development of the bioeconomy requires research with multiple scales, scopes and focusing on different needs. On the one hand, focused studies, starting from basic profitability, consumer attitudes, technology assessment and business model analysis are needed in specific bioeconomy sectors under development. On the other hand, holistic approaches are needed to ensure the understanding of the increasingly interwoven components of the bioeconomy and the integration of the bioeconomy in a territorialized economy, where the ecosystemic, social and public good dimensions of the bioeconomy are also perceived as relevant. Furthermore, both sector and multi-sector approaches are needed. These different views could feed each other with increased attention being devoted to new linkages between sectors and product chains, which are some of the qualifying features of the bioeconomy.

Moreover, bioeconomy research is developing at the crossroads of several disciplines, involving both biological and social sciences and, more importantly, their interconnections. The emphasis on knowledge associated with biological resources also brings directly into play the role of decision making in society, as well as responsibility and governance in ways that go beyond the simple recognition that there is a need for regulation and ethical considerations.

Having said that, at the end of this book, an attempt to depict the bioeconomy as a subject of research and a delimitation shaping some specific area for economic thought and applications appears to be a worthwhile challenge.

12.5 Further Research on the Economics of the Bioeconomy

Several of the areas of analysis identified in the book may involve suitable directions for further research. This final section concerns some perspective to research needs in the bioeconomy, to sum up those already illustrated at the end of each chapter. This also builds on a few previous papers (Viaggi *et al.*, 2012; Viaggi, 2015, 2016).

The conceptualization of the bioeconomy and the establishment of a holistic view of the bioeconomy as a subject of economic research is clearly of importance, notably to give the subject an identity of its own and to help find research pathways able to properly address the specificities of the bioeconomy. A number of related concepts are already present in the literature, using 'bio-' as a distinguishing feature and emphasizing different aspects of the bioeconomy. We have suggested the idea of socioecological value web (SEVW), or even of socio-ecological technological value-enhancing web (SETVEW), as a concept encompassing the whole view of a network of value-producing relationships and the detailed and specific connections among bioresources in time and space. This is probably just a bridging concept and further work in defining a solid conceptual basis, besides the inflation of bio-concepts, is required.

In this framework, linkages among bioeconomy sectors (in terms of both technology and markets) need to be better studied, as they are a key feature of the bioeconomy as a whole, and it is on the linkages, rather than on the individual sectors, that the bioeconomy is displaying most of its novelty (i.e. the circularity of the economy based on bio-based solutions or the convergence of sectors such as agriculture and bioenergy).

There is a need for an improved economic representation of bioeconomy technologies that break down biomass and recombine its compounds; one direction could be an explicit representation of input and output as bundles of attributes organized in goods (instead of just goods). An attempt has been made in this book, but a deeper understanding of this topic for the economics of production and for system coordination is needed. The second direction is a better representation of flexibility in processes and how this is connected with the input–output potential of plants and chains, and the system as a whole. Finally, this needs to be better linked to the properties of the underlying living organisms (such as population dynamics and genetic potential). This goes hand-in-hand with the further development of studies to better understand research and technology impact pathways in this changing context. This goes much beyond acceptability and adoption: inclusive innovation and participation in innovation is a well-defined issue in the bioeconomy, notably in relation to past difficulties and failures, and will probably need in-depth research in the future (Bryden *et al.*, 2017).

There is also a need to better investigate new institutional mechanisms and organizational forms able to manage properly a more and more divisible technology; of special interest is the role of entrepreneurship linked to research and innovation, as well as the social construction of successful technologies. This connects with the notion of business models and the 'shape' (or, better, 'non-shape') of enterprises, which, metaphorically, can be better represented as amoebas rather than any other constant-form concept. Enterprises are becoming increasingly a bundle of loosely connected rights and values, focusing on narrower parts of the production process and confined to limited lifespan and time frames, the components of which are linked by a well-defined mission. This trend is emphasized by the growing complexity of technology ownership, the articulation of the exploitation of innovation and the explicit link with individual know-how and attitudes through entrepreneurship, which make a specific connection between entrepreneurship and human resources.

A better study of the role and functioning of the demand side of the bioeconomy is also needed. On the one hand, the hybridization of the roles of citizens and consumers requires an understanding of the balance between political and economic behaviour; on the other hand, the interplay between consumers and producers, beyond simplified market mechanisms, also needs better interpretations. On this issue, new awareness building and marketing strategies (social networks) and thoroughly informed/aware communities (often limited in their interpretation ability) connecting supply and demand, boosted by new digital technologies, call for the need for collaboration and linkages between studies on production and consumption in order to directly deal with uncertainty about prices and market shares. This is particularly relevant for new technologies for which the market has yet to be developed, in which frequently changing or new policy measures are also often observed and human well-being, health and primary needs are involved. At the same time, promissory economies that are at the core of the bioeconomy point to the potential for the construction of future well-being through the building of preferences and values.

Digitalization and new communication technologies have a key role in several of the processes mentioned above. Although they are arguably playing a disruptive role in changing societies, their role as means and promoters of change in economic behaviour linked to innovation and exploitation of research has yet to be sufficiently studied. In the context of the bioeconomy, they can play an important role in linking individual consumer/citizens' information, social coordination, data intensive innovation processes and supply. Examples in this direction are now myriad, ranging from traceability to citizens' science, to customer genomics, and so on.

The management of uncertainty is a clear aspect of the future bioeconomy and related research; the main difficulty is related to the low level of formalization or analysis of the topic due to the lack of information about actual hazards. As a result, economics has mostly dealt with uncertainty by analysing *ex-post* how social systems have been managing uncertainty through political and regulatory processes. However, little progress can be envisaged without tangible progress with regard to risk and risk management. This is a basic problem in economics in the face of an increasingly complex and uncertain environment, but takes on a special role for the bioeconomy. It is linked to uncertainty with regard to emerging technologies, but also to their interplay with awareness and market building; trends are less and less a relevant indication for the future, while attention should be particularly focused on potential trend-breaking scenarios. Radical uncertainty is the emerging issue and new ways of understanding how individuals and society deal with it is extremely relevant for building a future context suitable for bioeconomy development.

In light of the decision-making challenges of the bioeconomy, an investigation of new potential tools to measure economic performance is needed. The existing tools are for the most part well established in the literature; sometimes they are seen to be novel when 'discovered' by different disciplinary fields or when new variants become available, but in the last couple of decades most of the innovations in this field have been incremental. Given the pressure from policy and society, it is time for more radical innovation.

Contaminations among existing tools are promising and are already under way, for example, studies proposing combinations of life-cycle analysis and multi-criteria analysis. Furthermore, modelling integration between territorial (ecosystem), chain and web approaches is a promising pathway still only marginally explored. Instruments to measure achievements and impact can have a major role in helping achieve some consensus and this role deserves further research (Bertrand *et al.*, 2016). In a more ambitious way, the integration between emerging large scale and dynamic conceptual views of the world and sustainability assessment is required (Bates and Saint-Pierre, 2018).

An investigation into the new role of policy and related mechanisms is also needed. Not only do regulation and policy-defined prices contribute to reveal and signal preferences in value chains, but the building of new markets is becoming a stronger policy issue, which has a broad range of implications. Indirect policy mechanisms, such as those helping to reveal preferences related to public goods or facilitating coordination, are becoming more important, yet continue to be poorly understood. An increased emphasis on policy measures related to awareness, information, education and knowledge management may also result in the need for different tools to study policies, e.g. more qualitative and systemic tools. Indeed, the traditionally clear distinction between private and public goods is weakening due to the 'marketization' of environmental and social values, the increasingly explicit socially (or policy) driven construction of values and preferences and the growing number of cases of goods that are somewhere between private and public. These trends expose the system to the instability of preferences over time, which may lead to difficulties in prediction, overlapping and double counting, yet also provides opportunities to guide the transition. It also highlights the challenges of studying the relevant mechanisms based on a mix of institutional solutions, policy and private (market) incentives. This situation needs to be cast in an explicitly dynamic framework, in which the transition perspective allows for a clearer interpretation of timing and sequencing of actions.

The interface between policy, political science and political economy is also key for a thorough understanding of the bioeconomy. This, however, requires a stronger interaction with other social sciences. In addition, the question of how to connect proper interpretation with a more normative or supporting role for economics remains largely open and is linked to the ability to understand and build a renewed world view.

12.6   Finally....

As the bioeconomy is becoming a major feature in the current technology era, policy makers and practitioners need an improved understanding of its revolutionary potential. Society requires improved ways of thinking about it and needs to perform *ex-ante* and *ex-post* analyses that are better connected with decision making and capable of managing interplays between different governance levels and bioeconomy web subsystems. This is also in line with a growing emphasis on incorporating sustainability objectives into both firms and consumer behaviour, while dealing with globalized markets and global environmental and social concerns in, and through the bioeconomy. The bioeconomy is embedded in the changes reshaping our society and economies and brings into question the role of people as a living organism/species, as well as the interlinked changes in communication options and social relationships.

The fact that the development of the bioeconomy is not a 'given', but rather dependent on uncertain political processes and even more so on uncertain context needs (i.e. population, climate) emphasizes the role of economics in providing better contributions to wise and pragmatic decisions. As some authors have already pointed out (Birch and Tyfield, 2012), the understanding of bioeconomy development requires the joint interpretation of techno-scientific and economic changes and their conscious elaboration into new concepts.

High costs are the most evident barrier to the market development of several bioeconomy technologies. However making it to the market requires much more than cost calculation and implies dealing with technology potential, uncertainties, the mechanisms of value creation and leadership in entrepreneurial and consumers communities. This highlights the need for joint theoretical/conceptual and methodological/practical developments, which may find their basis in the improved knowledge of new technologies, as well as in a better understanding of surrounding societal change.

Economists are expected to play their role in the innovation system and contribute to its improvements. At the same time, they are expected to guarantee robust independent judgements that are suitable for evidence-based support to decision making. This ability to strike a balance at the crossroads between detailed and holistic views, as well as between facts and social constructions of the future, is the real challenge for future economic research and, perhaps, for the bioeconomy as a whole.

References

Aguilar, A., Wohlgemuth, R. and Twardowski, T. (2017) Perspectives on bioeconomy. *New Biotechnology* 40(Pt A), 181–184. doi: 10.1016/j.nbt.2017.06.012.

Bates, S. and Saint-Pierre, P. (2018) Adaptive policy framework through the lens of the viability theory: A theoretical contribution to sustainability in the Anthropocene era. *Ecological Economics* 145, pp. 244–262. doi: 10.1016/j.ecolecon.2017.09.007.

Bertrand, E., Pradel, M. and Dussap, C.-G. (2016) Economic and environmental aspects of biofuels. In Soccol, C.R. and Brar, D.K., Faulds, C, Pereira Ramos, L. (eds) *Green Energy and Technology*. New York, USA, Springer, pp. 525–555. doi: 10.1007/978-3-319-30205-8_22.

Birch, K. and Tyfield, D. (2012) Theorizing the bioeconomy: Biovalue, biocapital, bioeconomics or … what?. *Science Technology & Human Values* 38(3), 299–327. doi: 10.1177/0162243912442398.

Bryden, J., Gezelius, S.S., Refsgaard, K. and Sutz, J. (2017) Inclusive innovation in the bioeconomy: Concepts and directions for research. *Innovation and Development* 7(1), 1–16. doi: 10.1080/2157930X. 2017.1281209.

Dias, R.F. and De Carvalho, C.A.A. (2017) Bioeconomia no Brasil e no mundo: panorama atual e perspectivas. *Revista Virtual de Quimica* 9(1), 410–430. doi: 10.21577/1984-6835.20170023.

Dupont-Inglis, J. and Borg, A. (2017) Destination bioeconomy – The path towards a smarter, more sustainable future. *New Biotechnology* 40(Pt A), 140–143. doi: 10.1016/j.nbt.2017.05.010.

Egelyng, H., Romsdal, A., Hansen, H.O., Slizyte, R., Carvajal, A. K., Jouvenot, L., Hebrok, M., Honkapää, K., Wold, J.P., Seljåsen, R. and Aursand, M. (2016) Cascading Norwegian co-streams for bioeconomic transition. *Journal of Cleaner Production* 172, 3864–3873. doi: 10.1016/j.jclepro.2017.05.099.

German Bioeconomy Council (2018) Future opportunities and developments in the bioeconomy – A global expert survey. Available at http://gbs2018.com/fileadmin/gbs2018/Downloads/Bioeconomy_Global_Expert_Survey.pdf. (accessed 26 May 2018).

Graichen, F.H.M., Grigsby, W.J., Hill, S.J., Raymond, L.G., Sanglard, M., Smith, D.A., Thorlby, G.J., Torr, K.M. and Warnes, J.M. (2016) Yes, we can make money out of lignin and other bio-based resources. *Industrial Crops and Products* 106, 74–85. doi: 10.1016/j.indcrop.2016.10.036.

Green, T.W. (2016) Embracing our bioeconomic future: Engendering political and economic sustainability for the energy and chemical sector of the bioeconomy. In Liaison Functions 2016 - Core Programming Area at the 2016 AIChE Annual Meeting, San Francisco, California, USA.

Hagemann, N., Gawel, E., Purkus, A., Pannicke, N. and Hauck, J. (2016) Possible futures towards a wood-based bioeconomy: A Scenario Analysis for Germany. *Sustainability* 8(1), 98. doi: 10.3390/su8010098.

Hodgson, E., Ruiz-Molina, M.-E., Marazza, D., Pogrebnyakova, E., Burns, C., Higson, A., Rehberger, M., Hiete, M., Gyalai-Korpos, M., Lucia, L.D., Noël, Y., Woods, J. and Gallagher, J. (2016) Horizon scanning the European bio-based economy: a novel approach to the identification of barriers and key policy interventions from stakeholders in multiple sectors and regions. *Biofuels, Bioproducts and Biorefining* 10(5), 508–522. doi: 10.1002/bbb.1665.

Koukios, E., Monteleone, M., Texeira Carrondo, M.J., Charalambous, A., Girio, F., Hernández, E.L., Mannelli, S., Parajó, J.C., Polycarpou, P. and Zabaniotou, A. (2016) Targeting sustainable bioeconomy: A new development strategy for Southern European countries. The Manifesto of the European Mezzogiorno. *Journal of Cleaner Production* doi: 10.1016/j.jclepro.2017.05.020.

Mabee, W.E. (2017) Refocusing the bioeconomy in an uncertain policy environment. *Biofuels, Bioproducts and Biorefining* 11(3), 401–402. doi: 10.1002/bbb.1774.

Mathews, J.A. (2008) Biofuels, climate change and industrial development: Can the tropical South build 2000 biorefineries in the next decade? *Biofuels, Bioproducts and Biorefining* 2(2), 103–125. doi: 10.1002/bbb.63.

Meyer, R. (2017) Bioeconomy strategies: Contexts, visions, guiding implementation principles and resulting debates. *Sustainability* 9(6), 1031. doi: 10.3390/su9061031.

Philp, J. (2017) The bioeconomy, the challenge of the century for policy makers. *New Biotechnology* 40 (Pt A), 11–19. doi: 10.1016/j.nbt.2017.04.004.

Ronzon, T., Lusser, M., Klinkenberg, M., Landa, L., Sanchez Lopez, J., M'Barek, R., Hadjamu, G., Belward, A., Camia, A., Giuntoli, J., Cristobal, J., Parisi, C., Ferrari, E., Marelli, L., Torres de Matos, C., Gomez Barbero, M. and Rodriguez Cerezo, E. (2017) Bioeconomy Report 2016. JRC Scientific and Policy Report. EUR 28468 EN. Available at https://biobs.jrc.ec.europa.eu/sites/default/files/files/JRC_Bioeconomy_Report2016.pdf (accessed on 25 May 2017).

Spiertz, H. (2014) Agricultural sciences in transition from 1800 to 2020: Exploring knowledge and creating impact. *European Journal of Agronomy* 59, 96–106. doi: 10.1016/j.eja.2014.06.001.

Viaggi, D. (2015) Research and innovation in agriculture: beyond productivity? *Bio-based and Applied Economics* 4(3), 279–300.

Viaggi, D. (2016) Towards an economics of the bioeconomy: four years later. *Bio-based and Applied Economics* 5(2), 101–112.

Viaggi, D., Mantino, F., Mazzocchi, M., Moro, D. and Stefani, G. (2012) From agricultural to bio-based economics? Context, state-of-the-art and challenges. *Bio-based and Applied Economics* 1(1), 3–11.

Withers, J., Quesada, H. and Smith, R.L. (2017) Bioeconomy survey results regarding barriers to the United States advanced biofuel industry. *BioResources* 12(2), 2846–2863. doi: 10.15376/biores.12.2.2846-2863.

Zilberman, D., Kaplan, S. and Wesseler, J. (2015) The loss from underutilizing GM technologies. *AgBioForum* 18(3), 312–319.

Index

Note: bold page numbers indicate figures; italic page numbers indicate tables.

CABI – who we are and what we do

This book is published by **CABI**, an international not-for-profit organisation that improves people's lives worldwide by providing information and applying scientific expertise to solve problems in agriculture and the environment.

CABI is also a global publisher producing key scientific publications, including world renowned databases, as well as compendia, books, ebooks and full text electronic resources. We publish content in a wide range of subject areas including: agriculture and crop science / animal and veterinary sciences / ecology and conservation / environmental science / horticulture and plant sciences / human health, food science and nutrition / international development / leisure and tourism.

The profits from CABI's publishing activities enable us to work with farming communities around the world, supporting them as they battle with poor soil, invasive species and pests and diseases, to improve their livelihoods and help provide food for an ever growing population.

CABI is an international intergovernmental organisation, and we gratefully acknowledge the core financial support from our member countries (and lead agencies) including:

Discover more

To read more about CABI's work, please visit: **www.cabi.org**

Browse our books at: **www.cabi.org/bookshop**, or explore our online products at: **www.cabi.org/publishing-products**

Interested in writing for CABI? Find our author guidelines here: **www.cabi.org/publishing-products/information-for-authors/**